Day

NIS

CONTRACTUAL PROCEDURES IN
THE CONSTRUCTION INDUSTRY

692. 8

lot 4 08

CONTRACTUAL PROCEDURES IN THE CONSTRUCTION INDUSTRY

FOURTH EDITION

ALLAN ASHWORTH

Adjunct Professor in Quantity Surveying,
UNITEC, New Zealand;
Visiting Professor in Construction Education,
Liverpool John Moores University, UK

An imprint of **Pearson Education**

Harlow, England · London · New York · Reading, Massachusetts · San Francisco · Toronto · Don Mills, Ontario · Sydney
Tokyo · Singapore · Hong Kong · Seoul · Taipei · Cape Town · Madrid · Mexico City · Amsterdam · Munich · Paris · Milan

Pearson Education
Pearson Education Limited,
Edinburgh Gate, Harlow,
Essex CM20 2JE, England
and associated companies throughout the world

First published 1985
Second edition 1991
Third edition 1996
Reprinted 1997
Fourth edition 2001

ISBN 0-582-43223-5

British Library Cataloguing in Publication Data

A catalogue entry for this title is available from the British Library

Set by 35 in 10/12pt Ehrhardt
Printed in Malaysia

To Margaret, Amanda, and Caroline

CONTENTS

LIST OF FIGURES

LIST OF TABLES

LIST OF BOXES

PREFACE

Contractual procedures is about procurement in the construction industry. The continued aim of this book is to provide a knowledge of the nature, purpose and legal requirements of the procedures that are used in the construction industry. This knowledge forms the basis of an understanding which can then be applied across the various types of project and procurement arrangements encountered in the industry. The procurement arrangements for any construction project, irrespective of the form of contract that is used, follow a similar profile: employer's requirements, contractor's obligations, quality, payments, time, subcontractors, insurances, etc.

This book has been written for people in the construction industry. It has been prepared largely to meet the needs of students on undergraduate and higher technician courses across the various professional disciplines that exist in the industry. It also provides a good basis for anyone wishing to gain an understanding of the contractual procedures that are used in the industry.

The construction industry continues to change and evolve, and will do so throughout the millennium. Procurement, a word that did not exist in this context some twenty years ago, has seen the traditional procedures being replaced with new ideas. Some of the supposed new ideas are really not so new, but are old practices in a new guise. There is also a sense that there is nothing new under the sun. History somehow wants to repeat itself. As historians suggest, this is because people did not listen the first time around! The industry long ago was comprised of different tradespeople. Now they are known as subcontractors. The architect was responsible for their management and coordination. Today many a project manager does a similar task. The different trade contractors recouped their costs for the work done. Today there is a dislike for quantifying construction work, although it is becoming popular in some surprising countries around the world. There is a desire in the UK to want to return cost-plus arrangements. This desire will be short-lived. Procurement is being influenced by other industries and from countries abroad in Europe, the USA and Japan. The dissatisfaction with our current arrangements is to some extent influenced by changes in society at large; a wish to seek out a panacea which is as illusive as the end of the rainbow. The dissatisfaction is also to some extent a result of a lack of empirical research that would provide us with a better understanding. There is also the added consideration that too few architects have experience on construction sites, and conversely, those involved in building manufacture and production have only a limited understanding of design procedures or a history of architecture. It will be fascinating to see how procurement continues to evolve in the construction industry during the early years of the twenty-first century.

A section of the book that has continued to increase in size with subsequent editions is that which deals with procurement issues. The Latham Report published in 1994 has already had a huge impact on procurement practices. This report has also spurred a number of other important documents and changes. I have added a chapter on lean construction techniques which has been borrowed from the manufacturing industry and is generating interest on a worldwide basis.

Finally, I would like to express my very real thanks to my wife, Margaret, for her continued patience, understanding and support during times of writing and revision.

Allan Ashworth
York 2000

ACKNOWLEDGEMENTS

- The first two case studies in Chapter 13 have been adapted from *Rethinking construction*, 1998.

- RIBA Publications for permission to quote from JCT 98.

- Standing Joint Committee for the Standard Method of Measurement of Building Works.

- Davis, Langdon and Everest for permission to quote from their Contracts in Use Survey.

- The Construction Industry Board for permission to use information from their leaflet on Lean Construction.

TABLE OF RELEVANT STATUTES (ACTS OF PARLIAMENT)

A complete list of statutes can be found in *Halsbury's Statutes of England* (fourth edition)

Administration of Justice (Appeals) 1934
Arbitration Acts 1934, 1950, 1975, 1979, 1996
Architects (Registration) Acts 1931, 1938
Architects Registration (Amendment) Act 1969
Banking Act 1987
Bankruptcy Acts 1869, 1914
Bills of Exchange Act 1982
Bills of Sale Act 1891
Building Act 1984
Building Societies Act 1986
Charities Act 1992, 1993
Civil Jurisdiction and Judgments Act 1982
Civil Liability (Contribution Act) 1978
Companies Acts 1948, 1985, 1989
Construction (Design and Management Regulations) 1994
Construction Act 1996
Consumer Credit Act 1974
Consumer Protection Act 1987
Contracts (Applicable Law) Act 1990
Copyright, Design and Patents Act 1988
Corporation Taxes Act 1988
County Court Act 1984
Courts and Legal Services Act 1990
Defective Premises Act 1972
Employer's Liability (Compulsory Insurance) Act 1969
Employment Act 1989, 1990
Employment Protection Acts 1975, 1978
Environmental Protection Act 1990
European Communities Act 1972

PART 1

CONTRACT LAW

CHAPTER 1

THE ENGLISH LEGAL SYSTEM

There are many individuals within the construction industry who will, at some time in their careers, become professionally involved in either litigation or arbitration or adjudication. The laws which are applied in the construction industry are both of a general and a specialist nature. They are general in the sense that they embrace the tenets of law appropriate to all legal decisions; and are special since the interpretation of construction contracts and documents requires a particular knowledge and understanding of the construction industry. Note, however, that the interpretation and application of law will not be contrary to or in opposition to the established legal principles and precedents found elsewhere. It is appropriate at this stage to consider briefly the framework of the English legal system.

THE NATURE OF LAW

Law, in its legal sense, may be distinguished from scientific law or the law of nature and from the rules of morality. In the first case, scientific laws are not man-made and are not therefore subject to change. In the case of morality it is less easy to draw a distinction between legal rules and moral precepts. It may be argued, for example, that the legal rules follow naturally from a correct moral concept. The difference between the two is, perhaps, that obedience to law is enforced by the state whereas morals are largely a matter of conscience and conduct. The laws of a country are, however, to some extent an expression of its current morality, since laws can generally only be enforced by a common consensus. Law, therefore, may be appropriately defined as a body of rules for the guidance of human conduct but which may be enforced by the authorities concerned.

CLASSIFICATION OF LAW

Law is an enormous subject and some specialization is therefore essential. A complete classification system would require a very detailed chart. Essentially the basic division in the English legal system is the distinction between criminal and civil law. Usually, the distinction will be obvious. It is the difference between being prosecuted for a criminal offence and being sued for a civil wrong. If the aim of the person bringing the case is to punish the defendant then it will probably be a criminal case. However, if the aim is to obtain some form of compensation or other benefit then it will generally be a civil case.

Alternative methods of classification are to subdivide the offences that are committed against persons, property or the state under these headings. Laws may also be classified as either public or private. Public law is primarily concerned with the state itself. Private law is that part of the English legal system which is concerned with the rights and obligations of the individuals.

SOURCES OF ENGLISH LAW

Every legal system has its roots, the original sources from which authority is drawn. The sources of English law can be categorized in the following ways.

Custom

In the development of the English legal system, the common law was derived from the different laws associated with the different parts of the country. These were adapted to form a national law common to the whole country. Since the difference between the regions stemmed from their different customary laws, it is no exaggeration to say that custom was the principal original source of the common law. The term 'custom' has three generally accepted meanings:

- *General custom*: accepted by the country at large
- *Mercantile custom*: principles established on an international basis
- *Local custom*: applicable only to certain areas within a country

The following conditions must be complied with before a local custom will be recognized as law:

- The custom must have existed from 'time immemorial'. The date for this has been fixed as 1189.
- The custom must be limited to a particular locality.
- The custom must have existed continuously.
- The custom must be a reasonable condition in the eyes of the law.
- The custom must have been exercised openly.
- The custom must be consistent with, and not in conflict with, existing laws.

In some countries the writings of legal authors can form an important source of law. In England, however, because of tradition, such writings have in the past been treated with comparatively little respect. They are therefore rarely cited in the courts. This general rule has always been subject to certain exceptions and there are 'books of authority' which are almost treated as equal to precedents. Many of these books are very old, and in some cases date back to the twelfth century.

Legislation

The majority of new laws are made in a documentary form by way of an Act of Parliament. Statute has always been a source of English law and by the nineteenth century it rivalled decided cases as a source. If statute and common law clash, then the former will always prevail, since the courts cannot question the validity of any Act. The acceptance by the courts of Parliament's supremacy is entirely a matter for history. Today it is the most important new source of law because:

■ The complex nature of commercial and industrial life has necessitated legislation to create the appropriate organizations and legal framework.
■ Modern developments such as drugs and the motor car have necessitated legislation to prevent their abuse.
■ There are frequent changes in the attitudes of modern society, such as that relating to females, and the law must thus keep in step with society.

Before a legislative measure can become law, it must undergo an extensive process:

■ The measure is first drafted by civil servants who present it to the House of Commons or the House of Lords as a bill.
■ The various clauses of the bill will already have been accepted and agreed by the appropriate government department prior to its presentation.

Before the bill can become an Act of Parliament, it must undergo five stages in each house:

■ *First reading*: the bill is introduced to the House.
■ *Second reading*: a general debate takes place upon the general principles of the measure.
■ *Committee stage*: each clause of the bill is examined in detail.
■ *Report stage*: the House is brought up to date with the changes that have been made.
■ *Third reading*: only matters of detail are allowed to be altered at this stage.

The length of time which is necessary for the bill to pass through these various stages depends upon the nature and length of the bill and how politically controversial it is. Once the bill has been approved and accepted by each House, it then needs the royal assent for it to become law.

A public bill is legislation which affects the public at large and applies throughout England and Wales. Scottish law is similar to English law but it is not exactly the same. A private bill is legislation affecting only a limited section of the population, for example, in a particular locality. A private member's bill is a public bill introduced by a back-bench Member of Parliament as distinct from a public bill, which is introduced by the government in power.

Delegated legislation arises when a subordinate body makes laws under specific powers from Parliament. These can take the form of:

■ Orders in council
■ Statutory instruments
■ By-laws

Whilst these are essential to the smooth running of the nation, the growth of delegated legislation can be criticized, because law-making is transferred from the elected representatives to the minister, effectively the civil servants. The validity of delegated legislation can be challenged in the courts as being *ultra vires*, i.e. beyond the powers of the party making it, and thus making it void. The judicial safeguard depends on the parent legislation, i.e. the Act giving the powers. Often this is extremely wide and such a restraint may therefore be almost ineffectual.

All legislation requires interpreting. The object of interpretation is to ascertain Parliament's will as expressed in the Act. The courts are thus at least in theory concerned with what is stated and not with what it believes Parliament intended. A large proportion of cases reported to the House of Lords and the High Court involve questions of statutory interpretation and in many of these the legislature's intention is impossible to ascertain because it never considered the question before the court. The judge must then do what he or she thinks Parliament would have done had they considered the question.

Since Britain's entry into the European Community on 1 January 1973 it has been bound by Community law. All existing and future Community law which is self-executing is immediately incorporated into English law. A self-executing law therefore takes immediate effect and does not require action by the UK legislature.

Case law

Case law is often referred to as judicial precedent. It is the result of the decisions made by judges who have laid down legal principles derived from circumstances of the particular disputes coming before them. Importance is attached to this form of law in order that some form of consistency in application in practice can be achieved. The doctrine of judicial precedent is known as *stare decisis*, which literally means 'to stand upon decisions'. In practice, therefore, a judge trying a case must always look back to see how previous judges dealt with similar cases. In looking back, the judge will expect to discover those principles of law which are relevant to the case now being decided. The decision made will therefore seek to be in accordance with the already established principles of law and may in turn develop those principles further. Note that the importance of case law is governed by the status of the court which decided the case. The cases decided in a higher court will take precedence over the judgments in a lower court.

Here are the main advantages claimed for judicial precedent:

- *Certainty*: because judges must follow previous decisions, a barrister can usually advise a client on the outcome of a case.
- *Flexibility*: it is claimed that case law can be extended to meet new situations thereby allowing the law to adjust to new social conditions.

A direct result of the application of case law is that these matters must be properly reported and published and should be readily available for all future users. Consequently, there is now available within the English legal system an enormous collection of law reports stretching back over many centuries. Within the construction professions a number of different firms and organizations now collate and publish law reports which are relevant to this industry. Computerized systems are also available to allow for rapid access and retrieval from such reports.

It is not the entire decision of a judge that creates a binding precedent. When a judgment is delivered, the judge will give the reason for the decision. This is known as *ratio decidendi*, and is a vital part of case law. It is the principle which is binding on subsequent cases that have similar facts in the same branch of law. The second aspect of judgments, *obiter dicta*, are things said 'by the way', and these do not have to be followed. Although the facts of a case appear similar to a binding precedent, a judge may consider that there is some aspect or fact which is not covered by the *ratio decidendi* of the earlier case. The judge will therefore 'distinguish' the present case from the earlier one which created the precedent.

A higher court may also consider that the *ratio decidendi* set in a lower court is not the correct law which should be followed. When another case is argued on similar facts, the higher court will overrule the previous precedent and set a new precedent to be followed in future cases. Such a decision does not affect the parties in the earlier case, unlike a decision that is reversed on appeal.

Finally a superior court may consider that there is some doubt as to the standing of a previous principle, and it may disapprove but not expressly overrule the earlier precedent.

Examples

Here are some examples of how sources of English law are appropriate to the construction industry:

- Custom
  - Right to light
  - Right of way
- Legislation
  - Highways Act 1980
  - Town and Country Planning Act 1990
  - Local Government Act 1988
  - Control of Pollution Act 1974
- Cases
  - *Hadley* v. *Baxendale* 1854
  - *Sutcliffe* v. *Thackrah and Others* 1974
  - *Dawber Williamson Roofing* v. *Humberside County Council* 1979

EU law

A completely new source of English law was created when Parliament passed the European Communities Act 1972. Section 2(1) of the Act provides that:

All such rights, powers, liabilities, obligations and restrictions from time to time created or arising by or under the Treaties, and all such remedies and procedures from time to time provided for by or under the Treaties, as in accordance with the treaties are without further enactment to be given legal effect or used in the United Kingdom shall be recognized and available in law, and be inforced, allowed and followed accordingly.

The effect of this section is that all UK courts have to recognize European Union (EU) law, whether it comes directly from treaties or other Community legislation. As soon as the 1972 Act became law, some aspects of English law were changed to bring them into line with European Union law.

There are several institutions to implement the work of the European Union. These include the European Parliament, the Council of Ministers, the European Commission and the European Court of Justice.

THE COURTS

There are a number of different courts in which civil actions may be tried (Figure 1.1). Cases are first heard in either the High Court or the County Court, and should an appeal be necessary, the matter is brought to the Court of Appeal. Where the matter is still not resolved then it is brought to the House of Lords. Technical cases may be heard in the official referee's court, this has been renamed the Technology and Construction Court. A further alternative is to allow the dispute to be settled through arbitration or alternative dispute resolution procedures (Chapter 5).

County court

There are approximately 340 districts in England and Wales in which a county court is held at least once a month. These are divided into 60 circuits, each with its own judge. A county court can hear almost all types of civil cases. Its jurisdiction is limited to actions in contract and tort up to sums of £5,000; actions for the recovery of land where the net annual value for rating does not exceed £1,000;

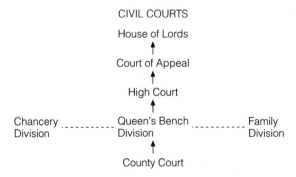

CIVIL COURTS

House of Lords
↑
Court of Appeal
↑
High Court
↑
Chancery ----------- Queen's Bench ----------- Family
Division   Division   Division
↑
County Court

Fig. 1.1　The court system

equity proceedings where the sum involved does not exceed £30,000 and some bankruptcy claims. These sums are kept under review. The main advantages claimed for county courts are their lower costs and shorter delays before coming to trial. These are also the two similar reasons claimed for arbitration. The court comprises a circuit judge assisted by a registrar who is a solicitor of seven years' standing. The latter mainly performs administrative tasks but, with the leave of the judge, he or she may hear actions in which the defendant does not appear, admits liability or where the claim does not exceed £2,000.

In order to reduce litigation coming before the county court, and enable those with very small claims to bring an action without fear of excessive court costs, a small claims procedure is now available.

The High Court

The High Court hears all the more important civil cases. It is the lower half of the Supreme Court of Judicature and was brought into being under the Judicature Acts 1873–1875. It comprises three divisions which all have equal competence to try any actions, according to the pressure of work, although certain specific matters are reserved for each of them:

- *The Queen's Bench Division* (QBD) deals with all types of common law work, such as contract and tort. This is the busiest division. Matters concerning the construction industry usually come to this High Court. The division is headed by the Lord Chief Justice and there are about 40–50 lesser judges. These are known as puisne (pronounced 'puny') judges. There are two specialist courts within QBD. The Commercial Court hears major commercial disputes, usually in private, with the judge hearing the case in the more informal role of an arbitrator. The Admiralty Court hears maritime disputes.
- *The Chancery Division* deals with such matters as trusts, mortgages, deeds, and land, taxation and partnership disputes. The division, whilst nominally headed by the Lord Chancellor and the Master of the Rolls, is actually run by a vice-chancellor, with the help of about ten to twelve lesser judges.
- *The Family Division* deals with matters of family disputes such as probate and divorce. This division is headed by a president and three lesser judges.

The High Court normally sits at the Strand in London but there are fifteen other towns to which judges of the High Court travel to hear common law claims.

The Court of Appeal

Once a case has been heard, either party may consider an appeal. This means the case is transferred to the Court of Appeal, where three judges usually sit to form a court. The High Court has the right to refuse an appeal. In civil appeals the appellant has six weeks from the date of judgment in which to give the Court of Appeal formal notice of appeal. The appellant must specify the exact grounds on which the appeal is based and on which the lower court reached an 'incorrect' decision.

The Civil Division hears appeals on questions of law and of fact, rehearsing the whole of the evidence presented to the court below relying on the notes made at the trial. If the appeal is allowed, the court may reverse the decision of the lower court, or amend it, or order a retrial. It can hear appeals about the exercise of discretion, for example, discretion as to costs.

Most appeals are heard by three judges, although some (e.g. appeals from county court decisions) can be heard by only two judges. Decisions need not be unanimous. The head of the court is the Master of the Rolls, perhaps the most influential appointment in our legal system.

The House of Lords

Appeals from decisions of the Court of Appeal are made to the House of Lords, although leave must first be obtained to do so and this is sparingly given. Permission is only given if the appeal is of general legal importance. The court used to sit in the Chamber of the House of Lords but since 1948 it has usually sat as an appellate committee in a committee room in the Palace of Westminster. The normal rule is that a case can only go to the House of Lords after it has been heard by the Court of Appeal, so the case progresses slowly up the judicial hierarchy. In exceptional circumstances it is possible to leap-frog over the Court of Appeal, but this is rarely done. There must be at least three judges for a committee to be quorate, although in practice appeals are heard by five judges.

All of the courts must apply statute law in reaching their decisions and, in general, the lower courts are bound by the decisions of the higher courts. In practice the law resulting from a case to the House of Lords can only be changed by an Act of Parliament.

Adjudication and arbitration

Adjudication and arbitration are alternatives to litigation in the courts, and are widely used for the settlement of disputes which involve technical or commercial elements. The tribunal is chosen by the parties concerned, and the powers of the arbitrator largely depend upon agreement between these parties. Adjudication and arbitration are more fully explained in Chapter 5.

THE LAWYERS

Solicitors

There are several UK legal professions; most lawyers are solicitors. There are over 80,000 solicitors in England and Wales. The solicitor's profession has a long history going back to the twelfth century, when the language of the courts was Norman French and litigants therefore needed a representative to act as a translator. This representative might also appear in the court on the client's behalf. They were known as attorneys, and are the forerunners of today's solicitors. Over the centuries a division between attorneys and court pleaders was established, today known as solicitors and barristers.

Solicitors are the lawyers that the public most frequently meets and as such are the general practitioners of the legal profession. A solicitor's work falls into two main categories of court work and non-court work. Non-court work accounts for about three-quarters of their business. A solicitor operates in many ways like a business person, with an office to run, clients to see and correspondence to be answered. Traditionally, property (conveyancing and probate) has been one of the main fee earners.

The Solicitors Act 1974 gave solicitors three monopolies: conveyancing, probate, and suing and starting court proceedings. The conveyancing monopoly was, however, significantly eroded by the Administration of Justice Act 1985. This allowed for licensed conveyancers to do this work.

The Law Society is the solicitor's professional body and it has two roles: (1) to act as the representative body of solicitors; (2) to ensure that proper professional standards are maintained and that defaulting solicitors are disciplined. It lays down the rules on professional conduct, the most important of which prohibits touting for business or unfair attraction of business. Advertising of solicitors' services is now allowed in the press and on the radio, but they must be very careful what they say about themselves. One firm of solicitors cannot claim that they do a better job than another.

Barristers

There are approximately 8,000 barristers in England and Wales. These are high quality, highly competitive specialist advocates who are experts in the conduct of the client's case or trial. For advocacy, advice or arbitration in matters involving English or European law the English Bar is a rich source of talent, expertise and specialization.

Barristers are the specialist advocates and the specialist advisers of the legal profession. About 10 per cent of the Bar is made up of Queen's Counsel, often referred to as silks. Whilst solicitors may appear in the lower courts, barristers have a monopoly over appearing in the higher courts. Some of their work is non-court work, such as advising on difficult points of law or on how a particular case should be conducted.

Barristers cannot form partnerships with other barristers; several barristers will, however, share a set of rooms, known as chambers, and employ clerical staff and a clerk between them. They will not, however, share their earnings in the way that a firm of solicitors would.

Judges

In contrast with many other European countries, the judiciary in England and Wales is not a separate career. Judges are appointed from both branches of the legal profession. They serve in the House of Lords (the final appellate court), the Court of Appeal, the High Court, Crown Court or as circuit or district judges.

The circuit judges sit either in Crown Courts to try criminal cases or in county courts to try civil cases. District judges sit in county courts. There also part-time judges appointed from both branches of the practising legal profession, who serve in Crown Court, county court or on various tribunals, for example, those dealing with unfair dismissal from employment.

However, the majority of cases are not dealt with by judges, but by laypersons who are appointed to various tribunals because of their special knowledge, experience and good standing. For example, the majority of minor criminal cases are judged by justices of the peace in magistrates' courts. They are not legally qualified or receive a remuneration but are respected members of the community who sit part-time.

All members of the judiciary are appointed by the Lord Chancellor – a member of the government and Speaker of the House of Lords. The Lord Chancellor holds a function similar to that of a minister of justice, although some matters concerning the administration of justice are the responsibility of the Home Secretary.

Once appointed, judges are completely independent of both the legislature and the executive, and so are free to administer justice without fear of any political interference.

FUTURE CONSIDERATIONS

Law is sometimes seen as a question of how far you can afford to go rather than how good your case is. A commission helps to remedy the huge problems of cost, delay, complexity and inequality in the civil justice system. The key is to recognize that justice is not an abstract quality. It has to be proportionate, it has to be within the means of the parties and it has to be expeditious. Many people are denied access to the courts because the costs involved are disproportionate to their claims. Judges should become trial managers able to dictate the pace of legal cases. Also attitudes must change but the powers to encourage settlement or to strike out unworthy cases must also be provided. Judges should also be able to encourage litigants to look at other ways of settling disputes such as mediation or alternative dispute resolution (Chapter 5). These procedures encourage early settlement. They should also be able to give summary judgment, leaving only the core of the dispute to go to trial. The following are some recommendations:

- Small claims court expanded to take claims up to £3,000.
- Fast-track cases up to £10,000 including personal injury claims with capped costs and fixed hearings.
- New multi-track for cases above £10,000 providing hands-on management teams by judges for heaviest cases.
- New post, head of civil justice, to run all civil courts as a single system; post to be filled by a senior judge.
- Better use of technology, with laptop computers for all judges and videoconferencing facilities.

- Incentives for early settlement, including 'plaintiff's offer' and referral to alternative dispute procedures.
- Solicitors to inform clients of charges as the bill for legal services mounts.
- Elimination of courtroom Latin.

The present consequences of the current system are that it is often excessive, disproportionate and unpredictable. The delay is frequently unreasonable. For example, in London, High Court cases take on average 160 weeks just to reach trial. Outside of London they take even longer, average 190 weeks. In the county courts the typical figure is 80 weeks.

A report chaired by Lord Justice Wolfe recommended a three-track civil system with a single entry point, headed by a senior judge in a new post called head of civil justice. A key point of the proposed system is to encourage settlement. Both parties will be able to make offers to settle at any stage, relating to the whole case or just one of the issues involved.

Note that annually almost 300,000 writs are lodged with the Lord Chancellor's Office, with less than 3 per cent ever coming to trial.

SOME LEGAL JARGON

abate	To reduce or make less.
actus reus	The guilty act.
ad valorem	According to the value. For example, stamp duty on sale of land is charged according to the price paid.
affidavit	A written statement to be used as evidence in court proceedings.
ancient lights	Windows which have had an uninterrupted access of light for at least twenty years. Buildings cannot be erected which interfere with this right of light.
attestation	The signature of a witness to the signing of a document by another person.
burden of proof	The obligation of proving the case.
caveat emptor	Let the buyer beware.
certiorari	An order of the High Court to review and quash the decision of the lower court which was based on an irregular procedure.
chattels	All property other than freehold real estate.
consideration	Where a person promises to do something for another, it can only be enforced if the other person gave or promised to give something of value in return. Every contract requires consideration.
costs	The expenses relating to legal services.
counterclaim	When a defendant is sued, any claim against the plaintiff may be included, even if it arises from a different matter.

custom	An unwritten law dating back to time immemorial.
defendant	A person who is sued or prosecuted, or who has any court proceedings brought against them.
enactment	An Act of Parliament, or part of an Act.
estoppel	A rule which prevents a person denying the truth of a statement or the existence of facts which another person has been led to believe.
ex parte	An application to the court by one party to the proceedings without the other party being present.
expert witness	One who is able to give an opinion on a subject. This is an exception to the rule that a witness must only tell the facts.
fieri facias	A court order to the sheriff requiring the seizure of a debtor's goods to pay off a creditor's judgment.
frustration	A contract is frustrated if it becomes impossible to perform because of a reason that is beyond the control of the parties. The contract is then cancelled.
good faith	Honesty.
goodwill	The whole advantage, wherever it may be, of the reputation and connection of the firm.
in camera	When evidence is not heard in open court.
injunction	A court order requiring someone to do, or to refrain from doing, something.
judicial review	An application made to the divisional court when a lower court or tribunal has behaved incorrectly.
limitation	Court proceedings must begin within a limitation period. Different periods exist for different types of claim.
liquidated sum	A specific sum, or a sum that can be worked out as a matter of arithmetic.
liquidator	A person who winds up a company.
moiety	One half.
obiter dictum	A statement of opinion by a judge which is not relevant to the case being tried. It is not of such authority as if it had been relevant to the case being tried.
official referee	A layperson appointed by the High Court to try complex matters in which he or she is a specialist.
plaintiff	Person who sues.
plc	A public limited company. Most used to call themselves Ltd but changed this when UK company law was brought into line with EC law in 1981.
pleadings	Formal written documents in a civil action. The plaintiff submits a statement of claim and the defendant a defence.
pre-trial review	Preliminary meeting of parties in a county court action to consider administrative matters and what agreement can be reached prior to the trial.

quantum meruit	As much as has been earned.
ratio decidendi	The reason for a judicial decision. A statement of legal principle in a *ratio decidendi* is more authoritative than if in an *obiter dictum*.
res ipsa loquitor	The matter speaks for itself.
retrospective legislation	An Act that applies to a period before the Act was passed.
seal	This used to be an impression of a piece of wax to a document. Now a small red sticky label is used instead. The absence of the seal will not invalidate the document, since to constitute a sealing requires neither wax, wafer, piece of paper nor even an impression.
sine die	Indefinitely.
special damage	Financial loss that can be proved.
specific performance	When a party to a contract is ordered to carry out their part of the bargain. Only ordered where monetary damages would be an inadequate remedy.
stare decisis	To stand by decided matters. Alternative name for the doctrine of precedent.
statute	An Act of Parliament.
statutory instrument	Subordinate legislation made by the Queen in Council or a minister, in exercise of a power granted by statute.
stay of proceedings	When a court action is stopped by the court.
subpoena	A court order that a person attends court, either to give evidence or to produce documents.
uberrimae fidei	Of utmost good faith.
unenforceable	A contract or other right that cannot be enforced because of a technical defect.
vicarious liability	When one person is responsible for the actions of another because of their relationship.
void	Of no legal effect.
voidable	Capable of being set aside.
with costs	The winner's cost will be paid by the loser.
writ	The document that commences many High Court actions.

CHAPTER 2

LEGAL ASPECTS OF CONTRACTS

A number of Acts of Parliament affect construction contracts, although it is only during this present century that they have begun to play any significant part. Historically the law of contract has evolved by judicial decisions, so that there now exists a body of principles which apply generally to all types of construction contract. These principles have been accepted on the basis of proven cases that have been brought before the courts.

Construction contracts are usually made in writing, using one of the standard forms available. The use of a standard form provides many advantages, and although standard forms are not mandatory, their use should be encouraged in all possible circumstances. It is important to remember, however, that the making of a contract does not require any special formality. A binding contract could be made by an exchange of letters between the parties, rather than signing an elaborate printed document. On some occasions a binding contract could be made by a gentlemen's agreement, i.e. by word of mouth. There are, however, many practical reasons why construction contracts for all but the simplest projects should be made using an approved and accepted form of contract.

DEFINITION OF A CONTRACT

A contract has been defined by Sir William Anson as 'a legally binding agreement made between two or more parties, by which rights are acquired by one or more to acts or forbearances on the part of the other or others'. The essential elements of this definition are as follows:

- *Legally binding*: not all agreements are legally binding; in particular, there are social or domestic arrangements which are made without any intention of creating legal arrangements.
- *Two or more parties*: in order to have an agreement there must be at least two parties; in law one cannot make bargains with oneself.
- *Rights are acquired*: an essential feature of a contract is that legal rights are acquired; one person agrees to complete part of a deal and the other person agrees to do something else in return.

- *Forbearances*: to forbear is to refrain from doing something; there may thus be a benefit to one party to have the other party promise not to do something.

AGREEMENT

The whole basis of the law of contract is agreement. Specifically, a contract is an agreement bringing with it obligations which are able to be enforced in the courts if this becomes necessary. Most of the principles of modern contract date from the eighteenth and nineteenth centuries. The concept of a contract at that time was of equals coming together to bargain and reach agreement, which they would wish to be upheld by the courts. Whilst it is still true that individuals come together to form agreements, note that many contracts are formed between parties who are not equals in any way, even where the law may pretend that they are. A major criticism of contract law in recent years has been that the wealthy, experienced and legally advised corporations have been able to make bargains with many people who are themselves of limited resources and poorly legally represented. Because of this the law of contract has gradually moved away from a total commitment to enforce, without qualification, any agreement which has the basic elements of a contract. In particular, Parliament has introduced statutes which are often designed to protect relatively weak consumers from business persons having greater bargaining power. Despite this, the courts are still reluctant to set aside an agreement having all of the elements of a contract and in this respect follow their nineteenth-century predecessors.

THE ELEMENTS OF A CONTRACT

Capacity

In general every person has full legal powers to enter into whatever contracts they might choose. There are, however, some broad exceptions to this general rule. Infants and minors, that is anyone under the age of 18 years as set out in the Family

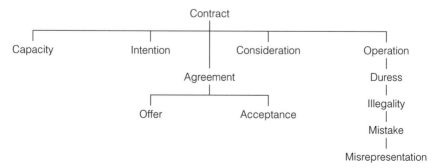

Fig. 2.1   The elements of a contract

Law Reform Act 1969, cannot contract other than in certain circumstances, such as for necessaries or benefits. Persons of an unsound mind, as defined in the Mental Health Act 1959, can never make a valid contract. Other persons of an unsound mind, and those unbalanced by intoxication are treated alike. Their contracts are divided into two types, those for necessary goods (the situation with minors above) and other contracts where the presumption is one of validity. Corporations are legal entities created by a process of law. A company can only contract on matters falling within its objects clause, and since the records of companies are matters of public record available for inspection at Companies House, it used to be the case that a company could not have a contract enforced against it if it lay outside its objects clause. The presumption was that those entering into contracts with a company knew or ought to have known the contents of the objects clause. Consequently, anyone making an *ultra vires* 'outside the powers' contract with a company only had themselves to blame. On entry into the European Community in 1973 this *ultra vires* doctrine had to be revised, since it was not followed in the other EU countries.

Intention to create legal relations

Merely because there is an agreement, it cannot be assumed that an enforceable contract exists. English law requires that the parties to a contract actually intended to enter into legal relations. These are relations actionable and enforceable in the courts. If it can be demonstrated that no such intention existed then the courts will not intervene, despite the presence of both agreement and consideration. In commercial agreements the courts presume that the parties do intend to enter into legal relations. (This is different from social, family and other domestic agreements, where the general rule is that the courts presume there is no intention to enter into legal relations.) These are the general rules and it is also possible to demonstrate the opposite intention.

In contract law we need to know what we have agreed. It is possible for two parties to use words which are susceptible to interpretation in different ways, so they do not have the same idea in mind when they agree. It is important that both parties have agreement to the same idea (*consensus ad idem*). The classic case to which students should refer is *Raffles* v. *Wichelhaus* (1864).

OFFER AND ACCEPTANCE

The basis of the contract is agreement and this is composed of two parts: offer and acceptance. In addition, conditions are generally required by law (in all but the simplest of contracts) to make the offer and acceptance legally binding.

The offer

An offer must be distinguished from a mere attempt to negotiate. An offer, if it is accepted, will become a binding contract. An invitation for contractors to submit tenders is inviting firms to submit offers for doing the work. The invitation often states that the employer is not bound to accept the lowest or any tender or to be responsible for the costs incurred.

An offer may be revoked by the person who made it, at any time before it is accepted. Thus a contractor may submit a successful tender in terms of winning the contract. However, the contractor may choose to revoke this offer prior to formal acceptance.

Tenders for building works do not remain on offer indefinitely. If they are not accepted within a reasonable time then the offer may lapse, or be subject to some monetary adjustment should it later be accepted. The building owner may stipulate in the invitation to the tenderers that the offer should remain open for a prescribed period of time.

Offers concerned with building projects are generally made on the basis of detailed terms and conditions. The parties to the contract will be bound by these conditions, as long as they know that such conditions were incorporated in the offer, even though they may never have read them or acquainted themselves with the details.

In many instances with building contracts the offer must follow a stipulated procedure. Such procedures often incorporate delivery of the offer by a certain date and time, in writing, on a special form, and in a particular envelope, and stipulate that the offer must not be disclosed to a third party. Failure to comply with these procedures will result in the offer being rejected.

The acceptance

Once an agreeable offer has been made, there must be an acceptance of it before a contract can be established. The acceptance of the offer must be unconditional and it must be communicated to the person who made the offer. The unconditional terms of acceptance must correspond precisely with the terms of the offer. In practice the parties may choose to negotiate on the basis of the offer. For example, the building owner may require the project to be completed one month earlier and this may result in the tenderer revising the tender sum. A fresh offer made in this way is known as a counter-offer, and is subject to the conditions now applied.

An offer or acceptance is sometimes made on the basis of 'subject to contract'. In practice the courts tend to view such an expression as of no legal effect. No binding contract will come into effect until the formal contract has been agreed.

FORM

The word *form* means some peculiar solemnity or procedure accompanying the expression of agreement. It is this formality which gives to the agreement its binding character. The formal contract in English law is the contract under seal – that is, one made by deed. The contract is executed – that is, it is made effective – by being signed, sealed and delivered.

- *Signature*: doubts have been expressed regarding the necessity for a signature; some statutes make a signature a necessity.
- *Sealing*: today this consists of affixing an adhesive wafer, or in the case of a corporate body an impression in the paper, and the party signs against this and acknowledges it as his or her seal.
- *Delivery*: this is not now necessary for legal effectiveness; as soon as the party acknowledges the document as the deed, it is immediately effective.
- *Witnesses*: these attest by signing the document; they are not usually a legal necessity.

CONSIDERATION

Another essential feature of a binding contract, other than a contract made under seal, is that the agreement must be supported by consideration. The most common forms of consideration are payment of money, provision of goods and the performance of work. Consideration has been judicially defined as 'some right, forbearance, detriment, loss or responsibility given, suffered or undertaken by the other in respect of the promise'. In building contracts the consideration of the contractor to carry out the works in accordance with the contract documents is matched by that of the building owner to pay the price. The following rules concerning consideration should be adhered to:

- Every simple contract requires consideration to make it valid.
- The consideration must be worth something in the eyes of the law. The courts are not concerned whether the bargain is a good one, but simply that there is a bargain.
- Each party must get something in return for the promise, other than something already entitled to, otherwise there is no consideration.
- The consideration must not be such that it conflicts with the established law.
- The consideration must not relate to some event in the past.

DURESS AND UNDUE INFLUENCE

Duress is actual or threatened violence to, or restraint of the person of, a contracting party. If a contract is made under duress it is at once suspect, because consent has not been freely given to the bargain supposedly made. The contract is voidable at the option of the party concerned. Duress is a common law doctrine which relates entirely to the person and has no relation to that person's goods. As such it is a very limited doctrine and is one where cases are rare.

UNENFORCEABLE CONTRACTS

Contracts may be described as void, or voidable or unenforceable. A void contract creates no legal rights and cannot therefore be sued upon. It may occur because of a mistake as to the nature of the contract, or because it involves the performance of something illegal that is prohibited by a statute. A void contract will also result because of the incapacity of the parties, as in the case of infants. Corporations cannot make contracts beyond their stated powers which are said to be *ultra vires* 'beyond one's powers'.

A contract is said to be voidable when only one of the parties may take advantage. In cases involving misrepresentation, only the party who has been misled has the right to a void in one of the ways previously described.

Unenforceable contracts are those that are valid, but owing to the neglect of the formalities involved, a party seeking to enforce it will be denied a remedy.

MISTAKE

The law recognizes that, in some circumstances, although a contract has been formed, one or both of the parties are unable to enforce the agreement. The parties are at variance with one another and this precludes the possibility of any agreement. Mistake may be classified as follows:

- *Identity of subject matter*: this occurs where one party intends to contract with regard to one thing and the other party with regard to another; the parties in this situation cannot be of the same mind and no contract is formed.
- *Identity of party*: if the identity of either party enters into consideration, this will negate the contract.
- *Basis of contract*: if two parties enter into a contract on the basis that certain facts exist, and they do not, then the contract is void.
- *Expressing the contract*: if a written contract fails to express the agreed intentions of the two parties, then it is not enforceable; courts may, however, express the true intention of the parties and enforce it as amended.

MISREPRESENTATION

Misrepresentation consists in the making of an untrue statement which induces the other party to enter into a contract. The statement must relate to fact rather than opinion. Furthermore, the injured party must have relied on the statement and it must have been a material cause of their entering into the contract. Where such a contract is voidable it may be renounced by the injured party, but until such time it is valid. Misrepresentation may be classified as:

- *Innocent misrepresentation*: an untrue statement is made in the belief that it is true.
- *Fraudulent misrepresentation*: an untrue statement is made with the knowledge that it is untrue or made recklessly without attempting to assess its validity.
- *Negligent misrepresentation*: a statement is made honestly but without reasonable grounds for belief that it is true. It is really a special case of innocent misrepresentation, for although the statement is made in the belief that it is true, insufficient care has been taken to check it.

Where misrepresentation occurs the injured party has several options:

- The injured party can affirm the contract, when it will then continue for both parties.
- The injured party can repudiate the contract and set up misrepresentation as a defence.
- An action can be brought for rescission and restitution:
  - Rescission involves cancelling the contract and the restoration of the parties to the state that they were in before the contract was made.
  - Restitution is the return of any money paid or transferred under the terms of the contract.
- The injured party can bring an action for damages. The claim for damages is only possible in circumstances of fraudulent misrepresentation.

DISCLOSURE OF INFORMATION

When entering into a contract it is not always necessary to disclose all the facts that are available. A party may observe silence in regard to certain facts, even though it may know that such facts would influence the other party. This is summed up in the maxim *caveat emptor* 'let the buyer beware'.

There are, however, circumstances where the non-disclosure of relevant information may affect the validity of the contract. This can occur where the relevant facts surrounding the contract are almost entirely within the knowledge of one of the parties, and the other has no means of discerning the facts. These contracts are said to be *uberrimae fidei* 'of the utmost good faith'.

PRIVITY OF CONTRACT

A contract creates something special for the parties who enter into it. The common law rule of privity is that only the persons who are party to the contract can be affected by it. A contract can neither impose obligations nor confer rights upon others who are not privy to it. For example, the clause in the standard form of building contract which allows the employer to pay money direct to the subcontractor may be used by the employer. It cannot be enforced by the subcontractor, who is not a party to the main contract. The subcontractor may seek to persuade the employer to adopt this course of action but cannot enforce the employer to do so in a court of law.

EXPRESS AND IMPLIED TERMS

The terms of contract can be classified as either express or implied. Those terms which are written into the contract documents or expressed orally by the parties are described as express terms. Those terms which were not mentioned by the parties at the time that they made the contract are implied terms, so long as they were in the minds of both parties. The courts will, where it becomes necessary, imply into building and engineering contracts a number of implied terms. Although the implied term is one which the parties probably never contemplated when making the contract, the courts justify this by saying that the implication is necessary in order to give business efficacy to the contract. This does not mean, however, that the courts will make a contract more workable or sensible. The courts will generally imply into a building contract the following terms:

- The contractor will be given possession of the site within a reasonable time, should nothing be stated in the contract documentation
- The employer will not unreasonably prevent the contractor from completing the work
- The contractor will carry out the work in a workmanlike manner

Implied terms which have evolved from decided cases often act as precedents for future events. In the majority of the standard forms of building contract all of these matters are normally express terms, since the contracts themselves are very comprehensive and hope to cover every eventuality.

There are some notable terms which are not normally to be implied into building and engineering contracts, such as the practicability of the design. However, there will be express or implied terms that the work will comply with the appropriate statutes.

Express terms

An express term is a clear stipulation in the contract which the parties intend should be binding upon them. Traditionally, the common law has divided terms into two categories of conditions and warranties:

- *Conditions* are terms which go to the root of the contract, and for breach of which the remedies of repudiation or rescission of the contract and damages are allowed
- *Warranties* are minor terms of the contract, for breach of which the only remedy is damages

Implied terms

Implied terms are those which, although not expressly stated by the parties by words or conduct, are by law deemed to be part of the contract. Terms may be implied into contracts by custom, statute or the courts:

- *By custom*: in law this means an established practice or usage in a trade, locality, type of transaction or between parties. If two or more people enter into a contract against a common background of business, it is considered that they intend the trade usage of that business to prevail unless they expressly exclude it.

- *By statute*: there are many areas in the civil law where Parliament has interfered with the right of parties to regulate their own affairs. This interference mainly occurs where one party has used a dominant bargaining position to abuse this freedom. Thus in the sale of goods, the general principle *caveat emptor* 'let the buyer beware' has been greatly modified, particularly in favour of the consumer by the provisions in the Sale of Goods Act 1979. In addition there have been important changes in relation to exemption clauses brought about by the Unfair Contract Terms Act 1977.
- *By the courts*: the court will imply a term into a contract under the doctrine of the implied term, if it was the presumed intention of the parties that there should have been a particular term, but they have omitted to expressly state it.

LIMITATIONS OF ACTIONS

Generally speaking, litigation is a costly and time-consuming process which becomes more difficult as the time between the disputed events and the litigation increases. Also rights of action cannot be allowed to endure forever. Parties to a contract must be made to prosecute their causes within a reasonable time. For this reason Parliament has enacted limitation acts which set a time limit on the commencement of litigation. The rules and procedures in respect of limitation of actions are contained in the Limitation Act 1980. The right to bring an action can be discharged in three ways:

- The parties to a contract might decide to discharge their rights
- Through the judgment of a court
- Through lapse of time

If an action is not commenced within a certain time then the right to sue is extinguished. Actions in a simple contract (not under seal) and tort become statute barred after six years. In the case of contracts under seal, this period is extended to twelve years. The Act does permit certain extensions to these time limits in very special circumstances, for example, where a person may be unconscious as the result of an accident. If damage is suffered at a later date then this will not affect the limitation period, although an action for negligence may be pursued. The period of limitation can be renewed if the debtor acknowledges the claim or makes a part payment at some time during this period. A number of international conventions, particularly with respect to the law of carriage, lay down shorter limitation periods for action than those specified in the Act. For example, in carriage by air under Warsaw Rules the period is two years.

THE UNFAIR CONTRACT TERMS ACT 1977

Contractual clauses designed to exonerate a party wholly or partly from liability from breaches of express or implied terms, first appeared in the nineteenth century.

The common law did not interfere but took the view that parties forming a contractual relationship were free to make a bargain within the limits of the law. This is still largely the rule today, although the growth of large trading organizations has led to an increase in both excluding and restricting clauses to the severe detriment of other parties.

The efforts of the courts to mitigate against the worst effects of objectionable exclusion clauses have been reinforced by the Unfair Contract Terms Act 1977. This Act restricts the extent to which liability can be avoided for breach of contract and negligence. The Act relates only to business liability, so transactions between private individuals are not covered. The reasonableness is further extended to situations where parties attempt to exclude liability for a fundamental breach, i.e. where performance is substantially different from that reasonably expected or there is no performance at all of the whole or any part of the contractual obligations. The following are subject to the Act's provisions:

- Making liability or its enforcement subject to restrictive or onerous conditions
- Excluding or restricting any right or remedy
- Excluding or restricting rules of evidence or procedure
- Evasion of the provisions of the Act by a secondary contract is prohibited

CONTRA PROFERENTEM

Any ambiguity in a clause in a contract will be interpreted against the party who put it forward. It is a general rule of construction of any document that it will be interpreted *contra proferentem*, that is, against the person who prepared the document. As an exclusion clause is invariably drafted by the imposer of it, this is an extremely useful weapon against exclusion clauses. The effect of the rule is to give the party who proposed the ambiguous clause only the lesser of the protections possible. The one who draws up the contract has the choice of words and must choose them to show clearly the intention.

AGENCY

Agency is a special relationship whereby one person (the agent) agrees on behalf of another (the principal) to conclude a contract between the principal and a third party. Providing that the agent acts only within the scope of the authority conferred upon them, those acts become the acts of the principal, and the principal must therefore accept the responsibility for them. The majority of contractual relationships involve some form of agency. For example, the architect in ordering extra work is acting in the capacity of the employer's agent. The contractor may presume, unless anything is known to the contrary, that the carrying out of these extras will result in future payment by the client.

A contract of agency may be established by:

- *Express authority*: authority that has been directly given to an agent by his or her principal
- *Implied authority*: because a person is engaged in a particular capacity, others dealing with that person are entitled, perhaps because of trade custom, to infer that that person has the necessary authority to contract within the limits usually associated with that capacity
- *Ratification*: a principal subsequently accepts an act done by the principal's agent, even where this exceeds the agent's authority. This becomes as effective as if the principal originally authorized it.

In contracts of agency a principal cannot delegate to an agent powers which the principal does not already possess. The capacity of the agent is therefore determined by the capacity of the agent's principal.

CHAPTER 3

DISCHARGE OF CONTRACTS

A contract is said to be discharged when the parties become released from their general contractual obligations. The discharge of a contract may be brought about in several ways.

DISCHARGE BY PERFORMANCE

In these circumstances the party has undertaken to do a certain task and nothing further remains to be done. In general, only the complete and exact performance of the contractual obligations can discharge the contract. In practice, where a contract has been substantially performed, payment can be made with an adjustment for the work that is incomplete.

A building contract is discharged by performance once the contractor has completed all the work, including the making good of defects under the terms of the contract. The architect must have issued all the appropriate certificates and the building owner paid the requisite sums. If, however, undisclosed defects occur beyond this period, the building owner can still sue for damages under the statute of limitations. The contract has not been properly performed if there are hidden defects.

Thus, A undertakes to sell to B 1,000 roof tiles. A will be discharged from the contract when they have delivered the tiles, and B when they have paid the price. A question sometimes arises whether performance by another party will discharge the contract. The general rule is that where personal qualifications are a factor of consideration, then that person must perform the contract. Where, however, personal considerations are unimportant it would not matter who supplied the goods.

DISCHARGE UNDER CONDITION

Contracts consist of a large number of stipulations or terms. In many types of contract there are conditions which, although not expressed, are intended because the parties must have contracted with these conditions in mind. It will, however, be obvious that the terms and conditions of a building contract are not of equal importance. Some of the terms are fundamental to the contract, and are so essential that if they are broken the whole purpose of the contract is defeated.

Thus, if a contractor agreed to design and build a building for a specific use, and it is incapable of such a use, the building owner would be able to reject the project and recover the costs from the contractor.

Building contracts frequently contain a number of terms forming a specification. The builder agrees to construct the project in accordance with this specification. If the builder deviates from this in some small way then this will give rise to an action for damages and the contract will not be discharged.

It is, however, well established that subject to an express or implied agreement to the contrary, a party who has received substantial benefit under a contract cannot repudiate it for breach of condition.

DISCHARGE BY RENUNCIATION

This is effected when one of the parties refuses to perform obligations. Thus, A employs an architect to design and supervise a proposed building project. On completion of the design, A decides not to continue with the project. This is renunciation of the contract and the architect can sue for fees on a *quantum meruit* basis – for as much as has been earned.

DISCHARGE BY FRESH AGREEMENT

A contract can be discharged by a fresh agreement being made between the parties, which is both subsequent to and independent of the original contract. Such a contract may, however, discharge the parties altogether. This is known as a rescission of the original contract. Where one party to a contract is released by a third party from undertaking obligations, then the original contract discharged is termed 'novation'. Any alteration to a contract made with the consent of the parties concerned has the effect of making a new contract.

FRUSTRATION

A contract formed between two parties will expressly or impliedly be subject to the condition that it will be capable of performance. If the contract becomes incapable of performance then the parties will be discharged from it. Impossibility of performance is usually called frustration of contract. It occurs whenever the law recognizes that, without default of either party, a contractual obligation has become incapable of execution.

For example, the event causing the impossibility may be due to a natural catastrophe. A agrees to carry out a contract for the repair of a road surface, but prior to starting the work the road subsides to such an extent that it disappears completely down the side of a hill. Impossibility may also occur because a government may introduce a law that makes the contract illegal. For example, A contracts to carry out some insulation work using asbestos. The government subsequently introduce legislation forbidding the use of this material in buildings, thus making the contract impossible to carry out. A contractor cannot, however, claim that a contract is frustrated if by deliberate actions the contractor causes a delay to avoid completion. For example, a contract for the building of a sea wall must be completed prior to winter weather setting in, otherwise practical performance will become impossible. The knowledge that this predictable event will occur, coupled with a deliberate delay, does not result in a frustration of the contract.

Examples of building contracts being frustrated are extremely rare. Note that hardship, inconvenience or loss do not decide whether an occurrence frustrates the contract, because these are accepted risks. The legal effect of frustration is that the contract is discharged and money prepaid in anticipation of performance should be returned. A party receiving benefit from a partly executed contract should reimburse the other party to the value of that benefit.

DETERMINATION OF CONTRACT

There are provisions in all the common forms of contract that allow either party to terminate the contract. There must, of course, be good reasons to support any party who decides to determine, otherwise a breach of contract can occur. Clause 27 of the standard form of building contract provides circumstances which allow the employer to terminate the contract with the contractor. In one sense they are fairly exceptional happenings, and so they should be, since they provide a final option to the employer. The decision to determine must be taken very carefully and reasonably. It generally occurs because the contractor is failing to take notice of instructions from the architect to remedy an already existing breach of contract. Clause 28 provides for determination because of either a refusal or interference with payments due to the contractor. It may also occur where the works are suspended for an unreasonable length of time. Again the action to determine must be carefully taken, and is a last-resort decision on the part of the contractor. The contractor takes this decision when it is felt it is no longer possible to continue, perhaps after a number of delays in payment, to work with the employer. Determination of contract by either party results in two losers and no winners. Although there are provisions in the contract for financial recompense to the aggrieved party, these rarely suffice.

ASSIGNMENT

It sometimes happens that one party to a contract wishes to dispose of its obligations under it, but the extent to which this is permitted is limited. The other party may have valid reasons why it prefers the obligations to be performed by the original contractor. The rule is that liabilities can only be assigned by novation. This is the formation of a new contract between the party who wishes performance and the new contractor, who is accepted as adequately qualified to perform as the original contractor. Liabilities can only be assigned by consent. By contrast, rights under a contract can usually be assigned without consent of the other party, except where the subject matter involves a personal service. Even, however, in those circumstances where the contract does not involve personal service, but specifically restricts the right to assign the contract or any interest in it to a third party then the assignment of rights will not be permitted.

CHAPTER 4

REMEDIES FOR BREACH OF CONTRACT

A breach of contract occurs when one party fails to perform an obligation under the terms of the contract. For example, a breach of contract of the JCT 98 conditions of contract occurs if:

- The contractor refuses to obey an architect's instruction (clause 4).
- The employer fails to honour an architect's certificate (clause 28).
- The contractor refuses to hand over certain antiquities found on the site (clause 34).
- The contractor fails to proceed regularly and diligently with the works (clause 23).

Defective work is not necessarily in breach of contract, as long as the contractor rectifies it in accordance with an architect's instruction. A breach will occur where the contractor either refuses to remove the defective work or ignores the remedial work required.

A breach of contract may have two principal consequences. The first is damages. Then if the breach is sufficiently serious, there may be determination by the aggrieved party under the terms of the contract. The aggrieved party may decide to terminate the contract with the other, and also sue for damages, or alternatively take only one of these courses of action. Damages may include both the loss resulting from the breach, the loss flowing from the termination and the additional costs of completing the contract.

There is, with few exceptions, the rule that no one who is not a party to the contract can sue or be sued in respect of it. There may be other remedies, for instance in tort, but these are beyond the scope of this book.

The legal remedy for breach of contract is damages. This consists of the award of a sum of money to the injured party, designed to compensate for the loss sustained. The basis of the award of damages is by way of compensation. The damages awarded are made in an attempt to recompense the actual loss sustained by reason of the breach of contract. The injured party is to be placed, as far as money can do it, in the same situation as if the contract had been performed. It should be clearly understood that not every breach automatically results in damages. In order for damages to be paid, the injured party must be able to prove that a loss resulted from the breach. Furthermore, the innocent party must take all reasonable steps to mitigate any loss. Damages may be classified as follows.

NOMINAL DAMAGES

Where a party can show a breach of contract, but cannot prove any sustained loss as a result, then nominal damages may be awarded. These comprise merely a small sum in recognition that a contractual right has been infringed.

SUBSTANTIAL DAMAGES

Substantial damages represent the measure of loss sustained by the injured party. Despite their name they might be quite small.

REMOTENESS OF DAMAGE

A breach of contract can, in some circumstances, create a chain of events resulting in considerable damage, and the question may arise whether the injured party or parties can claim for the whole of the damage sustained. In ordinary circumstances the only damages that can be claimed are those which arise immediately because of the breach. Damage is not considered too remote if the parties at the time the contract was entered into contemplated that it could occur. If the contract specifies the extent of the liability, then no question of consequential damages arises. Defects liability clauses, which require the builder to rectify defects, do not limit the extent of the remedial work, but may also require other work to be put right that has resulted from these defects.

SPECIAL DAMAGE

Damages resulting from special circumstances are recoverable if they flow from a breach of contract, and if the special circumstances were known to both parties at the time of making the agreement.

LIQUIDATED DAMAGES

Each of the standard forms of contract provide for payment of agreed damages by the contractor when completion of work is not within the stipulated time. These payments are known as liquidated damages, and their amount should be recorded on the appendix to the form of a contract. The sum stated should be a genuine estimate of the damage that the building owner may suffer. If, however, the sum stated is excessive and bears no relation to the actual damage, then it may be regarded as a penalty. In these circumstances the courts will not enforce the amount stated in the contract but will assess the damage incurred on an unliquidated basis. Liquidated damages may be estimated on the basis of loss of profit in the case of commercial projects, but they are much more difficult to determine in the case of public works projects such as roads.

In some circumstances a building owner may seek the occupation of a project even though it remains incomplete. This would normally deprive the owner of the right to enforce a claim for liquidated damages.

UNLIQUIDATED DAMAGES

When no liquidated damages have been detailed in the contract, the employer can still recover damages should the contractor fail to complete on time. The sum awarded in this case is that regarded as compensation for the loss actually sustained by the breach.

SPECIFIC PERFORMANCE

Specific performance was introduced by the courts for use in those cases where damages would not be an adequate remedy. In building contracts this remedy is only really available in exceptional circumstances. The courts will enforce a party to do what it has contracted to do, in preference to awarding damages to the aggrieved party. A decree of specific performance will not be granted where the court cannot effectively supervise or enforce the performance.

INJUNCTION

Injunction is a remedy for the enforcement of a negative undertaking. A party is prohibited from carrying out a certain action. It is often awarded by the courts as an effective remedy against nuisance.

RESCISSION

This is an equitable remedy, which endeavours to place the parties in the pre-contractual position by returning goods or money to the original owners. This means that the parties are no longer bound by the contract. This is granted at the discretion of the court but will not be awarded where:

- The injured party was aware of the misrepresentation and carried on with the contract.

- The parties cannot be returned to their original position.
- Another party has acquired an interest in the goods.
- The injured party waited too long before claiming this remedy.

QUANTUM MERUIT

A claim for damages is a claim for compensation for loss. Where, under the terms of the contract, one party undertakes its duty for the other, and the other party breaks the contract, the former can sue upon a *quantum meruit* basis; that is, to claim a reasonable price for the work carried out. It is payment earned for work carried out. If the two parties cannot agree, the question of what sum is reasonable is decided by the court. A claim on a *quantum meruit* basis is appropriate where there is an express agreement to pay a reasonable sum upon the completion of some work. In assessing a *quantum meruit* claim, the parties may choose to use the various means that are available. For example, it may be based upon the costs of labour and materials plus profit, or the measurement of the work using reasonable rates.

CHAPTER 5

SETTLEMENT OF DISPUTES

And God said unto Noah, the Ark shall be finished within seven days. And Noah saith, it shall be so. But it was not so. And the Lord saith, what seemeth to be the trouble this time? And Noah saith, mine subcontractor hath gone bankrupt. The pitch which thou commandest me to use has not arrived. The plumber hath gone on strike. And the glazier departeth on holiday to Majorca, even tho I did offer him double time.

Christopher Taylor, Engineering Manager, Shell UK

Construction contracts in the distant past consisted of a document of about five pages long. They generally concluded with a handshake, but underlying such agreements were an essential set of values of competence, fairness and honesty. Today things are different. We have developed a complex and onerous set of conditions that attempt to cover every eventuality and in so doing create loopholes that the legal professions can feast upon. It has often been suggested that the only individuals to make any money in the construction industry are the lawyers! Precious time and resources are thus drawn away from the main purpose of the industry: getting the project built on time, to the right design and use of construction technology at an agreed price and quality.

Everyone in the industry agrees that construction contracts need to become less adversarial and more simply constructed, emphasizing the positive needs of the project. Consultants and contractors should be allowed and encouraged to use their best endeavours and to work together as a team (Chapter 11) rather than watching their individual backs all of the time. When a project ends up in a protracted dispute, the project will fail to meet its original goals and expectations. In addition, clients will suffer from high legal fees, delayed completion and occupation and general dissatisfaction. The contractor's profits will diminish and to these will be added additional legal fees. There are no winners under these circumstances.

THE REASONS WHY DISPUTES ARISE

The construction industry is a risky business. It generally does not build many prototypes, with each different project being individual in many respects. Even identical buildings that have been constructed on different sites create their own special circumstances, are subject to the vagaries of different site and weather conditions and use labour that may have different trade practices even from one site to another. Even the identical building constructed on an adjacent site by a different contractor will have different costs and different problems associated with its construction. The introduction of new building materials and designs, changes to the procurement and organization of the project and the poor margins of profitability provide a good platform for disputes. Disputes are therefore likely to arise under the best circumstances, even where every possibility has been potentially eliminated. Disputes between parties, it should be remembered, are really in no one's best interests. Here are some of the main areas where disputes might occur:

General

- Adversarial nature of construction contracts
- Poor communication between the parties concerned
- Proliferation of forms of contract and warranties
- Fragmentation in the industry
- Tendering policies and procedures

Clients

- Poor briefing
- Changes and variation requirements
- Changes to standard conditions of contract
- Interference in the contractual duties of the contract administrator
- Late payments

Consultants

- Design inadequacies
- Lack of appropriate competence and experience
- Late and incomplete information
- Lack of coordination
- Unclear delegation of responsibilities

Contractors

- Inadequate site management
- Poor planning and programming
- Poor standards of work
- Disputes with subcontractors
- Delayed payments to subcontractors
- Coordination of subcontractors

Subcontractors

- Mismatch of subcontract conditions with main contract
- Failure to follow and adopt agreed procedures
- Poor standards of work

Manufacturers and suppliers

- Failure to define performance or purpose
- Failure of performance

ISSUES FOR THE RESOLUTION OF DISPUTES

The following matters need to be resolved in order to reduce the possibility of future disputes occurring:

- Clarification of responsibilities
- Need of single-point responsibility contracts
- Allocation of risk to the parties who are best able to control it
- Further investigation of insurance-based alternatives
- Need to develop and extend non-adversarial methods of dispute resolution
- Partnership sourcing (contractors and consultants working in a consortium)
- Quality management and quality assurance

CLAIMS

It is evident from society in general that as individuals we are becoming more claims conscious. Firms of lawyers are now touting their services, often on a no-win no-fee basis. Everyone wants their pound of flesh and what they rightfully believe belongs to them. Claims are seen by many to be a last resort issue. Even in the construction industry this is true, although some would want to argue that some contractors prepare their claims alongside their tender submissions.

Contractual claims arise where contractors assess that they are entitled to additional payments over and above that paid within the general terms and conditions of the contract. For example, the contractors may seek reimbursement for some alleged loss that has been suffered for reasons beyond their control. On many occasions the costs incurred lie where they fall and contractors will have recourse to recover them. Thus losses and delays arising from the intervention of third parties who are unconnected with the contract almost invariably fall with the contractor. The fact that a loss has been sustained, without fault on the part of the loser, may merit sympathy, but does not in itself demand compensation. Where a standard form of contract is used, many attempts may be made by contractors to invoke some of the compensatory provisions of the contract in order to secure further payment to cover the losses involved.

The details of such claims will be investigated by the quantity surveyors and a report made to the architect, engineer or other lead consultant. The report should summarize the arguments involved and set out the possible financial effect of each claim. Quantity surveyors frequently end up negotiating with contractors over such issues in an attempt to solve the financial problems and to arrive at an amicable solution, wherever this is possible. This is preferable to a lengthy legal dispute.

As with many issues in life, contractual claims are rarely the fault of one side only. If the claim cannot be resolved in this way then some form of legal proceedings may be initiated. Particular care therefore needs to be properly exercised in the conduct of the negotiations since they may have an effect upon the outcome of any subsequent legal proceedings. Claims may be classified in several different ways. They usually reflect a loss and expense to a contractor.

Contractual claims

Contractual claims have a direct reference to conditions of contract. When the contract is signed by the two parties, the contractor and the employer, there is a formal agreement to carry out and complete the works in accordance with the information supplied through the drawings, specification and contract bills. Where the works constructed are of a different character or executed under different conditions then it is obvious that different costs will be involved. Some of these additional costs may be recouped under the terms of the contract, through, for example, remeasurement and revaluation of the works, using the appropriate rules from the contract. Other additional costs that an experienced contractor had not allowed for within the tender may need to be recovered in a different way. This is usually under the heading of a contractual claim.

Ex gratia payments

Ex gratia payments are not based upon the terms or conditions of contract. However, the carrying out of the works has nevertheless resulted in some loss and expense to the contractor. The contractor has completed the project on time, to the required standards and conditions and at the price agreed. Perhaps, due to a variety of different reasons, and at no fault of the contractor, a loss has been sustained that cannot be related to the contractual conditions. On rare occasions a sympathetic employer may be prepared to make a discretionary payment to the contractor. Such payments are made out of grace and kindness. They may be made because of a long-standing relationship and trust between employer and contractor, or because of outstanding service and satisfaction provided by the contractor. Nevertheless, they are rare occasions.

JCT 98

The standard form of building contract (JCT 98) seeks to clarify the contractual relationship between the employer and the contractor. As far as possible, ambiguities have been removed, but some nevertheless remain. If such forms or conditions of contract were not available, then the uncertainty between the two parties would be even greater. This could have the likely effect of increasing tender totals. Under the present conditions of contract, the contractual risks involved are shared between the employer and the contractor. Claims may arise most commonly under clause 26, and these are known as loss and expense claims. They may also arise due to a breach of contract. The contractor must make a written application to the architect, in the first place, stating that a direct loss and expense has occurred or is likely to occur in the execution of the project. The contractor must further state that any reimbursement under the terms of the contract is unlikely to be sufficient (clause 26.1). This information should be given to the architect promptly in order to allow as much time as possible to plan for other contingencies.

As soon as reasonably possible, the contractor should provide the architect with a written interim account providing full details of the particular claim and the basis upon which it is made. This should be amended and updated when necessary or when required. If the contractor fails to comply with this procedure, this might prejudice the investigation of the claim by the architect and any subsequent payments by the employer to the contractor.

The contractor is entitled to have such amounts included in the payment of interim certificates under clause 30.2.2. However, in practice a large majority of claims are not agreed until the completion of the contract. In these circumstances the contractor is entitled to receive part of the claim included in an interim certificate where this can be substantiated.

Contractors

Many contractors have well-organized systems for dealing with claims on construction projects and the recovery of monies that are rightly due under the terms of the contract. They are likely to maintain good records of most events, but particularly those where difficulties have occurred in the execution of the work. However, some of the difficulties may be due to the manner in which the contractor has sought to carry out the work and thus remain the entire responsibility of the contractor.

Claims that are notified or submitted late will inevitably create problems in their approval. In these circumstances the architect might not have the opportunity to check the details of the contractor's submission. Such occurrences will not be favourably looked upon by the architect or the employer.

The contractor must prepare a report on why a particular aspect of the work has cost more than expected, substantiate this with appropriate calculations and support it with reference to architect's instructions, drawings, details, specifications, letters, etc. The contractor must also be able to show that as an experienced contractor they could not have foreseen the difficulties that occurred. They will also need to show that the work was carried out in an efficient, effective and economic manner.

Claims are for additional payments that cannot be recouped in the normal way simply through measurement and valuation. They are based on the assumption that the works constructed differed considerably from the works for which the contractor originally submitted a tender. The differences may have changed the contractor's preferred method of working and this in turn may have altered or influenced the costs involved. The rates inserted by the contractor in the contract bills are not now a true reflection of the work that has been executed.

Example

The construction of a major new factory on a greenfield site requires a large earth-moving contract. The quantities of excavation and its subsequent disposal have been included in the contract bills and priced by the contractor. It is found that the quantities of materials to be taken to tips has increased by 25 per cent by volume. This is due to variations to the contract and the unforeseen nature of some of the ground conditions of the construction site.

The contractor's pre-tender report indicates that a variety of tips at different distances from the site will be used for the disposal of the excavated materials. The contractor's tendering notes indicated that the tips nearer the site would be filled first. In this case they result in lower haulage costs, but can also be shown to have lower tipping charges. The disposal of the excavated material therefore includes two separate elements:

- The haul charges to the tip
- The costs of the tip

In the contract bills, the rate used by the contractor for disposal of excavated materials represents an average rate. This is based upon calculated average haul distances and average tipping charges based upon the weighted quantities in each tip. A revised average rate for the disposal of the excavated materials can be calculated similarly for the actual quantities of excavated materials that are involved. The data is shown in Box 5.1.

To simply continue to apply the contract bill rates in similar situations to that shown in Box 5.1 is erroneous. The rates no longer reflect the work to be carried out and the contractor's method of working, which have been changed by the variation to the contract. Other factors that might also need to be considered are the excavation of different types of construction materials that have been encountered, whether they bulk at different rates or whether they are more difficult to handle.

The increase in the amount of excavated materials, on this scale, may also have other repercussions such as an extension of time which might also need to be considered. The method of carrying out the works might also now be different from that originally envisaged by the contractor. Different types of mechanical excavators may be required or the plant originally selected to do the work might no longer be the most appropriate. This is especially so where cut and fill excavations are considered. The contractor may also be involved in hiring additional plant at higher charges and employing workpeople at overtime rates, in order to keep the project on schedule.

Box 5.1 Earthmoving claim

Contract bills

Excavated materials for disposal in the contract bills = 1,000,000 m³.

Contract bill rate is based upon:

Tip A	500,000 m³	distance = 1 km	tip charge = £0.10/m³
Tip B	300,000 m³	distance = 2 km	tip charge = £0.20/m³
Tip C	200,000 m³	distance = 4 km	tip charge = £0.30/m³

$$\text{Average distance} \;\; = \frac{500{,}000 \,+\, 600{,}000 \,+\, 800{,}000}{1{,}000{,}000} = 1.9 \, \text{km/m}^3$$

$$\text{Average tip charge} = \frac{50{,}000 \,+\, 60{,}000 \,+\, 60{,}000}{1{,}000{,}000} = £0.17/\text{m}^3$$

Final account

Actual quantities in the final account:

Tip A	500,000 m³	distance = 1 km	tip charge = £0.10/m³
Tip B	300,000 m³	distance = 2 km	tip charge = £0.20/m³
Tip C	300,000 m³	distance = 4 km	tip charge = £0.30/m³
Tip D	150,000 m³	distance = 6 km	tip charge = £0.50/m³

$$\text{Average distance} \;\; = \frac{500{,}000 \,+\, 600{,}000 \,+\, 1{,}200{,}000 \,+\, 900{,}000}{1{,}250{,}000} = 2.56 \, \text{km/m}^3$$

$$\text{Average tip charge} = \frac{50{,}000 \,+\, 60{,}000 \,+\, 90{,}000 \,+\, 75{,}000}{1{,}250{,}000} = £0.22/\text{m}^3$$

The contractor must, as a matter of good practice always put in writing:

- Applications for instructions, drawings, etc.
- Application for the nomination of subcontractors
- Progress of the works and any delays
- Notification of any claims under the contract in respect of:
  - variations
  - extensions of time
  - loss and expense
- Confirmation of any oral instructions from the architect

The contractor should also ensure that any certificates that are required under the terms of the contract are issued at the appropriate time. These may have some effect upon the validity or otherwise of a contractor's claim at a later stage.

ADJUDICATION

The dispute resolution techniques that are described in JCT 98 include adjudication, arbitration and litigation. The courts may also need to be called upon to enforce settlements that are reached by other methods. The parties may also decide to agree amongst themselves to use other alternative methods to settle their differences, such as alternative dispute resolution. This is considered towards the end of this chapter. It is claimed to be a non-adversarial technique, although its rise in popularity in the construction industry appears now to have waned in favour of more established techniques.

Adjudication was first introduced into the UK in the mid 1970s. Its application was restricted to disputes that occurred between the main contractor and the directly employed or domestic subcontractors. The process involved using an independent third party, an adjudicator, to help resolve a dispute that had arisen. The adjudicator could be appointed as part of the subcontract conditions, but invariably was only appointed after the dispute had occurred. The main advantage of using an adjudicator was the rapid response of the decision. The decision was binding, although as in all disagreements the parties had the right to take the dispute to a higher authority. Adjudication was subsequently introduced into the JCT form with contractor's design and more recently into JCT 98. This followed one of the principles of better practice recommended by the Latham Report.

Adjudication is described in clause 41A of JCT 98 and this clause is discussed in Chapter 25. The referral of a dispute to an adjudicator must be made within 7 days (clause 41A2) and a decision must be given to the parties concerned, in writing, within 28 days (clause 41A5). The period for the decision can be extended for a further 14 days if the parties to the dispute agree. The adjudicator, like anyone in hearing disputes, must act impartially to determine the facts and law that are applicable to the dispute.

The appendix to the form of contract seeks to identify who should nominate the adjudicator. Unless the parties agree to the contrary, this shall be the president or vice-president of the Royal Institute of British Architects. As an alternative the adjudicator may be a president or vice-president of the RICS, the Construction Confederation or the National Specialist Contractors Council. The adjudicator can be named in the contract in order to save time should a dispute occur. Whilst this is useful, the appointed adjudicator might not be suitable to review all disputes. The *JCT adjudication agreement* is a standard form that has been produced by JCT.

Within 7 days of the notice to refer a matter to adjudication, a referral document should be provided that includes the particulars of the dispute, a summary of the issues involved and the remedy and relief that the adjudicator should consider. The powers of the adjudicator are described in clause 41A5.5 as follows:

- Using the adjudicator's own knowledge and expertise
- Opening up, reviewing and revising certificates, opinions, decisions or notices
- Requiring the parties to provide additional information
- Requiring the parties to carry out tests or open up work
- Visiting the site and workshops
- Obtaining information from the employees of the parties concerned
- Obtaining information from other third parties
- Determining the payment of any interest within the terms of the contract

The parties are normally responsible for their own costs, but the adjudicator may direct that, in fairness, the unsuccessful party can recover their costs from the successful party. There are several identified advantages of using adjudication in preference to other methods of settling disputes:

- It seeks to eliminate conflicts as quickly as possible by resolving disputes as they arise.
- It is recognized that the adjudicator's decision will be provided in the fastest possible way.
- It is intended to be the least expensive process for settling disputes, by reducing lawyers' charges.
- It can act as a referral system, preceding arbitration or litigation where the dispute is not resolved at this stage.
- It is anticipated that many disputes will be resolved and terminated at this stage, rather than proceeding towards more litigious action

However, in practice the complexity that sometimes occurs with construction disputes cannot be dealt with effectively, and more rigorous methods of settlement may need to be employed. Sometimes adjudication may simply be considered as a temporary solution to a problem.

ARBITRATION

Arbitration is the main alternative to legal action in the courts, in order to settle an unresolved dispute. No one is compelled to submit a dispute to arbitration unless they have agreed to do so within the terms of the contract. Care should be taken, therefore, when considering the completion of the appendix in respect of clause 41. Once a person has agreed to this method of settling a disagreement, they cannot then take legal proceedings prior to arbitration. If they attempt to do so, the courts will stay such proceedings. All of the standard forms of contract used in the construction industry include an arbitration provision. It is therefore the procedure that is most commonly used for dealing with disputes that arise between the various parties concerned.

Arbitration is described as a private procedure for settling a wide range of disputes in a diverse range of industries. The dispute is settled by an impartial individual or panel of arbitrators appointed solely for that purpose. The decision of the arbitrator is described as an award. Arbitrators do not need to have any particular skills or qualifications. In the majority of cases, an individual is appointed to serve as an arbitrator on the basis of experience or expertise in the subject of the dispute. However, most practising arbitrators are fellows of the Chartered Institute of Arbitrators (FCIArb). There is a preference in the construction industry for arbitrators who have qualifications in construction and law.

The arbitration agreement in the JCT 98 is covered in article 7A of the articles of agreement. The parties, under article 5.2, agree that proceedings will not take place until after practical completion has been achieved, or unless the contractor's termination of the contract has been made or if the project is abandoned. There are exceptions, however, to these rules where matters can be taken to arbitration during the progress of the works. Of course, where the two parties otherwise agree, the guidelines can be amended as required. The following matters can therefore be dealt with prior to the issue of the certificate of practical completion:

- *Article 3*: contractor's objection to the appointment following the death of the architect
- *Article 4*: contractor's objection to the appointment following the death of the quantity surveyor
- *Clause 4*: dispute over the power to issue an instruction
- *Clause 30*: a certificate being improperly withheld or not being in accordance with the conditions
- *Clause 25*: dispute over the difference of an extension of time
- *Clauses 32/33*: disputes concerning outbreak of hostilities or war damage

An arbitrator's powers are very wide. They may review and revise certificates and valuations. They may also disregard opinions, decisions or notices that have already been given.

Here are the essential features of a valid arbitration agreement:

- The parties must be capable of entering into a legally binding contract.
- The agreement should whenever possible be in writing.
- It must be signed by the parties concerned.
- It must state clearly those matters which will be submitted to arbitration, and when the proceedings will be initiated.
- It must not contain anything that is illegal.

Arbitration Act 1996

The Arbitration Act 1996 extended the previous Act of 1979. It came into force in 1996 and it applies to England, Wales and Northern Ireland. Scotland is excluded because different laws apply in Scotland. The footnote in JCT 98 reminds the reader of this. The Act consists of a number of provisions, some of which will apply to all arbitrations, and others to arbitrations provided by the parties in their agreement. The majority of arbitration agreements adopted by the construction industry accept the Act in its entirety.

Arbitration versus litigation

Advantages

- Arbitration is generally less expensive than court proceedings.
- Arbitration is a more speedy process than an action at law; a year awaiting a case to come before the courts is not uncommon.
- Arbitration hearings are usually held in private; this therefore avoids any bad publicity that might be associated with a case in the courts.
- The time and place of the hearing can be arranged to suit the parties concerned; court proceedings take their place in turn amongst the other cases and at law courts concerned.

- Arbitrators are selected for their expert technical knowledge in the matter of dispute; judges do not generally have such knowledge.
- In cases of dispute which involve a building site or property, it can be insisted that the arbitrator visit the site concerned; although a judge may decide upon a visit, this cannot be enforced by the parties concerned.

Disadvantages

- The courts will generally always be able to offer a sound opinion on a point of law; the arbitrator may seek the opinion of the courts, but this could easily be overlooked and then a mistake could occur.
- An arbitrator does not have the power to bring into an arbitration a third party against their wishes; the courts are always able to do this.

Terminology

arbitrator	The person to whom the dispute is referred for settlement. Arbitrators are often appointed by, for example, the president of the RIBA, RICS, CC, NSCC or CIArb (JCT 98 appendix). In practice they may be selected because of their expert technical knowledge regarding the subject matter in dispute.
umpire	It may occasionally be preferable to appoint two arbitrators. In this event a third arbitrator, known as an umpire, is appointed to settle any dispute over which the two arbitrators cannot agree.
reference	The actual hearing of the dispute by the arbitrator.
award	The decision on the matter concerned made by the arbitrator.
respondent	This is the equivalent of the defendant in a law court.
claimant	The equivalent of the plaintiff.
expert witness	This is a special type of witness who plays an important part in arbitrations. Ordinary witnesses must confine their evidence strictly to the statement of facts. Expert witnesses may, however, forward their opinion based upon technical knowledge and practical experience. Prior to presenting evidence, they must show by experience and academic and professional qualifications that they can be recognized as an expert in the subject matter.

Appointment of the arbitrator

An arbitrator or umpire should be a disinterested person who is quite independent from the parties that will be involved in the proceedings. The person appointed should be someone who is sufficiently qualified and experienced in the matter of the dispute. However, it is for the parties concerned to choose the arbitrator, and the courts will not generally interfere even where the person appointed is not really the most appropriate person to settle the dispute. Arbitrators may, however, be disqualified if it can be shown that:

- They have a direct interest in the subject matter of the dispute (e.g. where the decision may have direct repercussions on their own professional work).
- They may fail to do justice to the arbitration by showing a bias towards one of the parties concerned (e.g. it could be argued that architects might show favour to an employer since they are normally employed on this 'side' of the industry).

Each of the parties to the arbitration agreement must be satisfied that the arbitrator who is appointed will give an impartial judgment on the matter of the dispute. The remarks expressed by the arbitrator during the conduct of the case will probably reveal any favouring of one of the parties in preference to the other. This may lead to removal by the courts of the arbitrator on the grounds of misconduct.

Once an arbitrator has been appointed, the general matters relating to the dispute will need to be established. It will also be necessary to determine:

- The facilities available for inspecting the works
- Whether the parties will be represented by counsel
- The matters the parties already agree upon
- The time and place of the proposed hearing

Outline of the procedure

The pleadings

Pleadings are the formal documents which may be prepared by counsel or a solicitor. The arbitrator will first require the claimant to set out the basis of the case. This will be included in a document, termed the *points of claim*, which will then be served upon the respondent. The respondent then submits a reply in answer to the points of claim, termed the *points of defence*. The respondent may also submit points of counterclaim, which will be served on the claimant at the same time. This may raise relevant matters that were not referred to in the points of claim. The claimant, in reply to the matters raised in the counterclaim, will submit points of reply and defence to counterclaim.

The purpose of the above documents is very important, since they will make clear to the arbitrator the matters that are in dispute. Furthermore, the parties involved cannot stray beyond the scope of the pleadings without leave from the arbitrator.

Discovery

Once the pleadings have been completed, the precise issues which the arbitrator is to decide should be clear. Every fact to be relied upon must be pleaded, but the manner in which it is to be proved need not be disclosed until the reference. The term *discovery* means the disclosure of all documents which are in the control of each party and which are in any way relevant to the issues of arbitration. Each party must allow the other to inspect and to take copies of all or any of the documents in their list, unless they can argue on the grounds that it is privileged. The most important of these types of document are the communications between a party

and their solicitors for the purpose of obtaining legal advice. A party who refuses
to allow inspection may be ordered to do so by the arbitrator.

In fixing the date and place of the hearing, arbitrators have the sole discretion,
subject to anything laid down in the arbitration agreement. They must, however,
be seen to act in a reasonable manner. A refusal to attend the hearing by either
party, after reasonable notice has been given, may empower the arbitrator to
proceed without that party, i.e. *ex parte*.

The hearing

The procedure of the hearing follows the rules of evidence used in a normal court
of law. The parties may or may not be represented by counsel or other
representatives.

The claimant sets out the case, and calls each of the witnesses in turn. Once
they have given their evidence they are cross-examined by the respondent
(or counsel/representative where appropriate). The claimant is then allowed
to ask these witnesses further questions on matters which have been raised
by the cross-examination.

A similar procedure is then adopted by the respondent, who sets out the details
of the counterclaim if this is necessary. Witnesses are examined and these in turn
are cross-examined by the claimant. The claimant then replies to the respondent's
defence and counterclaim, and presents a defence to the counterclaim. When this
has been completed, the respondent sums up the case in an address to the arbitrator
known as the respondent's closing speech. The claimant then has the right of reply
or the last word in the case.

The trial is now ended and both parties await the publication of the award. The
arbitrator may now decide to inspect the works, if this has not already been done,
or re-examine certain parts of the project in more detail. The arbitrator will usually
make the decision in private, and will set the decision out in the award. This is
served on the parties after the appropriate charges have been met. It is enforceable
in much the same way as a judgment debt, where the successful party can reclaim
such costs as the arbitrator has awarded.

Evidence

To enable the arbitrator to carry out justice between the parties, the evidence
must be carefully considered. This is submitted in turn by the claimant and the
respondent. Evidence is the means by which the facts are proved. There are rules
of evidence which the arbitrator must ensure are observed. These have been
designed to determine four main problems:

- Who is to assume the burden of proving facts?
  Generally speaking, the person who sets forth a statement has the burden of
  determining its proof. The maxim 'innocent until proved guilty' is appropriate
  in this context.

- What facts must be proved?
  A party must give proof of all material facts which were relied upon to establish the case, although there can be exceptions to this rule. For example, the parties may agree on formal 'admissions' in order to dispense with the necessity of proving facts which are not in dispute.
- What facts will be excluded from the cognizance of the court?
  In order to prevent a waste of time or to prevent certain facts from being put before juries, which might tend to lead them to unwarranted conclusions, English law permits proof of facts which are in issue and of facts which are relevant to the issue.
- How proof is to be effected?
  The law recognizes three kinds of proof. Oral proofs are statements made verbally by a witness in the witness box. Documentary proof is contained in the documents that are available. Real proof could include models or a visit to the site in order to view the subject matter.

It is usual in arbitration for the evidence to be given on oath. The giving of false evidence is perjury and is punishable accordingly by fine or imprisonment. The arbitrator has no right to call a witness, except with the consent of both parties. Witnesses will, however, usually answer favourably to the party by whom they are called, in order to further that party's case. They must not in general be asked leading questions which attempt to put the answer in the mouth of the witness. For example, this is a leading question: Did you notice that the scaffold was inadequately fixed? It must be rephrased: Did you notice anything about the scaffolding? Leading questions can, however, be used to the opposing party's witnesses. The arbitrator must also never receive evidence from one party without the knowledge of the other. Where communications are received from one party, the other party must immediately be informed. An arbitrator must always refuse to admit evidence on the grounds that:

- The witness is incompetent, e.g. refuses to take the oath, too young, a lunatic.
- The evidence is irrelevant, e.g. evidence which the arbitrator considers has no real bearing upon the facts.
- The evidence is inadmissible, e.g. hearsay evidence not made under oath and not capable of cross-examination. (There are, however, exceptions to this rule such as statements made on behalf of one party against their own interests.)

Where documents are used as evidence it is the arbitrator's responsibility to ensure their authenticity. Where one party produces a document, this must be proved unless the other side accepts it as valid. Documents under seal must be stamped, and generally the original document must be produced wherever possible.

Stating a case

A question of law may arise during the proceedings, and the arbitrator may deal with this in one of three ways:

- Decide the matter personally
- Consult counsel or a solicitor
- State a case to the courts

Having decided on the third course of action, a statement is prepared outlining clearly all the facts in order that the courts may decide a point of law. Once the court has given its decision, the arbitration proceedings can continue. The arbitrator must then proceed in accordance with the court's decision. Failure to do so results in misconduct on the part of the arbitrator. The arbitrator may voluntarily take this course of action or be required to do so by one of the parties.

The arbitrator may also state a case to the courts upon the completion of the arbitration. In this case the award will be based upon the alternative findings to be resolved by the courts. The decision that the courts reach will depend upon which alternative is to be followed. For example, the arbitrator may state that a particular sum should be paid by one of the parties to the other if the courts approve the arbitrator's view of the law. Where the courts do not confirm this opinion, the arbitrator will have also indicated the course of action to be taken.

The award

The arbitrator's award is the equivalent of the judgment of the courts. The award must be made within the terms of reference, otherwise it will be invalid and therefore unenforceable, The essentials of a valid award can be summarized as follows:

- It must be made *within the prescribed time limit* that has been set by the parties.
- It should *comply with any special agreements* regarding its form or method of publication that has been laid down in the arbitration agreement.
- It must be *legal and capable of enforcement* at law.
- It must *cover all the matters* which were referred to the proceedings.
- It must be *final* in that it settles all the disputes which were referred under the arbitration.
- It must be *consistent and not contradictory or ambiguous*; its meaning must be clear.
- It must be *confined solely to the matters in question*, and not matters which are outside the scope of dispute.
- The award should *generally be in writing*, in order to overcome any problems of enforcing it in practice.

Publication of the award

The usual practice is for the arbitrator to notify both the parties that the award is ready for collection upon the payment of the appropriate fees to the arbitrator. If the successful party pays the fees then they are able to sue the other party for that amount, assuming that the costs follow the event. If the award is defective or bad, or it can be shown that there have been irregularities in the proceedings, application may be made to the courts to have it referred back for reconsideration or set aside altogether.

Referring back the award

Here are the grounds on which the court is likely to refer back an award:

- Where the arbitrator makes a mistake so that it does not express the true intentions.
- Where it can be shown that the arbitrator has misconducted the proceedings, for example, in hearing the evidence of one of the parties in the absence of the other.
- Where new evidence, which was not known at the time of the hearing, comes to light and as such will affect the arbitrator's award.

The application to refer back an award should be made within six weeks after the award has been published. The courts have the full discretion regarding the costs of an abortive arbitration. There is, of course, the right of appeal against the court's decision. The arbitrator's duty in dealing with the referral will depend largely upon the order of the court. As a rule, fresh evidence will not be heard unless new evidence has come to light. The amended award should normally be made within three months of the date of the court's order.

Setting aside the award

When the courts set aside an award it becomes null and void. The situations where the courts will do this are similar to those for referring back an award, but much more serious:

- Where the award is void, for example, if the arbitrator directs an illegal action.
- The discovery of evidence that was not available at the time the arbitration proceedings were held.
- Where the arbitrator has made an error on some point of law.
- Misconduct on the part of the arbitrator by permitting irregularities in the proceedings.
- Where the award has been obtained improperly, for example, through fraud or bribery.
- Where the essentials of a valid award are lacking, for example, the award is inconsistent or impossible of performance.

Misconduct by the arbitrator

Arbitrators must carry out their duties in a professional manner. Where they are guilty of misconduct, the award can be referred back by the courts, and in a serious case the effects of setting aside an award are to make it null and void. Misconduct may be classified as *actual* or *technical*.

Actual misconduct would occur where arbitrators have been inspired in their decisions by some corrupt or improper motive or have shown bias to one of the parties involved. Technical misconduct is when some irregularities in the proceedings occur, and may include:

- The hearing of evidence of one party in the absence of the other
- The examination of witnesses in the absence of both parties
- The refusal to state a case when requested to do so by one of the parties
- Exceeding jurisdiction beyond the terms of the reference
- Failing to give adequate notice of the time of the proceedings
- Delegating authority
- An error of law

Although an arbitrator may be removed because of misconduct, this will not terminate the arbitration proceedings in favour of, say, litigation. The courts, on the application of any party to the arbitration agreement, may appoint another person to act as arbitrator. The courts may also remove an arbitrator who has failed to commence the proceedings within a reasonable time or has delayed the publication of the award.

If the arbitrator dies during the proceedings, the arbitration will not be revoked. The parties must agree upon a successor.

The parties may at any time, by mutual agreement, decide to terminate the arbitrator's appointment. Where duties have been commenced, there will be entitlement to some remuneration.

Costs

An important part of the arbitrator's award will be the directions regarding the payment of costs. These costs, which include the arbitrator's fees, can sometimes exceed the sum which is involved in the dispute. The arbitrator must exercise discretion regarding costs, but should follow the principles adopted by the courts.

The agreement generally provides for the 'costs to follow the event', which means that the loser will pay. Where an arbitration involves several issues, and the claimant succeeds on some but fails on others, the costs of the arbitration will be apportioned accordingly. If the award fails to deal with the matter of costs, then any party to the reference may apply to the arbitrator for an order directing by whom or to whom the costs shall be paid. This must be done within 14 days of the publication of the award. A provision in an arbitration agreement that a party shall bear their own costs or any part of them is void.

The entire costs of the reference, which will include not only the costs of the hearing but also the costs incurred in respect of preliminary meetings and matters of preparation, are subject to taxation by the courts. This involves the investigation of bills of costs with the objective of reducing excessive amounts and removing improper items. For example, if a party instructs an expensive counsel which the issues involved did not justify, then the fees paid will be substantially reduced. If witnesses have been placed in expensive hotels, then this may be struck from the claim as an unnecessary expense. Such items are known as 'solicitor and client charges' and have usually to be borne by the successful party.

JCT arbitration rules

The Joint Contracts Tribunal has published a set of arbitration rules for use with all of its various forms of building contracts. The rules contain stricter time limits than those prescribed by some arbitration rules or those frequently used in practice, largely in an attempt to avoid unnecessary costs.

The rules follow much of what has previously been described but also seek to capitalize on what might be termed good practice. There are twelve rules:

- *Rule 1*: arbitration agreements in each of the JCT forms of contract
- *Rule 2*: interpretation and provisions as to time
- *Rule 3*: service of statements, documents and notices – content of statements
- *Rule 4*: conduct of the arbitration – application of rule 5, rule 6 or rule 7
- *Rule 5*: procedure without hearing
- *Rule 6*: full procedure with hearing
- *Rule 7*: short procedure with hearing
- *Rule 8*: inspection by arbitrator
- *Rule 9*: arbitrator's fees and expenses – costs
- *Rule 10*: payment to trustee-stakeholder
- *Rule 11*: the award
- *Rule 12*: powers of arbitrator

LITIGATION

Litigation is a dispute procedure which takes place in the courts. It involves third parties who are trained in the law, usually solicitors and barristers, and a judge who is appointed by the courts. This method of solving disputes is often expensive and can be a very lengthy process before the matter is resolved, sometimes taking years to arrive at a decison. The process is frequently extended to higher courts involving additional expense and time (Chapter 1). Also since a case needs to be properly prepared prior to the trial, a considerable amount of time can elapse between the commencement of the proceedings and the trial, as noted above.

A typical action is started by the issuing of a writ. This places the matter on the official record. A copy of the writ must be served on the defendant, either by delivering it personally or by other means such as through the offices of a solicitor. The general rule is that the defendants must be made aware of the proceedings against them. The speed of a hearing in most cases depends upon the following:

- Availability of competent legal advisers to handle the case, i.e. its preparation and presentation
- Expeditious preparation of the case by the parties concerned
- Availability of courts and judges to hear the case

The amount of money involved in the case will determine whether it is heard in the county court or High Court. Where the matter is largely of a technical nature, the case may be referred in the first instance to the Official Referees Court (now referred to as the Technology and Construction Court). An official referee is a circuit judge whose court is used to hearing commercial cases, and hence handles most of the commercial and construction disputes. Under these circumstances a full hearing does not normally take place, but points of principle are established. The outcome of this hearing will determine whether the case then proceeds towards a full trial.

Under some circumstances the plaintiff may apply to the court for a judgment on the claim (or the defendant for a judgment on the counterclaim), on the ground that there is no sufficient defence. Provided that the court is satisfied that the defendant (or plaintiff) has no defence that warrants a full trial of the issues involved, judgment will be given, together with the costs involved.

Every fact in a dispute that is necessary to establish a claim must be proved to the judge by admissible evidence whether oral, documentary or of other kind. Oral evidence must normally be given from memory by a person who heard or saw what took place. Hearsay evidence is not normally permissible.

In a civil action the facts in the dispute must be proved on a balance of probabilities. This is unlike a criminal case where proof beyond reasonable doubt is required. The burden of proof usually lies upon the party asserting the fact.

ALTERNATIVE DISPUTE RESOLUTION

Alternative dispute resolution (ADR) is a non-adversarial technique which is aimed at resolving disputes without resorting to the traditional forms of either litigation or arbitration. The process was developed in the USA but has also been widely used elsewhere in the world. It is claimed to be less expensive, fast and effective. It is also less threatening and stressful. ADR offers the parties who are in dispute the opportunity to participate in a process that encourages them to solve their differences in the most amicable way that is possible. Table 5.1 illustrates a comparison between litigation, arbitration, adjudication and ADR.

Before the commencement of an ADR negotiation the parties who are in dispute should have genuine desire to settle their differences without recourse to either litigation or arbitration. They must therefore be prepared to compromise some of their rights in order to achieve a settlement. Proceedings are non-binding until a mutually agreed settlement is achieved. Either party can therefore resort to arbitration or litigation if the ADR procedure fails.

Common forms of ADR

Conciliation

Conciliation is a process where a neutral adviser listens to the disputed points of each party and then explains the views of one party to the other. An agreed solution may be found by encouraging each party to see the other's point of view. With this approach, the neutral adviser plays the passive role of a facilitator. Recommendations are not made by the adviser, any agreement is reached by

Table 5.1 Characteristics of dispute resolution

Characteristics	Litigation	Arbitration	Alternative dispute resolution
Place/conduct of hearing	Public court: unilateral initiation; compulsory	Private (with few exceptions); bilateral initiation; voluntary (subject to statutory provisions)	Private: bilateral initiation; voluntary
Hearing	Formal: before a judge	Formal: conforming to rules of arbitration; before an arbitrator	Informal: before a third party (a neutral)
Representation	Legal: lawyers influence settlement	Legal: lawyers influence settlement	Legal only if necessary; disputants negotiate settlement
Resolution/ disposal	Imposed by a judge after adjudication; limited right of appeal	Award imposed by an arbitrator; limited right of appeal	Mutually accepted agreements; option of arbitration if dissatisfied
Outcome	Unsatisfactory: legal win or lose	Unsatisfactory: legal win or lose	Satisfactory: business relationship maintained
Time	Time-consuming	Can be time-consuming	Fast
Cost	Expensive; uneconomic	Uneconomic	Economic

Source: Based upon *Alternative dispute resolution (ADR) in construction* by A A Kwayke, CIOB Construction Papers No. 21 (1993)

the parties agreeing to settle their differences. Where an agreement is achieved, the neutral adviser will put this in writing for each of the parties to sign.

Mediation

In mediation the neutral adviser listens to the representations from both parties and then helps them to agree upon an overall solution. An active role is played by the adviser by putting forward suggestions, encouraging discussions and persuading the parties to focus upon the key issues. Private discussions may be held with each party in order to explain the points concerned and to attempt to formulate a mutually acceptable solution to the problem. Where this is successful then the agreement is put in writing and signed by the parties concerned.

Executive tribunal

An executive tribunal is a more formal arrangement undertaken by a group comprising a neutral adviser together with representatives of the parties involved in the dispute. This group has the authority to settle the matters of dispute that created the conditions of ADR in the first place. At the hearing each party makes representations to the chairperson and each party can raise questions and seek points of clarification. Witnesses may be called upon to give evidence, although this is unusual. Each of the parties in dispute then attempts to settle the matter in private in order to achieve a negotiated settlement. Again the agreement is put in writing for endorsement by the parties concerned.

Combination

Whilst the above arrangements may appear to be separate approaches to ADR, in practice a combination of the different aspects may be used in order to solve the dispute.

ADR advisers

The neutral advisers are typically construction professionals or lawyers. Lawyers are likely to have some knowledge of the construction industry and will probably have been involved in settling such disputes in the traditional way. Persons appointed as ADR advisers will need to be trusted by the different parties and acceptable to both with known impartiality and fairness. They will seek to develop a process that suits the particular circumstances involved, seek out the truth of the dispute through questioning and debate and be the sorts of individuals who are able to solve problems. They will have good communication skills and be able to put each of the parties and the witnesses at ease in order to build their trust and confidence. They will in addition be able to appreciate all of the relevant issues involved and direct the parties towards the matters that are of crucial importance to the case.

Role of advisers in ADR

- Educating the parties involved of the procedures to be employed and the outcomes expected
- Arranging the meetings, setting agendas and outlining the protocol to be used
- Managing the process
- Clarifying the issues involved and seeking conciliation rather than encouraging confrontation
- Allowing each party to state and explain their case
- Encouraging each party to see the other party's point of view
- Preparing a report and assisting the parties to accept the agreement as binding

INSTITUTE OF COMMERCIAL LITIGATORS

Five firms of quantity surveyors (James R. Knowles, Beard Dove, Bucknall Austin, Cyril Sweet, and Tweeds) have banded together to form the Institute of Commercial Litigators (ICL). They have made an application for rights to conduct litigation to the Lord Chancellor. The expectation from both solicitors and quantity surveyors is that the decision will favour ICL's application.

The structure of the quantity surveyors' argument fundamentally rests in two areas, legal expertise and construction costs. They argue that they have the necessary construction knowledge which solicitors do not have. They are also able to carry out such services more economically. They therefore claim that clients and contractors in the construction industry will get a better deal than at present.

In the meantime, ICL has been gathering support from other interested quantity surveyors and this has already exceeded over 1,000 affiliate members. This overwhelming support indicates it has already become a serious organization in the construction industry.

PART 2

PROCUREMENT

CHAPTER 6

FORMS OF CONTRACT

The UK is fortunate to have the benefit of a wide variety of published institutional standard forms of contract available for use in the building and civil engineering industries. However, a number of government-sponsored reports have also highlighted that this has major disadvantages, identifying duplication of effort and a wasteful use of resources at almost every level of activity. It has been suggested that many of these forms also probably help to fuel the adversarial nature of the construction industry in which they are applied. It can also be argued that to write and interpret the clauses of the various forms alone represents an industry in itself.

The widespread use of standard forms within the construction industry is also partly accounted for by the practical impossibility of writing a set of new contract conditions for every project, even if this were in any way desirable.

The construction of building and civil engineering projects represents a major investment for any client. In some cases this will represent the single highest purchase ever made in an organization. For clients who undertake construction projects as a regular part of their activities, the correct choice of a form of contract is more important, since the application of the principles involved may be seen as precedents in the administration of their contracts. Since large sums of money are likely to be involved in these activities, it is important that the contractual arrangements should always be formal and legal from the outset of the project. Where a client allows consultants or contractors to begin their work on an informal basis, then the client's bargaining position is thereafter weakened or even eliminated. Under these circumstances, a worst-case scenario for clients is that they may expect to spend years, often at substantial legal expense, in arguing over the precise nature of the contractual arrangements which should have been clarified from the start of the project.

TYPES OF CONTRACT ENVISAGED

The building or civil engineering contract typically refers to the contractual arrangement between the client (employer in building forms of contract and promoter in civil engineering) and contractor. The large part of this book has been written with this in mind. However, it should be remembered that with the construction of most building and civil engineering projects, contracts also need to be formed between other individuals or firms. Besides the main contract, these include:

- Engagement of the different consultants
- Nominated subcontractors and suppliers
- Trade or domestic subcontractors

In addition, the different contractual arrangements also require collateral warranties and performance bonds and in some cases personal and parent company guarantees.

EMPLOYER AND CONTRACTOR

As long ago as 1964, the Banwell Report recommended the use of a single form of contract for the whole of the construction industry, this being both desirable and practicable. More recently the Latham Report (1994) has reiterated these comments. *Rethinking construction*, the Egan Report (1998), makes further comments. Chapter 11 considers some of the issues raised from these reports in more detail. Unfortunately, since 1964 this apparently good suggestion has been thwarted with just the opposite taking place. Since that time there has been a plethora of different forms of contract designed to suit the individual interests of particular clients and changes in the way that construction work is now often procured. The different forms of contract also have the vested interest of different parties and institutions, who for a variety of reasons whilst incorporating good practices will at the same time wish to retain their separate identity. The better (fairer) forms of contract have incorporated the views of the different interested groups within the construction industry, such as employers and contractors. Such forms have then been prepared as a joint effort between the various different parties involved. The JCT forms (see below) include wide representation from different organizations in the construction industry.

The widespread use of different forms of contract is of course exacerbated where international projects are concerned. Not only are further additional forms required, but the procurement methods used can also be considerably different from those used in the UK and different laws may exist. Note that aspects of the laws of England and Scotland themselves continue to remain at variance. There are common threads that run throughout all of the forms of contract regarding payments, variations, quality, time, etc. However, whilst the general layout and content of the various forms may appear somewhat similar, the details may vary considerably. The interpretation of the individual clauses will also differ. In some cases these have been clarified through the application of case law resulting from differences of opinion being settled in a court of law. The principles of the case law may also only apply to the form of contract in question and thus may not be applied universally across all of the different forms.

The selection of a particular form of contract depends upon several different circumstances, such as the following:

- *Type of work to be performed*: building, civil engineering, process plant engineering.
- *Size of project*: forms are available for major and minor works and those of an in-between nature.
- *Status of designer*: architects are more likely to prefer JCT or ACA, whereas civil engineers will opt for an ICE or NEC form.
- *Public or private sector*: different forms are available for use by private clients and local and central government. Large industrial corporations may in addition have their own forms of contract.
- *Procurement method to be used*

A major advantage of using a standard document is that those who use it regularly become familiar with its contents and can apply them more easily and more consistently in practice. Individuals become aware of the strengths and weaknesses of a form and are able to identify the potential areas where disputes may arise, and take corrective action where possible. They are also able to identify the form's suitability for projects with which they are concerned. The range of forms adopted by consultants is often more restricted than those faced by contractors.

MAIN CONTRACT FORMS

According to *Contracts in use* surveys (Chapter 7), the Joint Contracts Tribunal form of contract, referred to as JCT 98, remains the most popular form of contract for building contracts in the UK. This has commonly been referred to as The Standard Form of Building Contract (SFBC). The Institution of Civil Engineers (ICE) form remains the most popular form for civil engineering contracts. Whilst JCT 98 is in common use, its original introduction in 1980 (JCT 80) in the industry faced considerable opposition. It resulted in the reprinting of the older 1963 form of contract due to its more popular demand by those in practice. Whilst this form is still preferred by some practitioners, its general use has now all but disappeared. It could also be argued that the introduction of JCT 80 also encouraged the Association of Consultant Architects (ACA) to prepare their own form of contract. This had great similarities to the former 1963 RIBA (Royal Institute of British Architects) form. Due to the overt complexity of administering aspects of JCT 80, particularly in respect of nominated subcontractors, the Joint Contracts Tribunal introduced a new intermediate form of contract (IFC) in 1984. This has since been revised to bring it in line with JCT 98 and this form is referred to as IFC 98. The reason for introducing this form was to provide contract conditions that were more appropriate for use on 'medium-sized' building projects. In practice this form has received a more widespread use, often on the sort and size of projects that should have adopted JCT 98 as the preferred form of contract.

JOINT CONTRACTS TRIBUNAL (JCT) FORMS

The JCT is widely represented from within the construction industry. Its constituent bodies are:

- Royal Institute of British Architects
- Construction Confederation
- Royal Institution of Chartered Surveyors
- Association of County Councils
- Association of Metropolitan Authorities
- Association of District Councils
- Confederation of Associations of Specialist Engineering Contractors
- Federation of Associations of Specialists and Subcontractors
- Association of Consulting Engineers
- British Property Federation
- Scottish Building Contract Committee

Here are the different forms of contract available from JCT.

Joint Contracts Tribunal Standard Form of Building Contract (JCT 98)

There are separate editions for use with either local authorities or the private sector (Figure 6.1). There are also different versions available for each, to allow for with quantities, without quantities or with approximate quantities. However, the six different versions are very similar in their content throughout. Since its introduction in 1980, a number of amendments have been introduced. The incorporation of such amendments is cumbersome to use. It was therefore decided to publish a revised JCT 80 incorporating the various amendments as JCT 98. No doubt future amendments will be issued from time to time until a new edition is published in perhaps fifteen to twenty years.

Three amendments to JCT 98 are applicable:

- Construction industry scheme (CIS) published in June 1998
- Sundry amendments published in January 2000
- Amendment TC/94 introduced in April 1994 for terrorism cover

In addition there are also a number of additional supplements to cover:

- Fluctuations, two alternatives for private versions and local authority versions
- Sectional completion, relevant to with quantities, without quantities and with approximate quantities
- Optional clause 30.4A for contractor's bond in lieu of retention, for use with the private edition of the forms
- Contractor's designed portion; this should not be confused with design and build, which is an entirely different concept, with its own particular JCT form (CD 98)

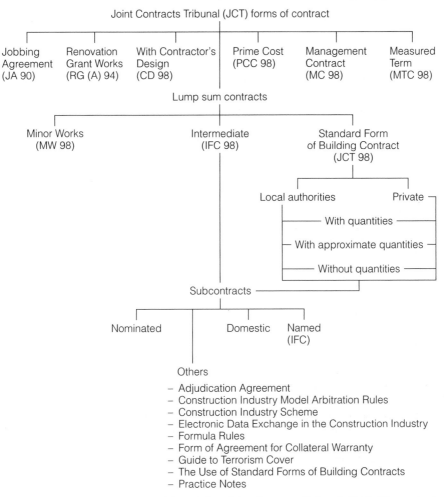

Fig. 6.1 JCT forms of contract (Adapted from Joint Contracts Tribunal, 1998, *The standard form of building contract*, Blackwell Science, Oxford)

JCT 98 Standard Form of Building Contract: Practice Notes

The Joint Contracts Tribunal have issued the following practice notes in respect of JCT 98:

- Practice Note 1 (Series 2): Construction Industry Scheme
- Practice Note 2 (Series 2): Adjudication in JCT Forms

In addition some of the JCT practice notes (series 1) on the previous JCT 80 form are still available. Where relevant to current practices they will be updated and reissued as practice notes (series 2). The rest will be withdrawn.

It is likely that additional practice notes will continue to be issued as the need arises in order to clarify matters arising from the use of the contract in practice.

Box 6.1   Practice notes issued under JCT 98

1	Sectional completion supplement	
2	Clauses 21, 22, and 30.1: insurance provisions	withdrawn
3	Clause 21.2: insurance – liability, etc., of employer	withdrawn
4	Clauses 5.2.2 and 5.4: drawings – additional copies	
5	Clause 30.3 and 16.2: payment for off-site materials	
6	Value added tax	withdrawn
7	Standard form of building contract for use with bills of approximate quantities	withdrawn
8	Statutory tax deduction scheme	out of print
9	Domestic subcontractors	out of print
10	Nomination of subcontractors	out of print
11	Employer/nominated subcontractor agreements	out of print
12	Direct and final payment to nominated subcontractors	out of print
13	Renomination of a subcontractor	out of print
14	Variations and provisional sum work	out of print
15	Nominated suppliers	out of print
16	Extensions and liquidated damages	out of print
17	Fluctuations	out of print
18	Payment and retention	out of print
19	Application and practice notes to contracts in Scotland	out of print
20	Deciding on the appropriate form of JCT main contract	
21	The employer's position under the 1980 edition of the standard form of building contract compared to the 1963 edition	
22	Amendments to the insurance and related liability provision: 1986	out of print
23	Contract sum analysis	out of print
24	Insolvency of contractor	
25	Performance-specified work	
26	Valuations and certification for interim payments including variations	
27	Application of construction (design and management) regulations	
28	Mediation on building contracts or NSC disputes	

Some of the practice notes originated with the 1963 form and have thus been adapted to suit this particular form. The practice notes are not finally authoritative; they often deal with a practical application of the form, and contain material which will be included in future revisions to the form. Such notes, although they express only an opinion in legal terms, are nevertheless considered to be based upon expert legal advice. However, they do not have the same binding effect as the decisions from courts of law.

Box 6.1 contains the full list of practice notes that have been issued under JCT 98, together with their status. There are also a series of JCT rules, warranties, etc.

- Guide to Terrorism Cover
- Electronic Data Interchange in the Construction Industry
- The Use Standard Forms of Building Contract

- JCT Adjudication Agreement
- JCT 1998 Edition of the Construction Industry Model Arbitration Rules
- Formula Rules
- Form of Agreement for Collateral Warranty
- Government Department Supplement

JCT Intermediate Form of Contract (IFC 98)

This form was first introduced in the mid 1980s in response to the overcomplexity of some of the provisions and procedures of the then JCT 80. This form has received similar revisions to JCT 80 and there are the same amendments that are applicable to JCT 98. The form is considered in further detail in Chapter 30.

JCT Agreement for Minor Building Works (MW 98)

The original minor works form was introduced at the same time as JCT 80. It now contains similar provisions to those of JCT 98, but in a much more simplified format. It is intended for use on minor building works schemes and small projects to be carried out on a lump sum basis and where an architect or supervising officer has been appointed on behalf of the employer (Chapter 29). The form cannot be used where bills of quantities have been prepared or where the employer wishes to nominate subcontractors or suppliers separately. The form includes the two amendments Construction Industry Scheme and Sundry Amendments.

JCT Standard Form of Building Contract with Contractor's Design (CD 98)

The use of this form is on projects of a design and build nature. It includes many of the clauses found in the 'parent' JCT 98. It also includes clauses especially designed to deal with this type of contractual procurement arrangement. It includes the same three amendments as JCT 98. (Chapter 11).

JCT Management Works Contract (MC 98)

The head contract to provide for contracts in which the contractor manages the works. It provides for the works to be subcontracted using the works contract documents, which correspond approximately to the NC series under JCT 98. It includes the two amendments Construction Industry Scheme and Sundry Amendments. The form is used on management contracts. It includes the following:

- Works Contract/1: Section 1: Invitation to Tender
- Works Contract/1: Section 2: Works Contractor's Tender
- Works Contract/1: Section 3: Articles of Agreement
- Works Contract/2: Works Contract Conditions

JCT Prime Cost Contract (PCC 98)

This form replaces the old fixed-fee contract. The main contract form consists of recitals, articles of agreement, attestation clause, conditions of contract and eight schedules. It includes the two amendments Construction Industry Scheme and Sundry Amendments.

JCT Measured Term Contract (MTC 98)

This form is designed for use by employers in the public and private sectors who need to undertake regular maintenance, building improvements and minor works programmes. The buildings concerned are likely to be within a defined geographical area. The appointment of a contractor to carry out these works is likely to be for a specified period of time. The contract is widely used with the National Schedule of Rates. It includes the two amendments Construction Industry Scheme and Sundry Amendments.

JCT Standard Form of Building Works of a Jobbing Character (JA 90)

This contract is designed for use by local authorities and other employers who place a number of small jobbing contracts with various contractors and who are experienced in ordering jobbing work and dealing with contractors' accounts. A new edition is currently in preparation. There are two alternatives:

- *JA/ T Jobbing Agreement Tender* for use by local authorities and other employers who place a small number of jobbing contracts with various contractors. It is used for work of small value where MW 98 is too detailed.
- *JA/ C Jobbing Agreement Conditions* for use with the above or by employers who do not wish to use JA/T but place a number of small jobbing contracts with various contractors by means of works orders.

JCT Agreement for Renovation Grant Works (RG/A 94)

There are two versions of this form for use where a grant is receivable under the Housing Act 1989. One is used where an architect is appointed and the other where no architect is engaged. A new edition is in preparation.

OTHER MAIN FORMS OF CONTRACT

British Property Federation (BPF)

This is more of a procurement arrangement than a form of contract. It is described in Chapter 8.

General Conditions of Government Contracts for Building and Civil Engineering Works (GC/Works/1 and GC/Works/2)

These are published by the Stationery Office. They are used almost exclusively by central government departments. GC/Works/1 is used on major projects and GC/Works/2 on minor building projects. There are in addition other related forms for both mechanical and electrical services in buildings. The body responsible for content, style and updating is the Department of the Environment, Transport and the Regions (DETR). They were revised in 1998 and are now compliant with the Latham Report and the Housing Grants Act. GC/Works/1 is available in three formats: lump sum with quantities, lump sum without quantities, and single-stage design and build.

Association of Consultant Architects Form of Building Agreement (ACA 99)

This form has been designed as an alternative to JCT 98. Its compilers were unhappy with the complexities of the JCT forms and claim that it is short, clear, legally precise and has been prepared with the advice of legal experts and in consultation with clients, contractors and the professions. However, the introduction of the IFC form, with ostensibly the same objectives, is considered by some to have eliminated the need for this form.

ICE Conditions of Contract and Form of Tender, Agreement and Bond for Use in Connection with Works of Civil Engineering Construction (ICE 99)

This is published by the Institution of Civil Engineers (ICE), the Association of Consulting Engineers (ACE) and the Federation of Civil Engineering Contractors (FCEC). The first edition was published in 1945. The sixth edition was published in 1991 and revised in 1998. A new seventh edition was published in 1999. In addition to the form, appendices cover an arbitration procedure, a conciliation procedure and contract price fluctuations.

ICE Conditions of Contract for Design and Construct (ICEDC/98)

This new contract was launched in 1992 (revised 1998) and has been developed from the ICE Conditions of Contract, sixth edition. The form has been designed to meet a widely felt need in the construction industry for a form of contract for use over the whole range of design and construct situations.

ICE Minor Works Contract (ICEM 98)

This has been designed to be used on minor works civil engineering contracts and is based on ICE 99.

Conditions of Contract (International) for Works of Civil Engineering Construction (FIDIC 87)

This form is prepared by the International Federation of Consulting Engineers. The fourth edition was published in 1987. It is approved by several other organizations representing the construction interests of various other countries.

The Engineering and Construction Contract (ECC 99)

The ECC provides a totally new approach in how engineering and construction contracts are structured and managed. It was formerly titled the New Engineering Contract (NEC). Its aim is to provide a variety of contract strategies which can be adopted as necessary to suit different project, client and contractor requirements. At the same time the ECC aims to provide a stimulus whereby all parties strive towards completion of a project without the disputes and adversarial approach inherent in some other contract strategies. The majority of its usage has so for been abroad, with relatively limited use in the UK. Those who have used the ECC have also generated widely differing responses.

The ECC is published as the ten documents listed below, available individually or all ten in a slip case. A professional service contract and an adjudicator's contract together with guidance notes is also available as a set.

- Option A: Priced Contract with Activity Schedule
- Option B: Priced Contract with Bill of Quantities
- Option C: Target Contract with Activity Schedule
- Option D: Target Contract with Bill of Quantities
- Option E: Cost Reimbursable Contract
- Option F: Management Contract
- The Engineering and Construction Subcontract
- Flow Charts
- Guidance Notes

Engineering and Construction Short Contract (ECSC 99)

The first edition of this form of contract was published in 1997 and intended for use on less complex projects than ECC 99. It is a contract for use internationally as well as in the UK. For simplicity there are no secondary options available. The method of payment is either based on a bill of quantities or a sum per item.

SUBCONTRACT FORMS

Nominated subcontracts for JCT 98

There are extensive arrangements and references in JCT 98 to the nomination procedures to be followed. These contemplate the use of the following documents and forms:

- NSC/T 98   Part 1 Invitation to tender to a subcontractor
            Part 2 Tender by a subcontractor
            Part 3 Particular conditions agreed by a contractor and
            a subcontractor
- NSC/A 98   Agreement between contractor and nominated subcontractor
- NSC/N 98   Standard form of nomination instruction for a subcontractor
- NSC/W 98   Standard form of employer and nominated subcontractor
            agreement
- NSC/C 98   Nominated subcontract conditions

Introduced in 1991 to replace the existing forms NSC1–4, they were revised in 1998. The revised procedure for nomination of a subcontractor can be purchased as a composite document covering all the above documents.

Nominated Suppliers Form of Tender for JCT 98

This form complies with JCT 98 clause 36, and although its use is not mandatory it is nevertheless advisable. The forms include:

- TNS/1   Standard form of tender by nominated supplier tender and schedules 1
          and 2
- TNS/2   Form of tender by nominated supplier schedule 3

Domestic forms of subcontract for JCT 98

These forms of subcontract have been prepared by the Construction Confederation for use with non–nominated subcontractors, i.e. those appointed directly by the contractor, commonly referred to as trade or domestic subcontractors. The documents include:

- DOM/1a   Domestic subcontract articles of agreement
- DOM/1c   Domestic subcontract conditions
- DOM/2a   Domestic subcontract articles in connection with contractor's design
           (CD 98)
- DOM/2c   Domestic subcontract conditions with CD 98

Named subcontractors for use with IFC 98

These include the following forms:

- NAM/T 98   Tender and agreement
- NAM/SC 98   Subcontract conditions
- NAM/FR   Formula rules for named subcontractors

Labour-only Subcontract Form

This form is used for work which is sublet by the main contractor under clause 19 of JCT 98. It came out in 1987 and is published by the Construction Confederation

(CC). CC also provide a shorter form for use where the labour-only subcontractor does not intend to use operatives for the execution of the subcontract works.

FCEC (Federation of Civil Engineering Contractors) Form of Subcontract

The sixth edition of this form was prepared in conjunction with the ICE form.

Form of Subcontract (International)

This form has been prepared for use with the FIDIC form of contract.

Engineering and Construction Subcontract

This is part of the Engineering and Construction Contract (see above).

OTHER FORMS OF CONTRACT

Other forms of contract are also available, most notably those prepared by the large industrial corporations who undertake a substantial amount of construction work. Even in cases where such forms are not used and one of the more conventional forms of contract is applied, it is not uncommon to find a few pages being added as supplementary conditions. These conditions generally place the standard forms in a more favourable position with the building owner or employer. A much greater risk is therefore placed with the contractor than is usually the case. Whether this offers any real advantages to the employer is open to speculation. Under 'normal' tendering circumstances, the contractors' prices will reflect the more onerous conditions.

It is much more unusual to tamper with the contract conditions themselves, since there is always the danger of making the entire contract null and void, or failing to alter the conditions consistently throughout the particular form being adapted.

OTHER CONTRACTUAL DOCUMENTATION

National Joint Consultative Committee

The National Joint Consultative Committee (NJCC) is an advisory body concerned with practice and procedures in building contracts. It has a good practice panel and a wide range of publications concerned with contracts. It provides a range of codes of procedure, the most common of which is *Code of procedure for single-stage selective tendering* (Chapter 8). In addition it provides codes on tendering procedure for industrialized building projects, two-stage selective tendering, selective tendering for design and build, management contracting and the letting and management of domestic subcontract works. The NJCC also offers good practice on preparing lists of approved contractors. It provides guidance notes on issues such as performance bonds, collateral warranties and alternative dispute resolution. It also produces a set of procedure notes covering a wide range of topics such as financial controls and cash flow, placing contracts with substantial building services engineering content, and the use and completion of nominated subcontracts.

Definition of Prime Cost for Daywork Carried out under a Building Contract

This is prepared jointly by the RICS and CEC. It applies to work described as daywork carried out under and incidental to a building contract. A separate document, *Definition of prime cost of building works of a jobbing and maintenance character*, covers jobbing and maintenance work carried out as a main or separate contract.

Formula Rules 1987

The consolidated set of formula rules covers all the main contract forms, such as JCT 98, CD 98 and IFC 98.

Standard Forms of Agreement for a Collateral Warranty by a Main Contractor

The JCT has published warranties by a contractor to a purchaser or tenant of building works and to a company providing finance for building works defined as a 'funder'. These warranties have been drafted to be compatible with warranties already issued by the BPF, RIBA, RICS and ACE which are given by a consultant to a purchaser or tenant or to a funder.

Guide to Terrorism Cover 1994

Provisions in JCT forms on insurance of the works of the existing structures and contents for physical loss or damage due to fire or explosion due to terrorism.

Construction Industry Model Arbitration Rules (CIMAR)

These have been introduced for the conduct of arbitration which will be contractually binding on the parties. The rules apply to the settlement of all disputes coming within the terms of the arbitration agreements in all forms of JCT contract and subcontract and other tribunal agreements.

Guidance on the Unfair Contract Terms Act 1977

This is published by CEC.

Appointment of consultants

Whilst 'the construction contract' is between the client (employer or promoter) and the contractor, the various consultants engaged on a project will also have a contract with the appointing client. Different forms of contract are used for the appointment of:

- *Architects*: Royal Institute of British Architects
- *Quantity surveyors*: Royal Institution of Chartered Surveyors
- *Structural engineers*: Association of Consulting Engineers
- *Building services engineers*: Association of Consulting Engineers
- *Project managers*: Royal Institution of Chartered Surveyors
- *Civil engineers*: Association of Consulting Engineers

The contracts in these circumstances are for the professional services relating to the provision of the construction project. The different forms cover duty of care, fees, copyright, collateral warranties and agreements, professional indemnity insurance, etc. The forms identify materials or substances that should not be used in the construction process. These are high alumina cement in structural elements, wood wool slabs in permanent formwork to concrete, calcium chloride in admixtures for use in reinforced concrete, asbestos products and naturally occurring aggregates for use in reinforced concrete which do not comply with British Standards 882 or 8110.

The fees charged by these consultants will be guided by the fee scales that are published by the different professional bodies. However, in common with the principles of competitive tendering, it is now usual to invite consultants to compete for work, with one of the criteria being the fees charged for the consultancy work.

CHAPTER 7

CONTRACT STRATEGY

In the early 1960s the clients of the construction industry had only a limited choice of procurement methods which they might wish to use for the construction of a new building. The RIBA form of contract, although even then not used universally, had yet to experience extensive rivalry from elsewhere, or even from within its own organization for a serious alternative. Bills of quantities were becoming the preferred document in place of the specification, and the mistrusted cost-plus type contracts which had been a necessity a few years earlier for the rapid repair of war-damaged property were already in decline.

Society is now a world of endless change. In recent years there have been many developments in the technologies that are used, in the types of societies in which we live, and amongst professional attitudes. The causes of these changes which have been experienced are not difficult to find. Society generally is on the lookout for something new, and the construction industry in a way mirrors that change to avoid being thought of as outdated or old-fashioned. The search is also for improvement, knowing that we have not as yet found the perfect system; even if one at all exists. In this context, largely due to improved geographical communications of the twentieth century, the construction industry has looked with interest to both East and West, to the way that others around the world arrange the procurement of capital works projects. Not surprisingly, it is the countries which appear to have been the most successful which have received most of the attention. The methods adopted in the USA or Japan for their construction industries have been in the forefront of our ideas for change.

The apparent failure of the construction industry to satisfy the perceived needs of its customers, particularly in the way it organizes and executes its projects, has been another catalyst for change. Various pressure groups too have evolved to champion the causes of the different organizations who perhaps have particular axes to grind. The 1970s oil price shocks, which had a massive influence on inflation and hence industries' borrowing requirements, also motivated the construction industry to improve its own efficiency through the way it managed and organized its work. The depression of the 1970s, like previous slumps, encouraged firms of all kinds to attempt to persuade employers to build, by using what were then innovative approaches to contract procurement.

INDUSTRY ANALYSIS

The construction industry can be measured in several different ways, e.g. turnover, profitability, number of firms, number of employees (see *The construction industry of Great Britain* by R C Harvey and A Ashworth, Butterworth-Heinemann, 1997). In 2000 its annual worth was about £60 billion. The construction industry at the start of the twenty-first century is buoyant. The construction industry is also diverse: about 60 per cent of the total output is repairs and maintenance, less than 25 per cent is public sector projects, and less than 20 per cent is civil engineering projects.

During the past decades, changes in the methods of construction procurement have been one of the most fundamental driving forces within the construction industry. Greater comparisons have been made with procurement methods used by other industries and in other countries around the world. Research has been undertaken to better inform clients, consultants and contractors. The different methods each have their own peculiar advantages and disadvantages and there is still no panacea that applies across the whole industry. The choice of method depends upon the following characteristics:

- Familiarity amongst parties
- Type of client
- Size of project
- Type of project
- Risk allocation
- Form of contract to be used
- Major objectives of the client
- Status of the designer
- Relationships with contractors and consultants
- Type of contract documentation

CONTRACTS IN USE SURVEY

There are several different organizations that monitor trends in procurement systems in the construction industry. The following analysis (Tables 7.1 to 7.5) is based on a survey carried out by Davis, Langdon and Everest every two years. The survey was first designed and published in 1984. The current data (Tables 7.1 to 7.5) has been extracted from their eighth survey. This survey is designed to determine trends in the use of standard forms and contract procurement methods in the construction industry. This is to assist employers, designers and contractors. The report has been prepared on behalf of the Royal Institution of Chartered Surveyors (RICS) and captures over one-fifth in value of new construction orders in the UK.

The scope of the survey excludes overseas work, civil and heavy engineering, term contracts, maintenance and repairs and subcontracts that formed a part of main contracts. Work done directly by contractors was excluded from the survey to avoid any possible double counting. This perhaps suggests that design and build represents a much larger proportion than indicated by this survey alone. This viewpoint is further considered in Chapter 11.

Table 7.1 Trends in methods of procurement: by value of contracts

	Percentage of contracts							
	1984	1985	1987	1989	1991	1993	1995	1998
Contract bills	58.7	59.3	52.1	52.3	48.3	41.6	43.7	28.4
Drawings/spec	13.1	10.2	17.7	10.2	7.0	8.3	12.2	10.0
Design and build	5.1	8.0	12.2	10.9	14.8	35.7	30.1	41.4
Approximate quantities	6.6	5.4	3.4	3.6	2.5	4.1	2.4	1.7
Prime cost	4.5	2.7	5.2	1.1	0.1	0.2	0.5	0.3
Management	12.0	14.4	9.4	15.0	7.9	6.2	6.9	10.4
Construction management	0.0	0.0	0.0	6.9	19.4	3.9	4.2	7.7
Total	100.0	100.0	100.0	100.0	100.0	100.0	100.0	100.0

Table 7.2 Trends in methods of procurement: by number of contracts

	Percentage of contracts							
	1984	1985	1987	1989	1991	1993	1995	1998
Contract bills	34.6	42.8	35.6	39.7	29.0	34.5	39.2	30.8
Drawings/spec	55.7	47.1	55.4	49.7	59.2	45.6	43.7	43.9
Design and build	2.4	3.6	3.6	5.2	9.1	16.0	11.8	20.7
Approximate quantities	3.2	2.7	1.9	2.9	1.5	2.3	2.1	1.9
Prime cost	2.3	2.1	2.3	0.9	0.2	0.3	0.7	0.3
Management	1.8	1.7	1.2	1.4	0.8	0.9	1.2	1.5
Construction management	0.0	0.0	0.0	0.2	0.2	0.4	1.3	0.8
Total	100.0	100.0	100.0	100.0	100.0	100.0	100.0	100.0

Trends in methods of procurement

Tables 7.1 and 7.2 show the value and number of contracts since 1984. The latest survey, based on 1998 data, was published in January 2000 and includes some projects that were valued in excess of £100 million. The survey is based on approximately 2,500 projects, of which 63 per cent of contracts were valued at below £500,000 and 30 per cent between £500,000 and £5 million. Thirty-five of the projects were valued in excess of £10 million. Tables 7.1 and 7.2 indicate the following characteristics:

■ A considerable decline in the value of projects using contract bills.
■ A much smaller trend in the number of projects not using contract bills.
■ A decline in the number of projects relying on drawings and specifications, although in terms of their value, little significant change is detected.

- The massive increase in the value of design and build projects, forecast for over forty years, appears to have taken place during the early 1990s, and has since been sustained.
- The increase in the number of projects using design and build is less pronounced, but now represents over 20 per cent of all projects, compared with just 2 per cent a decade earlier.
- Prime cost contracts have declined in both number and value, and represent a relatively insignificant amount in both cases.
- There is little trend in the use of management contracts, other than a decline in value on what represents a few per cent of projects by number.
- There is no trend in the use of construction management contracts, and their numbers are generally small.

Use of JCT standard forms

As expected, the JCT forms of contract continue to dominate and continue to be used on the majority of building projects (Table 7.3). The 91 per cent quoted in 1998 is the largest proportion, by number, since the survey was first introduced. In terms of the total value of projects, JCT is used on approximately three-quarters of all building projects. Table 7.4 indicates:

Table 7.3   Trend in use of JCT forms

	1984	1985	1987	1989	1991	1993	1995	1998
By number of contracts (%)	84	81	86	81	78	82	85	91
By value of contracts (%)	75	70	74	81	61	80	76	68

Table 7.4   Use of JCT standard forms

	By number of contracts (%)		By value of contracts (%)	
	1995	1998	1995	1998
JCT 80 with quantities	22.4	17.5	32.3	23.3
JCT 80 without quantities	5.2	7.8	3.6	6.2
JCT 80 with approx quantities	1.1	0.8	1.5	0.5
Minor works	22.7	20.1	2.3	1.0
IFC 80	21.9	25.2	10.5	5.3
Contractor's design	9.3	18.1	20.4	26.1
Management	0.8	0.7	4.5	4.9
Prime cost	0.5	0.3	0.3	0.3
Total	83.9	90.5	75.4	67.6

- JCT 80 (subsequently revised and replaced by JCT 98, see Part 4) in all of its variants remains the most commonly used form of contract for the procurement of building contracts.
- In terms of the number of contracts using any version of JCT 80 (now JCT 98), the percentage change between 1995 and 1998 is virtually unchanged (28.7 per cent and 26.1 per cent respectively).
- In terms of their value, there has been a significant shift from 37.4 per cent down to 30.0 per cent. The difference is probably made up with the increased number of contracts using contractor's design (plus 5.7 per cent).
- The use of the contractor's design form has increased substantially over this period. Had this been measured from 1993 (Tables 7.1 and 7.2) the increase would have been even greater than that indicated.
- The continued decline in the use of contract bills can be observed, although somewhere in the design and construction chain some form of measurement or quantification is required.
- The use of IFC 84 appears to have peaked in terms of the number of contracts on which it is used. In terms of the value of work using this form, the decline is more pronounced where contract bills are used as a part of the contract.

The survey also revealed a number of other aspects:

- A decline in the number of projects using one of the local authority variants.
- The average value scheme using IFC 84 was £408,000. This is higher than the recommended upper limit for this form of contract.
- Whilst the minor works form accounted for 20.1 per cent of all contracts, this represented only 1.0 per cent of the total value of work (Table 7.4). However, the trend in the use of this form is in decline, with the lowest recorded use on any of these surveys.
- The use of the management form has remained fairly constant.
- The number of instances of GC/Works forms identified a much lower use than previously.
- The use of any ICE (Institution of Civil Engineers) forms on building projects has always been low. The New (ICE) Construction Contract for building projects is negligible at the present time.
- The use of non-standard forms accounted for 5.4 per cent by number and 18.9 per cent by value. The 18.9 per cent figure represents a considerable slice of the procurement market.
- In line with the principles recommended by the Egan Report, there has been a significant increase in negotiated contracts. This is a desire by contractors to move away from low margin competitive tendering. Approximately 13 per cent were identified as having been negotiated.
- Partnership or alliance provisions account for a relatively small proportion of projects at the present time.
- Only a small number of projects included PFI arrangements, although this was increasing.

Table 7.5   Design and build

	By number of contracts (%)		By value of contracts (%)	
	1995	1998	1995	1998
JCT with contractor's design	10.0	18.9	21.0	27.1
GC/Works design and build	0.3	0.2	1.6	0.2
ICE design and build	0.1	0.4	0.3	9.6
Other design and build	1.4	1.3	7.2	4.6
Total	11.8	20.8	30.1	41.5

Design and build

Irrespective of the form of contract that is used, design and build as a procurement method has shown considerable increase in its use since the middle of the 1990s. Table 7.5 indicates the percentage number and value of projects using this method of procurement. One-fifth (20.8 per cent) of the projects surveyed used this method and two-fifths (41.5 per cent) of the value of new buildings were built using a variant of design and build. The increase in design and build contracts began in the mid 1990s. Also there has been a significant increase in the use of contractor's designed work within lump sum traditional contracts.

MAJOR ISSUES TO BE RESOLVED

Table 7.6   Developing a contract procurement straegy

- Consultant or contractor
- Price competition or negotiation
- Measurement or reimbursement
- Traditional or alternative procurement

Consultants versus contractors

The contracts survey referred to above underestimates the total amount of design and build (D&B) projects because it is almost entirely a survey of consultant organizations. The arguments for engaging a consultant rather than a contractor as the main employer's adviser are inconclusive. The respective advantages and disadvantages may be summarized as follows. Advantages of a contractor–centred approach are said to be:

- Better time management
- Single-point responsibility
- Inherent buildability
- Certainty of price
- Teamwork
- Inclusive design fees

Disadvantages may be:

- Problems of contractor proposals matching with employer requirements
- Payment clauses
- Emphasis may be away from design towards other factors
- Employers may still need to retain consultants for payments, inspections, etc.

Competition versus negotiation

There are a variety of ways in which a builder may seek to secure business. These include speculation, invitation, reputation, rotation arrangement, recommendation and selection. Irrespective, however, of the final contractual arrangements which are made by the employer, the method of choosing the contractor must first be established. The alternatives which are available for this purpose are either competition or negotiation.

Some form of competition on price, time or quality is desirable. All of the evidence which is available suggests that the employer under the normal circumstances of contract procurement is likely to strike a better bargain if an element of competition exists. There are, however, a number of circumstances which can arise in which a negotiated approach may be more beneficial to the employer. Some of these include:

- Business relationship
- Early start on site
- Continuation contract
- State of the market
- Contractor specialization
- Financial arrangements
- Geographical area

The above list is not exhaustive nor should it be assumed that negotiation would be preferable in all of these examples. Each individual project should be examined on its own merits, and a decision made bearing in mind the particular circumstances concerned and specific advantages to the employer.

Certain essential features are necessary if the negotiations are to proceed satisfactorily. These include equality of the negotiators in either party, parity of information, agreement as to the basis of negotiation and a decision on how the main items of work will be priced.

Measurement versus reimbursement

There are essentially only two ways of calculating the costs of construction work. Either the contractor adopts some form of measurement and is paid for the work on the basis of quantity multiplied by a rate, or the contractor is reimbursed the actual costs. A drawing and specification contract, for example, relies upon the contractor measuring and pricing the work, even though only a single sum is disclosed to the employer. The measurement contract allows for the payment for risk to the contractor, the cost reimbursement approach does not. Many of the measurement contracts may include for a small proportion of the work to be paid for under dayworks (a form of cost reimbursement) but it is more unusual to find cost reimbursement contracts with any measurement aspects. Here are some points to bear in mind when choosing between measurement or cost reimbursement contracts:

- *Contract sum*: this is not available with any form of cost reimbursement contract.
- *Final price forecast*: this is not possible with any of the cost reimbursement methods or with measurement contracts which rely extensively on approximate quantities.
- *Incentive for contractor efficiency*: cost reimbursement contracts can encourage wastage which must then be passed on to the employer.
- *Price risk*: measurement contracts allow for this, employers may therefore pay for such non-events.
- *Cost control*: the employer has little control over costs where any form of cost reimbursement contract is used.
- *Administration*: cost reimbursement contracts require a large amount of clerical work.

Traditional versus alternatives

Until recently the majority of the major building projects were constructed using single-stage selective tendering. This method of procurement has many flaws, so alternative procedures have been devised in an attempt to address them. The newer methods, or alternative procurement paths, overcame the failures of the traditional approach but they created their own particular problems. In fact, if a single method was able to be devised which addressed all of the problems then the remaining methods would quickly fall into disuse. In choosing a method of procurement, therefore, the following issues are of importance. They are more fully described in Chapter 9.

- Project size
- Costs inclusive of the design
- Time from brief to handover
- Accountability
- Design, function and aesthetics
- Quality assurance

- Organization and responsibility
- Project complexity
- Risk placing
- Market considerations
- Financial provisions

THE FRAMEWORK OF SOCIETY

The correct application of contractual procurement systems is influenced by a number of factors which are present within any society. Such factors, although external to.the construction industry, do have implications for the successful completion of each individual project. The appropriate recommendations today may also have different implications in the future, because the framework of society is a constantly evolving one. In selecting the right method, therefore, the following factors should be considered and evaluated:

- *Economic*: interest rates, inflation, land costs, investment policies, market levels, taxation opportunities, opportunity costs.
- *Legal*: contract law, case law, arbitration, discharge of contracts, remedies for breach of contract.
- *Technological*: new techniques, off-site manufacture, production processes, use of computers.
- *Political*: government systems, public expenditure, export regulations and guarantees, planning laws, trade union practice, policies.
- *Social*: demography, ageing population, availability of potential employees, retraining, the environment.

Many of these factors are unstable even in a well-run economy or for the duration of the contract period, due to the influences from other world markets. An employer's satisfaction with the project relies upon the procurement adviser's skill and to some extent the intuition which can be provided, in being able to reflect on how the above are likely to influence the method recommended for contract procurement.

EMPLOYERS' ESSENTIAL REQUIREMENTS

When buying a particular service, employers seek to ensure that it fully meets their needs. The consultant or contractor employed needs to identify with the employer's objectives within the context in which the employer has to operate, and particularly any constraints which may be present. *A study of quantity surveying practice and client demand* (RICS, 1984), identified some of the following criteria as important requirements for the majority of employers:

- Impartial and independent advice
- Trust and fairness in all dealings
- Timely information ahead of possible events
- Implications on the interreactions of time, cost and quality
- Options from which the employer can select the best possible route
- Recommendations for action
- Good value for any fees charged
- Advice based upon a skilled consideration of the project as a whole
- Sound ability and general competence
- Reliability of advice
- Enterprise and innovation

PROCUREMENT MANAGEMENT

It is of considerable importance to employers who wish to have buildings erected that the appropriate advice is provided on the method of procurement to be used. The advice offered must be relevant and reliable and based upon skill and expertise. There is, however, a dearth of objective and unbiased advice available. It is often difficult to elicit the relevant facts appropriate to a proposed building project. Construction employers will tend to rely on the advice from their chosen consultant or contractor. The advice provided is usually sound, and frequently successful according to the criteria set by the employer. It may, however, tend to be biased and even in some cases tainted with self-interest. It is perhaps sometimes given on the basis of 'who gets to the client first'. Methods and procedures have now become so complex, with a wide variety of options available, that an improvement in the management approach to the procurement process is now necessary to meet the employer's needs. The need to match, for instance, the employer's requirements with the industry's response is very important if customer care and satisfaction are to be achieved. The employer's procurement manager must consider the characteristics of the various methods that are available and recommend a solution which best suits the employer's needs and aspirations. The manager will need to discuss the level of risk involved for the procurement path recommended for the project under review.

The process of procurement management may be broadly defined to include the following:

- Determining the employer's requirements in terms of time, cost and quality.
- Assessing the viability of the project and providing advice in terms of funding and taxation.
- Advising on an organizational structure for the project as a whole.
- Advising on the appointment of consultants and contractors bearing in mind the criteria set by the employer.
- Managing the information and coordinating the activities of the consultants and contractor, through the design and construction phases.

The simplistic view is that architects design and contractors build; those are their strengths. It is important, however, that someone is especially responsible for the contractual matters surrounding the contract.

The NEDO publication *Thinking about building* makes the following suggestions on procurement:

- Do make one of your in-house executives responsible for the project.
- Do bring in an outside adviser if the in-house resources and skills are inappropriate.
- Do take special care to define the needs for the project.
- Do choose a procurement path to fit the defined priorities.
- Do go to some trouble to select the organizations and individuals concerned.
- Do ensure that a professional appraisal is done before the scheme becomes too advanced.

The effectiveness of a procurement path is a combination of three things:

- The correct advice and decision on which procurement path to use
- The correct implementation of the chosen path
- The evaluation during and after its execution.

COORDINATED PROJECT INFORMATION

Research at the Building Research Establishment on fifty representative building sites has shown that the biggest single cause of events which stop site managers, architects or tradespeople from working together is unclear or missing project information. Another significant cause is uncoordinated design, and at times the entire effort of site management can be directed to searching for missing information or reconciling inconsistencies in the information which is available. To overcome these weaknesses, the Coordinating Committee for Project Information (CCPI) was formed with the task of developing a common arrangement for work sections for building works that could be used throughout the various forms of documentation.

CCPI has consulted widely to ensure that the proposals are practicable and helpful, and its work has been performed by practising professionals who are themselves fully aware of the pressures involved. Is it really more economical to produce incomplete and inconsistent, if not contradictory, project information during the design process, and then to spend time wrestling with and trying to rectify the problems on site, with the attendant waste of time in dealing with claims? Is it not better to produce complete and coordinated information once and thus avoid such problems? If properly coordinated project documentation is produced, much abortive work will be avoided and costs will be so much less.

Regardless of the chosen procurement method, it is important that the principles of CCPI are adopted and put into practice. Procurement methods which are unable to incorporate such principles may be severely flawed in practice.

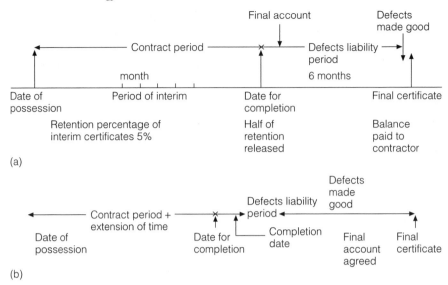

Fig. 7.1   Contractual terminology: (a) expected contractual position, (b) likely contractual position

ADDED VALUE

Employers are rightly concerned with obtaining value for money. Cheapness is in itself no virtue. It is well worthwhile to pay a little more if the gain in value exceeds the extra costs. Value for money, in any context, is a combination of subjective and objective viewpoints. There are some items which can be measured but there are other items which can only be left to opinion or at best expert judgement. Measurable items can largely be proven, or at least they could if our knowledge were fully comprehensive. But the professional's skill is largely in the area of judgment. The procurement method recommended to the employer needs to be the method which offers the best value for money. The need for careful assessment is required to obtain the desired results. Figure 7.1 illustrates the contractual terminology in both the expected and likely positions.

CONCLUSIONS

Procurement procedures today are a dynamic activity. They are evolutionary, to suit the changing needs of society and the considerations in which the industry finds itself operating. There are no standard solutions, but each individual project, which invariably represents a one-off, needs to be considered independently and analysed accordingly. A wide variety of factors have to be taken into account before any practical decisions can be made. The various influences concerned need to be weighed carefully, and always with the best interests of the employer in mind.

CHAPTER .8

CONTRACT PROCUREMENT

The execution of a construction project requires both design work and the carrying out of construction operations on site. If these are to be done successfully, resulting in a satisfactorily completed project, then some form of recognized procedure must be employed at the outset to deal with their organization, coordination and procedures. Procuring buildings is expensive. New buildings are one of the highest single-cost items for any firm or organization, be they developers or owner-occupiers. It is therefore important to get this aspect as right as possible.

Traditionally a client who wished to have a building constructed would invariably commission an architect to prepare drawings of the proposed scheme, and if the scheme was sufficiently large, employ a quantity surveyor to prepare appropriate contract documentation on which the building contractor could prepare a price. These would all be based upon the client's brief, and the information used as a basis for competitive tendering. This was the common system in use at the turn of the century and still continues to be widely used today.

However, particularly since the mid 1960s, a small revolution has occurred in the way designers and builders are employed for the construction of buildings. To some extent, this is the result of initiatives taken by the then Ministry of Works in the early 1960s. The Banwell Committee reported in 1967 (Chapter 9); it recommended several changes in the way that projects and contracts were organized, one of which was an attempt to try to bring the designers and the constructors closer together. The construction industry continues to examine and evaluate the methods available, and to devise new procedures which address the shortfalls and weaknesses of the current procedures. Some of these procedures remain fashionable for a short space of time then fall into disuse.

There is, however, no panacea that will suit all projects. In fact, it may be argued that change is occurring so fast that a present-day solution may be quite inappropriate for tomorrow. Each of the methods described has its own characteristics, advantages and disadvantages. All have been used in practice, some more than others, largely due to familiarity and ease of application. New methods will continue to evolve to meet the rapidly changing circumstances of the construction industry. These will occur in response to current deficiencies and to changes in the culture of the construction industry.

METHODS OF PRICE DETERMINATION

Building and civil engineering contractors are paid for the work they carry out, on the basis of one of two methods: measurement, i.e. payment against given criteria, or cost reimbursement, i.e. actual costs involved.

Measurement

The work is measured in place on the basis of its finished quantities. The contractor is paid for the work on the basis of quantity multiplied by a rate. Measurement may be undertaken by the employer's quantity surveyor, or by the contractor's surveyor or estimator. In the first example, an accurate and detailed contract document can be prepared. In the second example, the document prepared will be sufficient only to satisfy the particular builder concerned. The work may be measured as accurately as the drawings allow prior to the contract being awarded, in which case it is known as a lump sum contract. Alternatively, the work may be measured or remeasured after it has been carried out. In the latter case it is referred to as a remeasurement contract. All contracts envisage some form of remeasurement to take into account variations to the original plans. In such cases both the employer's and the contractor's surveyors measure the work together. Measurement contracts, unless they are entirely of a remeasurement type, allow for some sort of final cost to be recalculated. This offers advantages to the client for budgeting and cost control. Building contracts are more often lump sum contracts, whereas civil engineering projects are typically of the remeasurement type. Measurement is usually done against agreed criteria, such as a method of measurement. In the UK, the Standard Method of Measurement of Building Works (SMM7) has been designed and compiled for this purpose.

Cost reimbursement

Cost reimbursement arrangements allow the contractor to recoup the actual costs of the materials which have been purchased, the actual time spent on the work by the operatives, and the actual time used by mechanical plant. An agreed additional amount, often expressed as a percentage, is added to cover the contractor's overheads and profit. Daywork accounts, used on many contracts in conjunction with measurement, are costed and valued on much the same basis.

MEASUREMENT CONTRACTS

Drawings and specification

This is the simplest type of measurement contract and is really only suitable for small works or simple projects. In recent years there has been a trend towards using this approach on schemes much larger than it was originally intended for. Each contractor must then measure the quantities from the drawings and interpret the specification during pricing in order to calculate the tender sum. The method is wasteful of the contractor's estimating resources, since all contractor's need to prepare their own measurements. It does not easily allow for a fair comparison of the tender sums received by the employer. Interpreting the specification can be a difficult job, even for the more experienced estimator. The contractor has also to accept a greater proportion of the risk, being responsible for the prices as well as for the measurements that have been based upon interpretation of the contract information. In order to compensate for possible errors or omissions, evidence suggests that sensible contractors will tend to overprice this type of work.

Performance specification

This method is a much more vague and imprecise approach to tendering and the evaluation of the contractor's bids. The use of a performance specification requires the contractor to prepare a price for the work based only upon the employer's brief and user requirements alone. The contractor is then left to select the materials to be used and to determine the method of construction that suit these broad requirements. In practice the contractor will select materials and methods of construction which satisfy the prescribed performance standards in the cheapest possible way. Great precision is required in formulating a performance specification, if the desired results of the employer are to be achieved. Performance specifications are sometimes used with mechanical equipment that is used in buildings.

Schedule of rates

With some projects it is not possible to predetermine the nature and full extent of the proposed building works. In these circumstances, where it is desirable to form some direct link between quantity and price, a schedule of rates may be used. This schedule is similar to a bill of quantities, but without any actual quantities being included. It should be prepared using the rules of a recognized method of measurement. Contractors are invited to insert their rates against these items, and these are then used in comparison against other contractors' schedules in selecting the best tender. Upon completion of the work, this is remeasured and the rates that have been provided by the contractor are used to calculate the final cost. This method does not allow either the prediction of a contract sum, or an indication of the probable final cost of the project from the outset. Contractors also find it difficult realistically to price the schedule in the absence of any quantities, since the amount of work to be performed has a direct influence upon the costs incurred.

Schedule of prices

An alternative to the schedule of rates is to provide the contractors with a ready-priced schedule, similar to the former Property Services Agency's schedule of rates. The contractors in this case adjust each rate by the addition or deduction of a percentage. In practice a single percentage adjustment is normally made to all of the rates. This standard adjustment is unsatisfactory, since the contractor will view some of the prices in the schedule as being high, and others as being too low in terms of covering the contractor's own costs. However, a schedule of prices does have the advantage of producing fewer pricing errors in the tender documents when compared with the contractor's own price analysis of the work.

Bill of quantities

Even with all the new forms of contract arrangement, the bill of quantities continues to remain the most common form of measurement contract. It also remains the most common contractual arrangement for major construction projects in the UK (Chapter 7). The use of a schedule of quantities has many advantages over the alternative systems that are presently available (Chapter 10). The tender sums that are received from contractors are able to be judged almost against the criteria of cost, since all contractors are using the same qualitative and quantitative information. This type of documentation is still recommended as the most appropriate for all but the smallest and simplest types of building projects. Bills of quantities are referred to in JCT 98 as the contract bills.

Bill of approximate quantities

In some circumstances it is not possible to premeasure the work accurately, because parts of the design remain incomplete. In these circumstances a bill of approximate quantities can be prepared with the entire project being remeasured upon completion of the works. The JCT standard form of building contract (JCT 98) with approximate quantities would then be used. Whilst an approximate cost of the project can be obtained, the uncertainty in the design information makes any reliable forecast of cost impossible.

COST REIMBURSEMENT CONTRACTS

Cost reimbursement contracts are not favoured by many of the industry's employers, since there is an absence of a tender sum and a forecasted final account cost. Some of these types of contract also provide little incentive for contractors to control their costs, although different varieties of cost reimbursement have attempted to build in incentives for the contractor to keep costs as low as possible. Evidence from industry also shows them to be an unpopular method (Chapter 7). Cost reimbursement contracts are therefore used only in special or unusual circumstances. These might be as follows:

■ Emergency work projects, where time is not available to allow the traditional process to be used.

- When the character and scope of the works cannot be readily or easily determined.
- Where new technology is being introduced.
- Where a special relationship exists between the employer and the building contractor.

For example, part of a major highway between two important towns collapses and makes the route impassable. Diversions are put in place that increase journey times for passengers and freight traffic. It is important to reopen the highway as soon as possible. The design is started and a contractor is employed to commence preliminary works on site. Early completion of the project is the important criteria. There is insufficient time available to fully complete the design prior to construction work commencing on site. There is limited tender information and no time to obtain tenders in the usual way. This could further delay the work by anything up to three months even for a modest scheme (documentation preparation and pricing). The project is thus awarded to a firm who has successfully worked with the highways agency on previous occasions. On this project the cost reimbursement is a reasonable solution.

Cost reimbursement contracts can take many different forms but the following are three of the more popular types in use. Each of the methods repay the contractor's costs with an addition to cover profits. Prior to embarking on this type of contract it is especially important that all the parties involved are clearly aware of the definition of contractor's costs as used in this context.

Cost plus percentage

The contractor receives the costs of labour, materials, plant, subcontractors and overheads and to this sum is added a percentage to cover profits. This percentage is agreed at the outset of the project. A major disadvantage of this type of cost reimbursement is that the contractor's profits are directly related and geared to the contractor's own expenditure. Therefore, the more the contractor spends on the building works, the greater will be the contractor's profitability. Because it is an easy method to operate, this tends to be the selected method when using cost reimbursement. In the highway example described above, cost plus percentage would be the chosen method of cost reimbursement.

Cost plus fixed fee

With this method the contractor's profit is predetermined by the agreement of a fee for the work, before the commencement of the project. There is therefore some incentive for the contractor to attempt to control the costs, since the fee (contractor's profit) remains the same regardless of the actual costs involved. However, it is sometimes difficult to predict the costs of building with sufficient accuracy. Disagreement between the contractor and the employer's own professional advisers will occur where the predicted cost is widely at variance with the actual costs. The result is that the so-called *fixed fee* may thus need to be revised on completion of the project.

Cost plus variable fee

The use of this method requires a target fee to be set for the project prior to the signing of the contract. The contractor's fee is then composed of two parts, a fixed amount and a variable amount. The total fee charged then depends upon the relationship between the target cost and the actual cost. This method provides a supposedly even greater incentive to the contractor to control the construction costs. It has the disadvantage of requiring a reasonably accurate target cost to be fixed on the basis of a very *rough* estimate of the proposed project.

Examples

COST PLUS PERCENTAGE

Estimated cost	£500,000	Final cost	£511,964
Agreed 10% profit	£50,000	10% profit	£51,196
Tender	£550,000	Final account	£563,160

COST PLUS FIXED FEE

Estimated cost	£500,000	Final cost	£511,964
Fixed fee	£60,000	Fixed fee	£60,000
Tender sum	£555,000	Final account	£571,964

COST PLUS VARIABLE FEE

Estimated cost	£500,000	Final cost	£511,964
Fixed fee	£50,000	Fixed fee	£50,000
Variable fee		Variable fee	
10% ± £500,000		10% × 11,964	−1,196
Tender sum	£560,000	Final account	£560,768

The examples are for illustration only. Do not assume that the final accounts would necessarily follow these patterns. These are the simple explanations of the different calculations expected. In practice the financial adjustments are often very complex and time-consuming. And, as explained earlier, the preferred choice with cost reimbursement contracts is frequently the *cost plus percentage* version for ease of administration.

CONTRACTOR SELECTION

There are essentially two ways of choosing a contractor, either by competition or negotiation. Competition may be restricted to a few selected firms or open to almost any firm who wishes to submit a tender. The options described later are used in conjunction with one of these methods of contractor selection.

Selective competition

Selective competition is the traditional and most popular method of awarding construction contracts. It is also believed to be the fairest to all the parties concerned. In essence a number of firms of known reputation are selected by the employer on the advice of the design team to submit a price. The firm who submits the lowest tender is then awarded the contract. The *Code of procedure for single-stage selective tendering* is a useful document of good practice on guidance about awarding of construction contracts. The code takes into account the changes in buildings procurement and the general principles of current recommendations. Here are some of the more important points from this code:

- The code assumes the use of a standard form of building contract with which the parties in the construction industry are familiar. If other forms of contract are used, some modification of detail may be necessary. There are clear advantages to all parties in the knowledge that a standard procedure will be followed in inviting and accepting tenders. Figure 8.1 illustrates the traditional contractual relationships.
- The code recommends that the number of tenderers should be limited to six. The number of tenderers is restricted because the cost of preparing abortive tenders will be reflected in prices generally throughout the building industry. There are no such things as *free estimates*.
- In preparing a shortlist of tenderers, the following must be borne in mind:
  - the firm's financial standing and record
  - recent experience of building over similar contract periods
  - the general experience and reputation of the firm for similar building types
  - adequacy of management
  - adequacy of capacity

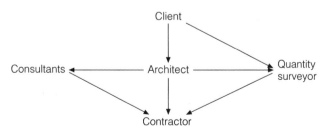

Fig. 8.1 Traditional relationships

- Each firm on the shortlist should be sent a preliminary enquiry to determine its willingness to tender. The enquiry should contain:
  - job title
  - names of employer and consultants
  - location of site and general description of the works
  - approximate cost range
  - principal nominated subcontractors
  - form of contract and any amendments
  - procedure for correction of priced bill
  - contract under seal or under hand
  - anticipated date for possession
  - contract period
  - anticipated date for dispatch of tender documents
  - length of tender period
  - length of time tender must remain open for acceptance
  - amount of liquidated damages
  - bond
  - special conditions
- Once a contractor has confirmed an intention to tender, that tender should be made. If circumstances arise which make it necessary to withdraw, the architect should be notified before the tender documents are issued or, at the latest, within 2 days thereafter.
- A contractor who has expressed a willingness to tender should be informed if not chosen for the final shortlist.
- All tenderers must submit their tenders on the same basis.
  - Tender documents should be dispatched on the stated day.
  - Alternative offers based on alternative contract periods may be admitted if requested on the date of dispatch of the documents.
  - Standard forms of contract should not be amended.
  - A time of day should be stated for receipt of tenders and tenders received late should be returned unopened.
  - The tender period will depend on the size and complexity of the job, but be not less than four working weeks, i.e. 20 days.
- If a tenderer requires any clarification, the architect must be notified and the architect should then inform all tenderers of this decision.
- If a tenderer submits a qualified tender, opportunity should be given to withdraw the qualification without amending the tender figure, otherwise the tender should normally be rejected.
- Under English law, a tender may be withdrawn at any time before acceptance. Under Scottish law, it cannot be withdrawn unless the words 'unless previously withdrawn' are inserted in the tender after the stated period of time the tender is to remain open for acceptance.
- After tenders are opened, all but the lowest three tenderers should be informed immediately. The lowest tenderer should be asked to submit a priced bill within 4 days. The other two contractors are informed that they might be approached again.

- After the contract has been signed, each tenderer should be supplied with a list of tender prices.
- The quantity surveyor must keep the priced bills strictly confidential.
- If there are any errors in pricing, the code sets out alternative ways of dealing with the situation:
  - The tenderer should be notified and given the opportunity to confirm or withdraw the offer. If it is withdrawn, the next lowest tenderer is considered. Where the offer is confirmed, an endorsement should be added to the priced bills that all rates, except preliminary items, contingencies, prime cost and provisional sums are to be deemed reduced or increased, as appropriate, by the same proportion as the corrected total exceeds or falls short of the original price.
  - The tenderer should be given the opportunity of confirming the offer or correcting the errors. Where it is corrected and is no longer the lowest tender, then the next tender should be examined. If it is not corrected then an endorsement is added to the tender.
- Corrections must be initialled or confirmed in writing and the letter of acceptance must include a reference to this. The lowest tender should be accepted, after correction or confirmation, in accordance with the alternative chosen. Problems sometimes occur because the employer will see that a tender will still be the lowest even after correction. If the first alternative has been agreed upon and notified to all tenderers at the time of invitation to tender, the choice facing the tenderer should clearly be to confirm or withdraw. The employer may require a great deal of persuading to stand by the initial agreement in such circumstances. The answer to the problem is to discuss the use of the alternatives thoroughly with the employer before the tendering process begins. The employer must be made aware that the agreement to use the code and one of the alternatives is binding on all parties. It is possible that an employer who stipulated the first alternative and subsequently allowed price correction could be sued by, at least, the next lowest tenderer for the abortive costs of tendering.
- The employer is not bound to accept the lowest or any tender and is not responsible for the costs of their preparation. There may be reasons why a decision is taken not to accept the lowest tender. Although the employer is entitled to do so, it will not please the other tenderers. The code is devised to remove such practices.
- If the tender under consideration exceeds the estimated cost, negotiations should take place with the tenderer to reduce the price. The quantity surveyor then normally produces what is called 'reduction or addendum bills'. They are priced and signed by both parties as part of the contract bills.
- The provisions of the code should be qualified by the supplementary procedures specified in EU directives which provide for a 'restrictive tendering procedure' in respect of public sector construction contracts above a specified value. Guidance on the operation of this procedure is given in *DOE Circular 59/73 (England and Wales)* and *SDD Circular 47/73 (Scotland)*, both of which are obtainable from the Stationery Office.

This method of contractor selection is appropriate for almost any type of construction project, where a suitable supply of contractors are available.

Open competition

With open competition the details of the proposed project are often advertised in the local and trade publications, or through the local branch of the *Construction Confederation*. Any contractor who then feels willing and able to carry out such a project can request the contract documentation. This method has the advantage of allowing new contractors or contractors who are unknown to the design team the possibility of submitting a tender for consideration. In theory any number of firms are able to submit a price. In practice there is usually a limit on the number of firms who will be supplied with the tender documents. Unsuitable firms are removed from the list where the number of firms becomes too large. The preparation of tenders is both expensive and time-consuming. The use of open tendering may relieve the employer of a moral obligation of accepting the lowest price, because firms are not generally vetted before tenders are submitted. Factors other than price must also be considered when assessing these tender bids, such as the capability of the firm who has submitted the lowest tender. There is no obligation on the part of the employer to accept any tender, should the employer consider that none of the contractors' offers is suitable. Prior to the Second World War, many major building contracts, particularly those in the public sector, were awarded using this method. Although it is still commonly used on minor works, its use on larger projects has been curtailed. There are no known records of the numbers of firms on an open tender list, although over fifty firms tendering for a single project have been known.

Negotiated contract

This method of contractor selection involves the agreement of a tender sum with a single contracting organization. Once the documents have been prepared, the contractor prices them in the usual way. The priced documents are then passed to the quantity surveyor, who will check the reasonableness of the contractor's rates and prices. The two parties then arrange a meeting to discuss the queries raised by the quantity surveyor and the negotiation process begins. It is important for both parties to arrive at reasonable rates and prices. Eventually a tender price is agreed which is acceptable to both parties. There is an absence of any competition or other restriction, other than the social acceptability of the price. This frequently results in tender sums that are higher than might have been obtained by using one of the previous competitive procurement methods. It is believed that, under normal circumstances, negotiated tenders are approximately 5 per cent higher in price than competitive tenders. Negotiation does, however, have particular applications as follows:

- Where a business relationship exists between the client and the building contractor.

- Where only one firm is capable of undertaking the work satisfactorily.
- On a continuation contract where the building contractor is already established on site.
- Where an early start on site is required by the client.
- Where it is beneficial to bring the contractor in during the design stage, to advise on constructional difficulties and how they might best be avoided.
- To make use of the contractor's buildability expertise.

Competitive tendering can create financial problems for both the industry and its clients, due to too keen pricing which in the end benefits no one. Negotiated contracts do result in fewer errors in pricing since all of the rates are carefully examined and agreed by each party. This will result in fewer exaggerated claims in an attempt to recoup losses on cut-throat pricing. This type of procurement can also involve the contractor in some participation during the design stage, which can result in on-site time and cost savings, thus improving value for money. It should also be possible to achieve greater cooperation during the construction process between the design team and the contractor. However, public sector clients tend not to favour negotiated contracts because of:

- Higher tender sums incurred
- Public accountability
- Suggestion of possible favouritism

CONTRACTUAL OPTIONS

The following contractual options are an attempt to address the employer's objectives associated with the cost, time and quality of construction. They are not mutually exclusive. For example, a serial contract can be awarded using in the first instance a design and build (D&B) arrangement. Fast tracking may be used in conjunction with management contracting. All of these options will also need to include either a selective or negotiated arrangement in the selection of a contractor.

Early selection

Early selection is also known as two-stage tendering. Its main aim is to involve the chosen contractor for the project as soon as possible, so that the contractor can have an input at the design stage. It therefore seeks to succeed in getting the firm who knows what to build (the designer) in touch with the firm who knows how to build it (the constructor), before the design is finalized. The contractor's expertise in construction methods can thus be harnessed with that of the designer to improve, for example, the buildability criteria of the project. A further advantage is that the contractor may be able to start work on site sooner than where the more traditional methods of procurement are used. In the first instance, an appropriate contractor must be selected for the project. This is often done through some form of competition and can be achieved by selecting suitable firms to price the major items of work connected with the project. A simplified bill of quantities can be prepared which might include the following items:

- Site-on costs on a time-related basis
- Major items of measured works
- Specialist items

Specialist items allow the main contractor the opportunity of pricing the profit and attendance sums. The contractors should also be required to state their overhead and profit percentages. The prices of these items will then form the basis for the subsequent and more detailed price agreement as the project gets under way.

The NJCC have recommended a *Code of procedure for two-stage selective tendering*. This code is not concerned with aspects of the design, which may in this process involve the contractor. The code assumes the use of a standard form of building contract after the second stage has been completed. During the first stage it is important to:

- Provide a competitive basis for selection
- Establish the layout and design
- Provide clear pricing documents
- State the respective obligations and rights of the parties
- Determine a programme for the second stage

Many of the conditions already outlined for single-stage selective tendering apply equally to two-stage tendering. Acceptance of the first-stage tender is a particularly delicate operation. The employer does not wish to be in the position of having accepted a contract sum at this stage. The terms of the letter of acceptance must be carefully worded to avoid such an eventuality. Depending upon the circumstances, it may be that a contract has been entered into. The question may be, what are the terms of the contract? There are two pitfalls:

- No contract exists: this is likely in many cases.
- A contract binds the employer to pay and the contractor to build.

Existence of a binding contract could be the far worse situation if insufficient care is given to the drafting of the invitation to tender, the tender and the acceptance. After a contractor has been appointed, all unsuccessful tenderers should be notified and, if feasible, a list of first-stage tender offers should be provided. If cost was not the sole reason for acceptance, the reasons for this should also be stated.

Design and build

Design and build projects aim to overcome the problem of having separate design and construction processes by providing for them within a single organization. The method has gained in popularity (Chapter 7). This single firm is generally the building contractor, who may employ the designers in-house or be responsible for employing consultants directly under their control. The major difference is that instead of approaching the designer for a building, the employer briefs the

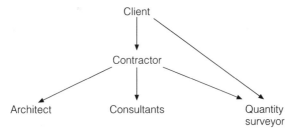

Fig. 8.2 Design and build relationships

contractor direct (Figure 8.2). The employer may choose to retain the services of an architect or quantity surveyor to assess the contractor's design or to monitor the work on site. The prudent employer will always want some form of independent advice.

The design evolved by the contractor is more likely to be suited to the needs of the contractor's organization and construction methodology, and this should save construction time and construction costs. Some argue that the design will be more attuned to the contractor's construction capabilities, rather than the design requirements of the employer. The final building should result in lower production costs on site and an overall shorter design and construction period, both of which should provide price savings to the employer. There should also be some supposed savings on the design fees, even after taking into account the necessary costs of any independent architectural advice. A further advantage to the employer is in the implied warranty of suitability, because the contractor has provided the design as a part of the all-in service.

Where CD 98 is used, the contractor has only a duty to use proper skills and care. An apparent disadvantage to the employer is the financial disincentive relating to possible changes to the design by the employer during construction. Where an employer considers these to be important to keep abreast of changing technologies or needs, excessive costs may be required to discourage them in the first place, or to allow for their incorporation within the partially completed building.

Design and build projects usually result in the employer obtaining a single tender from a selected contractor. Where some form of competition in price is desirable then both the type and quality of the design will need to be taken into account. This can present difficulties in evaluation and comparison of the different schemes. Many of these comments relate generally to design and build projects, not specifically to those which may use the JCT CD 98 form of contract with contractor's design.

Here are some of the advantages claimed for the design and build approach:

- The contractor is involved with the project from inception and is thus aware of all of the employer's requirements.
- The contractor is able to use their specialized knowledge and methods of construction in evolving the design.

- The time from inception to completion should be able to be reduced due to the telescoping of the various parts of the design and construction processes.
- There can be no claims for delays due to a lack of design information, since the contractor is in total control of it.
- There is direct contact between the employer and the contractor.

The major criticisms against design and build have tended to be of an architectural or aesthetic nature. After all most builders are not designers but constructors. In an attempt to combat this criticism, the building companies who are extensively involved in design and build work will have an architectural in-house office or employ an outside firm of architects. The rapid increase in this form of procurement indicates that architectural considerations are now given due recognition by the best design and build contractors.

Clients used to employ architects who would then employ building constructors. The trend at present is for clients to employ building firms who in turn employ designers. Design and build has changed the emphasis, in order to reduce the possible waste in resources and to increase or add value on behalf of the client.

Package deal

In practice the terms 'package deal' and 'design and build' are interchangeable. Design and build normally refers to a bespoke arrangement for a one-off project. But the package deal is strictly a special type of design and build project where the employer chooses a suitable building, often from a catalogue. The employer may also be able to view completed buildings of a similar design and type that have been constructed elsewhere. This type of contract procurement has been used extensively in the past for the closed systems of industrialized system buildings of timber or concrete. Multi-storey office blocks and flats, low-rise housing, workshop premises, farm buildings, etc., have been constructed on this basis. The building owner typically provides the package deal contractor with a site and supplies the user requirements or brief. An architect may be independently employed to advise on the building type selected, to inspect the works during construction or to deal with the contract administration. This architect's role is useful to the employer for those items which are outside the scope of the system superstructure. The type of building selected is an off-the-peg structure and often it can be erected very quickly. There is even less scope, however, for variations than with the more usual design and build approach, should the employer want to change aspects of the constructional detailing. It cannot be automatically assumed that this type of procurement will be a more economic solution to the employer's needs, either initially or even in the long term. Some system buildings constructed in the 1960s are now very costly to maintain. Initial costs can also be high but they have the advantage that they can be constructed quickly on site. This is due to the relative completeness of the design, the availability of standard components and the speed of construction on site.

Design and manage

Design and manage is really the consultant's counterpart to contractor's design and build. In this case the design manager, who may be an architect, engineer or surveyor, has full control not just of the design phase but also of the construction phase. The design and manage firm effectively replaces the main contractor in this role, which is now largely management and organization, administration and coordination of subcontractors. The design manager is responsible for all aspects of construction, including the programming and progressing and the rectification of any defects which may arise. The building contract is between the design and manage firms and the employer, giving the employer a single point of contractual responsibility. The work is generally let through competition in work packages to individual subcontracting firms. This method of procurement therefore offers many of the advantages of traditional tendering coupled with design and build. The design and manage firm will of course need to engage its own construction managers or develop existing staff who have potential in this direction. It will also need to consider continuity with this type of work.

It is suitable for all types and sizes of project, but employers undertaking large projects may, due to past experience, prefer to use a more traditional form of procurement using one of the larger contractors. A major disadvantage is with regard to the site facilities which will need to be provided by the design and manage consultant, who may need to be hired in a similar way to the subcontractors.

This type of procurement method should be able to offer comparative completion times when compared with the other methods that are available. Since there is the traditional independent control of the subcontractor firms, this should ensure a standard of quality at least as good as that provided by the other contracting methods. In terms of cost, since the work packages will be sought through competition, this will be no more expensive.

Turnkey method

Turnkey contracting is still somewhat unusual in the UK and has thus not been used to any large extent. It has had, however, certain notable successes in the Middle and Far East. The true turnkey contract includes everything that is required and necessary. This normally means everything from inception up to occupation of the finished building. The method receives its title from the *turning-the-key* concept whereby the employer, when the project is completed, can immediately start using the project since it will have been fully equipped by the turnkey contractor, including furnishings. Some turnkey contracts also require the contractor to find a suitable site for development. An all-embracing agreement is therefore formed with the one single administrative company for the entire project procurement process. It is therefore an extension of the traditional design and build arrangements, and in some cases it may even include a long-term repair and maintenance agreement.

On industrial projects the appointed contractor is also likely to be responsible for the design and installation of the equipment required for the employer's manufacturing process. This type of procurement method can therefore be appropriate for use on highly specialized types of industrial and commercial construction projects.

The entire project procurement and maintenance needs can therefore be handled by a single firm accepting sole responsibility for all events. However, it has been suggested that the employer's ability to control costs, quality, performance, aesthetics and constructional details will be very variable and severely restricted by using this procurement method. A contractor who undertakes such an all-embracing project will require a variety of strengths and may well have definite ideas about the importance of the different aspects and outcomes of the scheme.

Management contract

Management contracting evolved at the beginning of the 1970s in the UK with an aim of building more complex projects in a shorter period of time and for a lower cost. It may therefore be argued that the more complex the project, the more suitable management contracting may be. This method is also appropriate to a wide range of medium-sized projects. In many ways management contracting was ahead of its time.

The term *management contract* is used to describe a method of organizing the building team and operating the building process. The main contractor provides the management expertise required on a construction project in return for a fee to cover the overheads and profit. The intention is to place the main contractor in a professional capacity to be able to provide the management skills and practical building ability for a fee. The contractor does not therefore, in theory, participate in the profitability of the construction work. The construction work itself is not undertaken by the contractor, nor does the contractor employ any of the labour or plant directly, except with the possibility of setting up the site and those items normally associated with the preliminary works. Because the management contractor is employed on a fee basis, the appointment can be made early on in the design process. The contractor is therefore able to provide a substantial input into the design, particularly those associated with the practical aspects of constructing the building. Each trade section required for the project is normally tendered for separately by subcontractors, either on the basis of measurement or a lump sum. This should result in the least expensive cost for each of the trades and thus for the construction works as a whole. The work on site needs a considerable amount of planning and coordination, more so than a traditional procurement arrangement. This is the responsibility of the management contractor and an inherent part of the acquired skills.

In common with all procurement methods, there are advantages and disadvantages. It is somewhat open-ended, since the price can only be firmed up after the final works package quotation has been received. The later in the contract the work is let, the less time there will be for negotiating price reductions overall without seriously impairing a section of the works.

Management fee

Management fee contracting is a system whereby a contractor agrees to carry out building works at cost. In addition a fee is paid by the employer to cover the overheads and profit. Some contractors are prepared to enter into an agreement to offer an incentive on the basis of a target cost. This type of procurement is a similar approach to management contracting and cost-plus contracts and has therefore similar advantages and disadvantages.

An alternative approach can also be used where a bill of quantities can be prepared and priced net of the contractor's overheads and profit, or just the profit. These items are then recovered by means of a fee. The system can be as flexible and adaptable as the parties wish. Invariably, as with cost-plus contracting, the fees are percentage related unless some reasonably accurate forecast of cost can be made. Different contracting firms who use management fee contracting have different ways of determining the fee addition. In any case the total cost is largely unknown until completion is achieved and the records agreed. Overspending is therefore much more difficult to control, and any savings in cost which are needed tend to be required and made on the later sections of the project.

Construction management contracting

This offers a further alternative procedure to the management contract. The main difference is that the employer chooses to appoint a construction management contractor who is then responsible for appointing a design team (Figure 8.3) with the approval of the employer. The employer chooses to instruct the constructor rather than the designer. The construction management contractor is thus in overall project control of both the design and the construction phases. There are similarities with this method and design and build. A major difference is that the contractor would invariably appoint outside design consultants rather than choosing to employ an in-house service. The employer may thus form some contractual relationship with these consultants. In reality it is a reversal of the traditional arrangements.

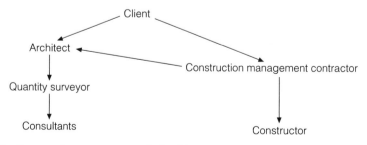

Fig. 8.3   Construction management relationships

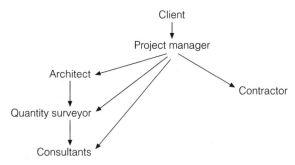

Fig. 8.4   Project management relationships

Project management

Descriptions of the different procurement arrangements may mean different things in different parts of the world, different industries or even in different sectors of the construction industry. They may even have different interpretations within a single country, such as the UK. Project management in this context is a function that is normally undertaken by the employer's consultants rather than by a contractor. Contractors do undertake project management but in a different context from that associated with procurement. The title of project coordinator is similar terminology and perhaps better describes the role of a project manager. The employer appoints the project manager who in turn appoints the various design consultants and selects the contractor (Figure 8.4). It is a more appropriate method for the medium- to large-sized project that requires an extensive amount of coordination, owing to its complexity. The function of the project manager is therefore one of organizing and coordinating the design and construction programmes. Any person who is professionally involved in the construction industry can become a project manager; it is the individual rather than the profession which is important.

In general terms, the need for a project manager is to provide a balance between function, aesthetics, quality control, economics, and the time available for constructing. The project manager's aims are to achieve an efficient, effective and economic deployment of the available resources to meet the employer's requirements. The tasks to be performed include identifying those requirements, interpreting them as necessary, and communicating them clearly to the various members of the design team and through them to the constructor. Programming and coordinating all of the activities and monitoring the work up to satisfactory completion are also a part of this role. A significant difference between this system and the majority of the others described is that the employer's principal contact with the project is through the project manager rather than through the designer.

The British Property Federation (BPF) system

The British Property Federation (BPF) represents substantial commercial property interests in the UK, and is thus able to exert some considerable influence on the building industry and its associated professions. The introduction of this system of building procurement to the industry created considerable interest due largely to the fact that it proposed very radical changes to the status quo and practices and procedures that were being used. Some members of the design team felt threatened as their traditional roles and importance were questioned. The BPF system unashamedly makes the interests of the employer paramount. It attempts to devise a more efficient and cooperative method of organizing the whole building process from inception up to completion by making genuine use of all who are involved with the design and construction process. Its development was due to employer dissatisfaction with the existing arrangements, where it was claimed that buildings on the whole cost too much, took too long to construct (when compared with other countries around the world) and did not always produce appropriate or credible results. The BPF manual is the only document which sets out in detail the operation of a system. Whilst the system may appear to be a rigid set of rules and procedures, its originators claim that it can be used in a flexible way and in conjunction with other methods of procurement. Since the system has been devised almost entirely by only one party to the building contract, it lacks the considerations and compromises that are inherent in agreements devised by joint employer and contractor bodies such as JCT. The system was designed as an attempt to help change outdated attitudes and to alter the way in which the various members of the professions and contractors dealt with one another. It seeks to create better cooperation between the various members of the building team, encourages motivation, and removes any possible overlap of effort amongst the design team. Innovations, which were post-Latham, include single-point responsibility for the employer, utilization of the contractor's buildability skills, reductions in the pre-tender period, redefinition of risks, and the preference for specifications rather than bills of quantities. This is despite the desire of contractors for measured quantities of work.

Fast tracking

Fast tracking results in the letting and administration of multiple construction contracts for the same project at the same time, as outlined in Figure 8.5. It is appropriate to large construction projects where the employer's needs are to complete the project in the shortest possible time. The process results in the overlapping of the various design and construction operations of a single project. These various stages can therefore result in the creation of separate contracts or a series of phased starts and completions. When the design for a complete section of the works, such as foundations, are completed the work is let to a contractor who will commence with this part of the construction work on site, whilst the remainder of the project is still being designed. The contractor for this stage or section of the project will then see this work through to completion. At the same time as this is being done another work section may be let, continuing and building upon what the first contractor has already completed. The second contractor may or may not be the same as the first contractor, depending upon the amount of specialization

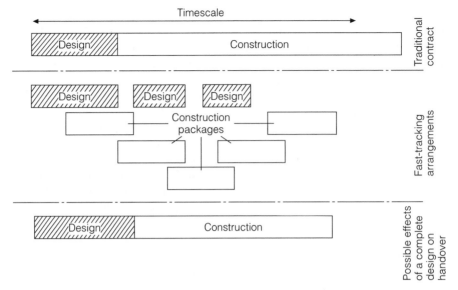

Fig. 8.5   Different contractual arrangements and their effects upon time

and competition that is involved. This staggered letting of the work has the objective of shortening the overall design and construction period from inception through to handover as shown in Figure 8.5. This type of procurement arrangement requires a large amount of foresight, since the later stages of the design must take into account what has by now already been completed on site. This type of arrangement also requires considerably more organization and planning, particularly from the members of the design team. In practice, where an efficient application of this procurement procedure is envisaged, then a project coordinator will need to be engaged. Although the handover date of the project to the employer should be much earlier than any of the other methods which might be used, this might be at the expense of the other facets of cost and performance. These aspects may be much inferior to those achieved by the use of the more traditional methods of procurement and arrangement.

Measured term

A measured term contract is suitable for major maintenance work projects. It is often awarded not just for a single building project, but to cover a number of different buildings belonging to the same client. A contractor will usually be appointed for a specified period of time, although this may be extended, depending upon the necessity of maintenance standards and requirements and the acceptability of the contractor's performance. The contractor is initially offered the maintenance work for a number of trades, such as painting and decorating, plumbing, etc. The work, when completed, will then be paid for using rates from an agreed priced schedule. This schedule may have been prepared specifically for the project concerned, or it may be a standard document similar to the Property Services Agency schedule of rates. Where the employer has provided the rates for the work, the contractor is normally given the opportunity of quoting a percentage addition or reduction to these rates. The contractor offering the most advantageous percentage will usually be awarded the contract. An indication of the amount of work involved over a defined period of time would therefore need to be provided for the contractor's better assessment of the quoted prices.

Serial tender

Serial tendering is a development of the system of negotiating further contracts, where a firm has already successfully completed a project, for work of a similar type. Initially the contractors tender against each other, possibly on a selective basis, for the first project. However, there is a contractual mechanism that several additional projects will automatically be awarded using the same contract rates. Some allowances are normally made for inflation, or perhaps more commonly increased costs are added to the final accounts. Contractors are therefore aware at the tender stage for the first project that they can expect to receive a further number of contracts in due course. This helps to provide some continuity in their workloads and provides an incentive for investment. Conditions would be written into the documents to allow further contracts to be withheld where the contractor's performance was less than satisfactory.

Serial contracting should result in lower costs to contractors since they are able to gear themselves up for such work, perhaps purchasing mechanical plant and equipment that might otherwise be too speculative. They should therefore be able to operate to greater levels of efficiency. The employer will also achieve some financial gain since some of these lower production costs will find their way into the contractor's tenders.

A further advantage claimed for the use of serial tendering is that the same design and construction teams can remain involved for each of the projects which form part of the serial arrangement. This is a form of partnering where the skills of a team are utilized over a period of time. This regular and close working association aids and develops an expertise, which accelerates the production of the work and eases any anticipated problems. In turn the operatives on site improve their efficiency as they progress from project to project.

Serial projects are particularly appropriate for buildings of a similar nature, such as housing and school buildings in the public sector where a large number of comparable schemes are constructed. This method may also be usefully employed in the private sector in the construction of industrial factory and warehouse units. It has also been successfully used with industrialized system building, where local authorities have worked together in a form of building consortium to gain marketing advantages from the construction industry and its material and component manufacturers.

Continuation contract

A continuation contract is an ad hoc arrangement to take advantage of an existing situation. Projects in an expanding economy are sometimes of insufficient size to cope with projected demands even before they become operational. During their construction, an employer may choose to provide additional similar facilities which may be constructed upon the completion of the original scheme. Such additions, often because of their size and scope, are beyond the definition of extra work or variations. For example, a housing project may be awarded to a contractor to build 300 houses. Due to perhaps an underestimation in demand, and where additional space permits, a continuation contract may be awarded to the same contractor to build an extra 100 houses on an adjacent site. Another example can occur where during the construction of a factory building it becomes apparent that it will be insufficient to meet the increased demands for the factory's product. Extensions to the factory building may therefore be agreed on the basis of a continuation contract. This will use the same designers and constructors, and the same contract prices adjusted only to allow for increased costs caused by inflation.

A continuation contract can be awarded as an add-on to the majority of the contractual arrangements. Continuing with the same building team, where this has already proved to be satisfactory to the employer, is a sensible arrangement under these circumstances. Each of the parties will already be familiar with each other's working methods and will be able to offer some cost savings since they are already established on the site. Both the employer and the contractor may wish to review the contractual conditions and the new market factors which may have changed since the original contract was signed. It is also not unreasonable to expect that the contractor will want to share in the monetary savings that will be available from the already established site organization from the initial project.

Speculative work (develop and construct)

Because of the market demand, many building firms have organized a part of their function into a speculative department. Some contractors are entirely speculative builders. These firms or divisions purchase land, obtain planning approvals, design and construct buildings for sale, rent or lease. It is also described as develop and construct. Housing is the common type of speculative unit, but factories, whole industrial estates, new office buildings and refurbishment projects can also be constructed on a speculative basis to meet further and expected demand. The contractors may be in partnership agreements with financial institutions and developers who have already identified that a demand for a particular type of development exists within a certain geographical area. Such assumptions are based upon detailed market research of the area, which is undertaken prior to the development taking place. The developer or speculative builder may employ their own in-house design staff, or may employ consultants to do the work for them. Whilst design and build is done for a particular employer, speculative building is carried out on the basis that there is a need for certain types of buildings and that these will be purchased, leased or rented at some later date.

Direct labour

Some employers, notably local authorities and large industrial firms, have departments within their organizations which undertake building work. In some cases they may restrict themselves to repair and maintenance works alone, or to the construction of minor projects. Other direct labour departments are much larger and are capable of carrying out any size of building project. Direct labour organizations within the public sector were at one time numerous. Successive governments, of different political persuasions, have for the time being abandoned many of these direct departments.

The employer, in order to execute construction projects in this way, needs to have either an extensive building programme or to be responsible for a large amount of repair and maintenance work. In-house services of this type allow the employer, at least in theory, much greater control than using some of the other methods which have been described. In practice the level of control achieved may be less than satisfactory, due to the accounting procedures which may be employed. Employers who choose to undertake their building works by this method often do so for either political or financial reasons.

REVIEW

At the inception of any construction project, both the designers and the contractors should enquire about the employer's real objectives. Almost without exception the employer's needs will be an amalgam of functional and aesthetic requirements, built to a good standard of construction and finish, completed when required, at an appropriate cost, offering value for money and easily and economically maintained (see early in this chapter). The contractual procurement arrangements must aim to satisfy all of these requirements, some in part and others as a whole, trading between them to arrive at the best possible solution.

The traditional approach has been to appoint a team of consultants to produce a design and estimate, and to select a constructor. The constructor would calculate the actual project costs, develop a construction programme to fit within the contract period, organize the workforce and material deliveries and build to the standards described in the contract documentation.

In practice the quality of the work was often poor, costs were higher than expected, and even where projects did not overrun the contract period, they took much longer to complete than similar projects in other countries around the world. Traditionally little attention or consideration was given towards future repairs or maintenance aspects.

The obvious deficiencies in the above, supposedly due to the separation of the design role from the construction effort, have been known for some time. However, many employers have generally wished to retain the services of an independent designer, believing they would serve their needs better than the contractors with their own vested interests. Employers also wished to retain the competition element in order to keep costs to a minimum and to help improve the efficiency of the contractor's own organization. In more recent years they have seen some benefits from buildability, benefits which have been achieved through the early and integrated involvement of the contractor during the design stage. This has also had a spin-off effect in reducing costs and the time the contractor is on site. Employers were, however, still loath to commit themselves to a single contractor at the design stage when costs might still be imprecise, and where their bargaining position might be seriously affected before the final deal could be struck.

Increasingly, however, contractors attempted to market their design and build methods as possible solutions to the employer's increasing dissatisfaction and frustration with existing procedures. The benefits of a truly fixed price, a single point of responsibility and having the contractor assume full responsibility for design and construction of the building, all were attractive propositions. Remedies open to the employer in the case of default or delay were now no longer a matter for discussion between the designer and the constructor. If necessary the employer could always employ professional advisers to oversee the technical and financial aspects of the project.

The entrepreneurs of the construction industry are always seeking out new ways of satisfying the increasing demands of clients. Some are busy examining the methods that are in use in the USA, and thus seek to import systems which they believe are superior and thus help to improve the falling image of the construction industry. Others feel that the employers' interests would be better served by an overarching project manager or construction manager. Contractors are themselves continuing to evolve changes in the way in which they employ and organize their workforce. An increase in subcontracting has occurred, as fewer and fewer operatives are directly employed by main contractors. Management contracting recognizes and encourages these developments as trends which are desirable.

The construction industry will continue to evolve and adapt its systems and procedures to meet the new demands. New procurement methods will be developed which utilize new technologies and new ways of working. Construction projects are different from the majority of manufactured goods, because they are procured in advance of their manufacture. The majority of the projects are also different from the previous ones and often incorporate some new characteristics. Some projects are almost wholly original and all depend upon their peculiar site characteristics. In order to achieve value for money and client satisfaction, the appropriate procurement procedures must be selected or adapted to suit the individual needs of the employer and the project.

CHAPTER 9

CONTRACT SELECTION

The selection of appropriate contract arrangements for any but the simplest type of project is difficult owing to the diverse range of options and professional advice. Much of the advice is in conflict. For example, professional design and build contractors are unlikely to recommend the use of an independent designer. Such organizations are also likely to believe that the integration of design and construction will result in better buildings, at a more economical cost and improve overall client satisfaction. Many of the design professions hold the opposite viewpoint, and believe that their independent approach will produce the best solution, particularly in the long term. In the past two decades there has been a significant shift in the way that construction projects are procured. This is partially in response to the structural changes which have occurred within the construction industry. Personal experiences, prejudices, vested interests, the general desire of many to improve the system and the familiarity with particular methods are factors that have influenced capital project procurement recommendations. Those who have had a bad experience with a particular procurement method will be reluctant at worst and cautious at best in recommending such an approach again. When this is suggested they will resist such a course of action and, where the results are then as expected, this further reinforces the view that such a method of procurement has but limited use.

In reality too little is known and understood. Too little research has been undertaken properly to evaluate the various procurement options. In practice this is difficult and complex because there are both successful and unsuccessful projects that have used identical means and methods of contract procurement. It is now believed that personal factors of all those involved have a major influence upon the possible outcomes. Nevertheless, it is essential to have more reliable information, based on empirical evidence rather than on subjective intuition and bias. It is also desirable to at least attempt to eliminate hearsay and folklore opinion.

The advice, wherever it originates, should always recommend a course of action which best serves the client's interests, not the advice giver's. This is true professionalism even though, for the party or organization concerned, it may mean losing a commission.

The employer may need to be convinced that a particular method and procedures which are recommended should be adopted as the most appropriate approach. This may at times be against the employer's own better judgment or preferences, particularly where there is some familiarity with the other options that are being suggested. Some clients are resistant to change, and may require persuasion to adopt what might appear to be radical proposals. It may be necessary also to accept that because of the mechanisms employed, a better course of action cannot easily be explained until after the event, and that such a recommendation may be rejected at the outset as being speculative.

Overseas clients working in different cultural environments may find that the traditional contractual procedures are wholly alien to their needs and expectations. They may require even greater assurances that their project should be constructed using a particular procurement method. Individuals working within the same team will also have different expectations and views on the optimum procurement path.

Procurement advisers should offer advice in the absence of any vested interests or personal gains. Any associated interests which might influence the judgments should be declared, in order to avoid possible repercussions at a later stage. Whilst the professional institutions retain codes of conduct and disciplinary powers governing their members' activities, there has been a blurring of commercial and professional attitudes with the former vying for position with the latter. Outmoded approaches and procedures should be removed when they no longer serve any really useful purpose. Sound, reliable and impartial advice is necessary from those who have the proficient skills, knowledge and expertise.

EMPLOYERS' REQUIREMENTS

A confirmed cynic once described the construction industry as 'the design and erection of buildings that satisfy the architect and constructor alone and not the employer'. In addition, buildings often failed to function correctly, were too expensive and took too long to construct. As in most cynicism the case is vastly overstated, yet there is enough truth not to dismiss it immediately.

One of the procurement adviser's main problems is in separating the employer's wants and needs. The employer's main trio of requirements are shown in Figure 9.1. The numerical values result from research and attempt to weigh these broad objectives in order of their relative importance. They reflect the 'average' or typical client, and offer some guidance on the main issues associated with a construction project. The weightings depend upon the employer's own aims and objectives. The attempted optimization of these factors may not necessarily achieve the desired solution. For example, an employer may desire exceptional quality standards and future performance, and they may be prepared to incur extra costs or a longer contract period. Alternatively a client may require the early use of the project as the main priority and be prepared to sacrifice a bespoke design or even high quality standards in order to achieve this. Rarely are clients prepared to pay for the highest known quality. Clients may also choose to set contrasting objectives

Performance (45)	Client (100) Cost (35)	Time (20)
Function (25)	Initial (20)	
Technology (15)	Life cycle (15)	
Aesthetics (5)		

Factors to consider

Quality	Budgets	Design length
Standards	Estimates	Start dates
Layout	Tender sums	Handover
Appearance	Cash flow	Completion
Maintenance	Final Account	
	Costs in use	

Fig. 9.1   Client's main requirements (%)

which are difficult to achieve using conventional procurement procedures. This may require the adviser to devise ingenious methods of procurement in order to satisfy such objectives.

In broad terms the employer's objectives and the procurement path will need to identify the following priorities:

- An acceptable layout for the project in terms of function and use.
- An aesthetically pleasing design not only to the employer, but also as viewed by others.
- The final cost of the building should closely resemble the estimate.
- The quality should be in line with current expectations.
- The future performance of the building and the associated costs-in-use should fit within specified criteria.
- The project is available for handover and occupation on the date specified for completion of the works.

FACTORS IN THE DECISION PROCESS

The following need to be considered when choosing the most appropriate procurement path for a proposed project. Their importance and emphasis will differ on individual projects, types of clients and the relevant market conditions at the time of procurement.

- Size
- Cost
- Time
- Accountability
- Design
- Quality assurance
- Organization
- Complexity
- Risk
- Market
- Finance

Project size

Projects of a small size are not really suited to the more elaborate forms of contractual arrangements, since such procedures are likely to be too cumbersome and not cost-effective. Smaller-sized schemes therefore rely upon the traditional and established forms of procurement, such as a form of competitive tendering or design and build. The medium- to large-scale schemes can use the whole range of options which are available. On the very large schemes a combination of different arrangements may be required to suit the project as a whole. On some of the very largest projects, bespoke arrangements might be introduced to meet particular circumstances.

Cost

Open tendering will generally secure for the client the lowest possible price from a contractor. This is the evidence that competition helps to reduce costs through efficiency, and lowers the price to the customer. There are limits to how far this theory can be applied in practice. If a large number of firms, for example, have to prepare detailed tenders then this increases industry's costs, which must be either absorbed or passed on to the industry's employers through successful tenders. There is no such thing as a free estimate! Negotiated tenders supposedly add around 5 per cent to the contract price. In the absence of competition, contractors will price the work *up to what the traffic will bear*. Some will argue that this is a major factor in determining project costs. Projects which require unusually short contract periods incur cost penalties, largely to reflect the demands placed upon the constructor for overtime working and rapid response management. The imposition of conditions of contract which favour the client or insist upon higher standards of work than are usual also increase building costs. The employer under these circumstances may end up paying more for the stricter conditions which might not be necessary or for a quality of work which adds little to the overall ambience of the project and is thus not required.

Employers in the past have been overly concerned with a lowest tender sum often at the expense of other factors, such as the principles behind the indeterminate life cycle cost. Cash flow projections which might have the effect of reducing the timing of expenditures are often ignored except perhaps by the most enlightened employers. A cash flow analysis, for example, may be able to show that the lowest tender is not always the least expensive solution. The timing of cash flows on a large project may have a significant effect upon the real costs to the employer. With projects that include options on construction method then the cash flow analysis becomes even more important. Procurement and contractual arrangements which take into account these factors should be considered more frequently.

The cost of the project is a combination of land, construction, fees and finance, and the employer will need to balance these against the various procurement systems that are available. In terms of cost savings a less expensive site elsewhere, rather than reducing the costs of quality, may be more appropriate.

Design and build projects show some form of cost savings in terms of professional fees, although the precise amount is difficult to calculate since fees are absorbed within other charges submitted by the contractor in the bid price. Where the building is of a relatively straightforward design such as a standard warehouse unit or farm building, then it can be more cost-effective to use a building contractor who has already completed similar projects rather than to opt for a separate design service. Cost reimbursement appears to be a fair and reasonable way of dealing with construction costs in a fair world. Society is not, and such an arrangement is often too open-ended for many of the industry's employers. Some clients may choose to use it out of necessity, setting objectives other than cost more highly. The majority of employers, however, will need to know as a minimum the approximate cost of the scheme before they begin to build, and there is nothing new in this (Luke 14:28).

Where a firm price is required before the contract is signed then one of several procurement arrangements can be used (Chapter 8). A firm price arrangement relies to a large extent on a relatively complete design being available. Where more price flexibility is possible then one of the more advanced forms of procurement might be used whilst at the same time achieving a measure of cost control.

Provisions exist under most forms of contractual arrangement for a fixed or fluctuating price agreement. The choice is influenced by the length of the contract period and the current and forecasted rates of inflation. Where the inflation rate is small in percentage terms and falling, then a fixed-price arrangement is preferable. When the rate of inflation is small and stable then it is common to find projects of up to 36 months' duration set up on this basis. When the rate of inflation is high, and particularly when it is rising then contractors will be reluctant to submit fixed-price tenders for more than about 12 months' duration. Although a fixed price is attractive to the employer, it cannot be assumed that this will necessarily be less expensive than a fluctuating type arrangement. Some contract conditions limit fluctuation reclaims to increases caused directly by changes in government legislation.

It is difficult to make cost–procurement comparisons, even where similar projects are being constructed under different contractual arrangements. It can be argued that competition reduces price, price certainty in the case of premeasurement or fixed price might not be the most economic, and that using serial tendering, bulk purchase agreements and similar concepts are ways of reducing building costs. Cash flows and projected life cycle costs on larger projects should be considered in any overall cost evaluation.

There have been suggestions in recent years that the lowest price might not be to the employer's best advantage. It is difficult to argue this case where the specification has been correctly prepared and contractors are simply competing on cost alone. However, it is important to realize that the least expensive design is necessarily always in the employer's best interests. It is frequently a sensible approach to spend more initially on the basis of recouping savings later during the project's life. Sometimes access to funds initially prohibits this option. Here are the cost factors to consider:

- Price competition/negotiation
- Fixed-price arrangements
- Price certainty
- Price forecasting
- Contract sum
- Bulk purchase agreements
- Payments and cash flows
- Life cycle costs
- Cost penalties
- Variations
- Final cost

Time

The majority of employers, once they have made the decision to build, want the project to be completed as quickly as possible. The design and construction phases in the UK are known to be lengthy and protracted. Some of the apparent delays are linked to the drawn-out planning processes rather than the actual design or the construction phases. It is difficult to make comparisons on a global basis since there are a wide variety of influences to be considered such as methods of construction, safety, organization of labour and quality assurance. This is evident from studies comparing UK and US performances. Construction techniques also vary and these in turn produce different qualities and costs. The time available will also influence the type of construction techniques which might be used. A need for rapid completion may force the employer to consider using an *off-the-peg* or package deal type building which can be constructed quickly.

In order to secure the early completion of the project, several different methods of procurement have been devised with this objective in mind. Such approaches have implications for the other factors under consideration such as design, quality or cost. There is an optimum time solution depending upon the importance which is attached to these other considerations. For example, in terms of cost, shorter or longer periods of time on site tend to increase building costs. The former is due in part to overtime costs and the latter to extended site on-costs. Some clients are prepared to pay extra costs in order to achieve earlier occupation. Different types of project also have different time concepts. A shopping centre redevelopment may require rapid completion, since earlier revenue may easily account for the extra costs. Different techniques of design and construction may also need to be used to achieve early completion. New educational building completions are linked to school term times, whereas housing starts and completions are at the rate at which they can be either let or sold.

Research has shown that construction work should not commence on site until the project is fully designed and ready for construction with only the minimal involvement thereafter by the designer. Variations will still be allowed, but are not encouraged. Such an approach helps to eliminate a large amount of construction uncertainty which is common on many supposedly designed projects today.

This approach allows the contractor to plan the organization and management of the project better and to spend less time awaiting drawings, details and other information. Overall the design and construction period is shortened and earlier completion is achieved. The contractor is on site for a shorter period of time, which saves on building costs. The mitigating difficulty of adopting this approach is that the client wants to see work beginning on site as soon as possible.

Many of the newer methods of contract procurement have been devised specifically to find a quicker route through the design and construction processes. In some ways they have been assisted by changes in the techniques used for the construction and assembly of the building products and materials.

Early selection was developed to allow the construction work to start on site whilst some of the scheme was still at the design stage. It also allowed the contractor an early involvement with the project. Some forms of cost forecasting and control can be used but they will be much less precise than with some of the more conventional methods of procurement. Critical path analysis can be used to find the quickest way through the construction programme, and the American fast-track system was imported for the sole purpose of securing an early completion of the project.

Inaccurate design information coupled with a need for speed in completion of the works has often resulted in abortive parts of the project, poor quality control and higher costs to both the contractor and the employer. The later forms of management contracting offer some solution to the time delay problem. Project management contracts, where an independent organization controls both the design and the construction teams, has provided some good examples in terms of project coordination.

Faster building for commerce (NEDO 1998) was a study aimed at helping the industry achieve earlier completions of its projects. The study claimed that the major influences on projects' time performance were customer participation, design quality and information, contractor's control over site operations and the integration of the subcontractors with design and construction. The study showed that fast times were 20–25 per cent shorter than the average, and the slowest times could take twice as long as the average. Overall, one-third of commercial projects finish on time, the rest overrun by a month or more, often because of extensive design alteration. These were claimed as the largest single factor affecting delays. Common ingredients in fast projects are organization to promote unity of purpose, competent management at all stages of the project and working practices which interlock smoothly and leave no gaps between activities and responsibilities. Many of the delays result from a lack of information, an underestimating of the supervision of subcontractors and late changes in requirements by employers. In some cases quality control was a problem where work needed to be redone and hence caused delays. Slow reactions by the utilities companies can cause extensive delays, especially on sites with high services contents. Here are the time factors to consider:

- Completion dates
- Delays and extensions of time
- Phased completions
- Early commencement
- Optimum time
- Complete information
- Fast tracks
- Coordination

Accountability

According to the dictionary definition, *accountability* is the 'responsibility for giving reasons why a particular course of action has been taken'. In essence it is not simply having to do the right things, but having to explain why a particular choice was made in preference to others that were available. It is of greater significance when dealing with public employers where it is necessary to justify why a particular course of action was taken. It has also become increasingly important with all types of employers where an emphasis is placed on achieving value for money on capital works projects. The documentation used for construction works is often complex and the technical and financial implications are considerable.

Employers need the assurance that they have obtained the best possible procurement method against their list of objectives. The possible trade-offs between competing proposals will need to be evaluated. Where tenders are sought in the absence of any form of competition, it is difficult to satisfy the accountability criteria in respect of price. There is also the difficulty of justifying subjective judgments where these appear to be in conflict with common practice, but the process of selection will never solely be a mechanistic process. The elimination of procedural loopholes should be such as to provide the employer with as much peace of mind as is possible.

Accountability is interlinked with finance and an emphasis on paying the smallest price for the completed project. It may be easy to demonstrate to some employers that to pay more for a perceived higher quality or earlier completion is worthwhile. Other employers may need more convincing and some will feel doubtful about non-monetary gains.

The procedure for the selection, award and administration of contracts must be as precise as possible. Auditing plays a useful role in the tightening up of the procedures used, with ad hoc arrangements that breach these procedures being discouraged. Systems that require huge amounts of documentation and the subsequent checking and cross-checking of invoices, time sheets, etc., are not favoured because of loopholes that can occur. Prime cost contracts are an example of this. Open-ended arrangements which are unable to provide a realistic estimate of cost are fraught with difficulties in terms of accountability. They provide difficulties in demonstrating value for money at the tender stage, since any forecast of cost will be too imprecise for reliability purposes. Here are the accountability factors to consider:

- Contractor selection
- Ad hoc arrangements
- Contractual procedures
- Loopholes
- Simplicity
- Value for money

Design

There is a good argument that the best design will be obtained from someone who is a professional designer. The design will then not be limited by the capabilities of the constructor or restrained to those designs which might be the most profitable in terms of their construction to such a firm. However, a design and build contractor is more likely to be able to achieve a solution which takes into account buildability, and produces a design which is sound in terms of its construction. There are examples where the employer, having been provided with an unsatisfactory building, has to wait impatiently whilst the consultant and constructor argue about their respective liability. There are also examples where design and build projects, some using industrialized components, have had to be demolished after only a few years of life because of their poor design concept, impossible and costly maintenance problems and an unacceptable user environment which they helped to create.

Contractors have on the whole been better at marketing their services and these have reaped benefits in the growing increase in design and build schemes (Chapter 7). Designers have to some extent been thwarted in their response to this upsurge in activity due to the restrictions imposed on advertising by their profession and a failure to respond adequately to the changes that have taken place in the industry and throughout many aspects of society. Only in recent years has the design and manage approach to the procurement of buildings been introduced.

The traditional methods of contract procurement fail in many aspects of building design, not least because of the absence of any constructor input. This is not a common feature in other comparable industries. However, they continue to remain a common method of procurement, despite their publicized drawbacks. Some of the forms of management contracting, which still largely retain the independence of the designer, did gain in their popularity, although bad publicity has curtailed their development. Where the employer has been encouraged to form a contractual relationship with a single organization, then on the whole this appears to have been beneficial.

Designs which evolve and develop only marginally ahead of the construction works on site must be of questionable worth in design solution terms, unless they are working within the constraints of either previously completed schemes or the confines of an existing structure. Here are the design factors to consider:

- Aesthetics
- Function
- Maintenance
- Buildability
- Contractor involvement
- Standard design
- Design before build
- Design prototypes

Quality assurance

Open tendering can result in a lower standard of work than might have been achieved by using a building contractor who submitted a higher price. You only get what you pay for is certainly true of construction quality. Where a building contractor has had to submit an uneconomical price at cost, then quality may suffer unless improved supervision is provided. Consistent and good quality control procedures are frequently lacking in the construction industry and are often relegated in favour of other criteria in the list of objectives. The quality of buildings has not improved in line with the quality of, for example, motor vehicles.

The quality of buildings depends upon a whole range of inputs: the soundness of the design, a correct choice of specification, efficient working details, adequate supervision and the ability of the builder. The skills of the operatives are also important, perhaps even more so today than a decade ago, now that the designs are tending to become more complex in their detailing and a higher level of skill is expected. The choice of a contractor who has a good reputation for the type and quality of work envisaged is important in achieving this objective. The use of labour-only operatives and the general subcontracting phenomena have sometimes contributed towards a deterioration in quality and performance, due to poor site coordination and supervision by site management.

The quality of design and build schemes depends largely upon the reputation of the building firm selected, particularly where the employer chooses not to involve any professional advisers. The quality of the materials and the work will be regulated entirely by the contractor alone. Speculative building schemes where the quality assurance is determined solely by the builder are not necessarily renowned for their high quality work, although they are market driven, relying upon customers to purchase their speculative units. Certainly, in the past, speculative house building was deficient in terms of quality. This has now improved to meet house buyers' expectations. Government regulations to improve building structures have been monitored for some time by external agencies, requiring defects to be corrected for periods of up to ten years.

Fast-track procurement methods which may involve a number of contractors on the same project can, without adequate supervision, result in widely varying standards of quality and work. There is also the added difficulty of coordinating different contractors on the site, such as occurs with work packages on management contracts. The use of selective tendering, properly managed, continues to offer a good solution in terms of quality control.

Quality standards cannot be judged at the building's completion alone, but need to be considered in the longer term. A virtually complete design prior to the commencement of work on site is likely to be beneficial in improving the qualitative aspects of the project rather than the more ad hoc design approach to problems as they occur on site.

The turnkey method, where the designer–contractor has a contractual responsibility for the long-term repair and maintenance of the project, does offer advantages, particularly in terms of quality assurance. Under this method of procurement there is the incentive that the designer–contractor will wish to reduce the likelihood of future defects arising by a more careful design and effective site management during construction. This may result in less innovation but it is preferable to inconvenience and costly failures in the future. Progress in construction is necessary, but not at the expense of prototype designs which result in poor quality assurance.

Although serial and continuation contracts, using the same design and detailing, should improve the quality aspects, in practice this has not always been achieved. The learning curves of the operatives in these cases have been an aid towards improved quality. However, poor design, detailing and construction methods have in too many cases been unfortunately repeated. The following are quality assurance factors to consider:

- Quality control
- Independent inspection
- Teamworking
- Coordination
- Subcontracting
- Buildability
- Future maintenance
- Design and detailing
- Reputation of craftspeople

Organization

Allowing the contractor total control of the building project, as in design and build or management contracting, removes a layer of organization and eliminates dual responsibility. This should result in fewer things being overlooked or forgotten, work left undone or subcontractors being unable to complete their work on time owing to a lack of information. An additional tier of organization also has the disadvantage of the parties blaming each other when disputes arise. The traditional methods of contract procurement appear to set the lines of demarcation between designer and constructor clearly, although the large number of disputes in practice might suggest that this is not the case. In practice the designer probably relies too heavily on the constructor. Elaborate conditions of contract, to cope with the organization of construction work, have had to be drawn up to anticipate most of the eventualities that might arise.

The employment of a single firm, such as a design and build contractor, allows for quick response management, the ability to deal with problems as they occur, and more freedom in the execution of the works. The extent of such freedom will vary with the conditions of contract being used. Where a separate designer is used, the response time is often much longer and this can result in delays to the contract. Where complex or difficult contract arrangements are employed, they can have the effect of removing the initiative from the contractor.

The more parties involved with a building project, the more complex will be the organization and the contractual arrangements. The employment of a group design practice should therefore result in fewer organizational difficulties than where individual firms are used. Management contracting is based upon awarding individual work packages to a range of specialist and general subcontractors. This can create problems of organization and coordination. There is much less control over such firms than when directly employed operatives of the main contractor are used. General contracting is, however, now unusual since about 90 per cent of construction work is subcontracted regardless of the method of procurement being used. These individual firms need to be programmed for precise periods of time, and a delay in allowing them to proceed with their work or a failure by them to complete on time can have a knock-on effect for the whole project. This presents even greater difficulties with tightly scheduled construction programmes. Here are the organizational factors to consider:

- Complexity of arrangements
- Single responsibility
- Levels of responsibility
- Number of individual firms involved
- Lines of management

Complexity

Projects which are complex in design or construction require more precise and comprehensive contractual arrangements. Complexity may be the result of an innovative design, the utilization of new construction methods, the phasing of the site operations or the necessity for highly specialized work. It can also be the result of employing several contractors on the same site at one time in order to achieve rapid progress or the complicated refurbishment of an existing building whilst still in use by its occupants. It is often necessary in circumstances like these to devise new contractual arrangements and to apply different types of procedures to the varying parts of the construction work. Where work can be reasonably well defined and forecasted, then traditional estimating processes can be used and the work paid for on the usual basis. Where the work is indeterminate, of an experimental type or requiring a solution from the contractor, then a lump sum or cost reimbursement approach with contractor design may need to be employed. In the latter case the contractor is given the opportunity to offer an acceptable solution to the problem as a part of the contract.

Where the project is very complex then the employer is likely to choose a separate designer with the skills required to produce the right solution. It is, however, important to involve the constructor as soon as possible with the project, particularly where this might influence the sequencing of site operations. A form of two-stage tendering might therefore be appropriate against this eventuality. Here are the organizational points to consider:

- Nature of complexity
- Capabilities of parties
- Main objectives of employer

Risk

Risk is inherent in the design and construction of a building. The employer's intentions will be to transfer as much of this as possible to either the consultants or the contractor. Risk may be defined as possible loss resulting from the difference between what was anticipated and what actually occurred. Risk is not entirely monetary. An unsatisfactory design, although completed successfully, can result in a weakening of the designer's reputation with a consequent loss of future commissions. Risk can be reduced but it is difficult to eliminate it entirely. For example, the risk associated with a very specialized form of construction can be reduced by selecting a contractor with the appropriate experience.

The transfer of risk from the employer to others involved with the project may appear to satisfy the accountability criteria. It may be argued as an appropriate course to follow for the employer, but it may not be a fair and reasonable approach. It may also not be the best route for the employer to follow, since the risk needs to be evaluated. All contractors' tenders contain a premium to cover contractual risk. Where the risk does not materialize then this becomes a part of the contractor's profit. The employer may thus be better advised to assess and accept some of the risk involved and thereby reduce the contractor's tender sum and also costs accordingly. This is a more common way of dealing with risk on construction projects.

The lump sum contract with a single price, which is not subject to any variation is at one extreme, and at the other is cost reimbursement, where risk and financial predictability are uncertain. In the former the employer is paying for eventualities which might not occur. In the latter the client is accepting the risk but only pays for events that happen. A balance has to be drawn. Risk should always be placed with the party to the contract who is in the best position to control it. Where this is not possible then it should at least be shared, although it may be difficult to convince the employer that this course of action is the most financially appropriate. Some projects involve a large amount of risk in their execution. In some cases the risk may be so high that it is impossible to get a contractor even to consider tendering under conventional arrangements. Some form of risk sharing may then become essential in order for the project to proceed. Here are some risk factors to consider:

- Risk evaluation
- Risk sharing
- Risk transfer
- Risk control

Market

The selection of a method of procurement will be influenced by the state of a country's economy and the industry's workloads. An appropriate recommendation for today may have different implications for some time in the future. When there is an ample amount of work available, contractors are able to choose those schemes which are the most financially lucrative. Under these circumstances employers will be unable to insist upon onerous contractual arrangements and conditions. Where the risk involved is high it will be even more difficult to persuade contractors to tender for the work. Employers may need to be advised to delay their building projects at such times, and to wait until the economy is more favourable. Many employers will, however, be unable or unwilling to follow this advice.

When construction prices are low, then a form of cost reimbursement or management fee approach can be expensive. During times like this, contractors are sometimes prepared to do work at cost, and take a gamble with the risk factors that nothing of financial significance will go wrong. In times of full order books the opposite is true and paying contractors their actual costs plus an agreed amount for profit can be a better proposition. When work is plentiful, contractors often have difficulty in recruiting a competent workforce of skilled operatives and this, coupled with similar restrictions in the availability of good supervision, can result in a deterioration in quality standards. When the available amount of construction work is restricted, the standards of work coupled with more intensive inspection are likely to enhance the overall quality of the project. Here are the market factors to consider:

- Availability of work
- Availability of contractors
- Economy affects procurement advice

Finance

The usual way of paying the contractor for the building work is through monthly or stage payments. These payments help the contractor to offset the financial borrowing that is required to pay for wages and salaries, goods and materials. Two alternatives to this can be used. The first is a delayed payment system similar to that used on speculative developments. The employer effectively pays for the work using a single payment upon completion of the project. The employer has to accept the design as it is built, but acquires immediate occupation of the project. The financial borrowing requirements of the contractor are higher, but the employer makes savings by paying for the work at the end of the project. The other alternative is for the employer to fund the work in advance and thereby reduce the contractor's interest charges that are otherwise included in the tender. In these situations there are cautions of which the client needs to be aware. The industry is notorious for the number of insolvencies, and the employer would thus want to ensure the financial soundness of the appointed contractor. Contractors also tend to be less interested in a project once they have received payment for work. The employer can devise remedies to deal with these factors. With the former, a performance bond can be adopted; with the latter, liquidated damages can be applied. Here are some financial factors to consider:

- Payment systems
- Financial soundness of parties
- Financial remedies
- Contract funding

CONCLUSIONS

The choice of a particular method of contract procurement for a construction project involves identifying the employer's objectives, balancing them with the procurement methods which are available and taking into account the considerations outlined above. Figure 9.2 provides a comparison of some of the payment methods.

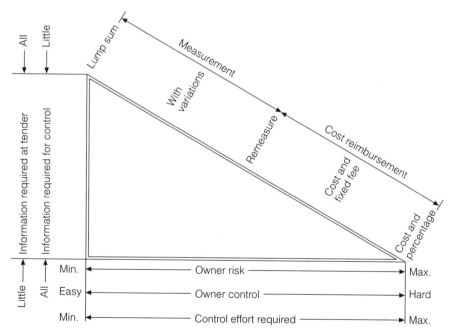

Fig. 9.2   Range of contract types (Adapted from Burgess, R A, ed., 1980, *Construction projects, their financial policy and control*, Longman, London, p. 149)

Table 9.1   Identifying the client's priorities

			Traditional selective tendering	Early selection	Design and build	Construction management	Management fee	Design and manage
TIMING	Is early completion important to the success of the project?	Yes:		*	*	*	*	*
		Average:		*	*	*	*	*
		No:	*					
VARIATIONS	Are variations to the contract important?	Yes:	*	*		*	*	*
		No:			*			
COMPLEXITY	Is the building technically complex or highly serviced?	Yes:	*	*		*	*	*
		Average:		*	*	*	*	*
QUALITY	What level of quality is required?	High:	*	*	*	*	*	*
		Average:	*	*	*	*	*	
		Basic			*			
PRICE CERTAINTY	Is a firm price necessary before the contracts are signed?	Yes:	*	*	*	*		
		No:					*	*
RESPONSIBILITY	Do you wish to deal only with one firm?	Yes:		*	*	*	*	*
		No:	*					
PROFESSIONAL	Do you require direct professional consultant involvement?	Yes:	*	*		*	*	
		No:			*			*
RISK AVOIDANCE	Do you want someone to take the risk from you?	Yes:	*		*			
		Shared:		*		*		*
		No:					*	

Source: Adapted from *Thinking about building*, EDC report

It compares factors such as the information requirements with the control and risk considerations. If an employer's main priority, for example, is for a lump sum price from a contractor, then the full information at the tender stage must be provided for an accurate price to be prepared. Risk and control effort throughout the duration of the contract will then be minimal to the employer. On large and complex projects the ability to provide this detailed information is difficult. The quality and reliability of the design information will determine how precisely the building's costs can be forecasted and controlled. The poorer and more imprecise it is, the greater will be the risks to the employer. The risk, of course, may never materialize and hence there will be no loss to the employer.

The three broad areas of concern to the employer, and the expectations that are required from any building project have already been identified in Figure 9.1. The balance of these will vary but their analysis will help to influence and select the most appropriate procurement method.

Table 9.1 offers a checklist of questions to help determine an appropriate contract strategy. It is based upon an EDC report entitled *Thinking about building*, but has been adapted accordingly. It provides examples of some of the more usual contractual arrangements which are available. It does, however, need to be emphasized that the solutions recommended are based largely upon judgment rather than objective analysis. In answering the questions, users can be provided with appropriate solutions. Under differing circumstances, or where other factors need to be taken into account, the solutions must be adjusted accordingly. The questions themselves are not weighted and users will need to do this in order of importance. Some employers may wish to emphasize only a single aspect such as quality, and choose a method and contractor which are capable of securing this. The majority of employers are, however, interested in an amalgam view, trading off the various factors against each other. It is inappropriate to use the chart in an incremental fashion by adding the various answers together.

The correct choice of a procurement method is a difficult task owing to the wide variety of options which are available. Some of the changes in methods of procurement are the result of a move away from the craft base to the introduction of off-site manufacture, the use of industrialized components and the wider application of mechanical plant and equipment. Improved knowledge of production techniques, coupled with the way in which the workforce is organized, has enabled the contractor to analyse the resources involved and move towards their greater optimization. Contractors also have a much greater influence upon the design of the project, and the recognition of buildability has influenced the design and how the work is carried out on site, hence it has influenced the quality of the finished work. The time available for construction and the subsequent costs involved have also been affected by these changes.

CHAPTER 10

CONTRACT DOCUMENTS

The contract documents under any building project should include the following information as a minimum:

- *The work to be performed*: this usually requires some form of drawn information. It assists the client's own understanding of the project through the use of schematic layouts and elevations. Even the non-technical employer is usually able to grasp a basic idea of the architect's design intentions. Drawings will also be necessary for planning permission and building regulations approval. Finally, they will be needed by the prospective builder in order that the architect's intentions can be carried out during estimating, planning and construction. On all but the simplest types of project, therefore, some drawings will be necessary. On some projects these will be supported by additional information such as three-dimensional models and computer-aided design.
- *The quality of work required*: it is not easy on drawings alone to describe fully the quality and performance of the materials and the standards of work expected. The usual way is to describe the quality and standards by a specification or a bill of quantities.
- *The contractual conditions*: in order to avoid possible misunderstandings, it is desirable to have a written agreement between the employer and the builder. For simple or small-scale projects, conditions of contract such as the minor works agreement may be sufficient. On more complex projects, one of the more comprehensive forms of contract should be used (Chapter 6). It is preferable to adopt the use of a standard form of contract rather than to devise separate conditions of contract for each project.
- *The cost of the finished work*: this should be predetermined, wherever possible, by an estimate of costs from the builder. This is best achieved through the use of some form of measured quantities of the work. On some projects it may only be feasible to assess the cost once the work has been carried out. In these circumstances the method of calculating this cost should be clearly agreed.
- *The construction programme*: the length of time available for the construction work on site will be important to both the client and the contractor. The client will need to have some idea of how long the project will take to complete in order to

plan arrangements for the handover of the project. The contractor's costs will, to some extent, be affected by the time available for construction. The programme should include progress schedules to assess whether the project is on time, ahead or behind the programme.

The contract documents on a building contract therefore will normally comprise the following:

- Form of contract, including
  - articles of agreement
  - appendix
- Contract drawings
- Bill of quantities, specifications or measured schedules

On a civil engineering project it is more usual to include both a bill of quantities and a specification. The JCT forms used on building contracts allow only one of these documents, depending on with or without quantities.

FORM OF CONTRACT

The form of contract is the principal contract document and will generally comprise one of the preprinted forms described in Chapter 6; that is, forms of contract that have been developed and agreed by the various parties in the construction industry. The form of contract, under JCT 98, takes precedence over the other contract documents (clause 2.2). The conditions of contract seek to establish the legal framework under which the work is to be undertaken. Although the clauses aim to be precise and explicit, and to cover any eventuality, disagreement in their interpretation does occur. When disputes arise an attempt in the first instance is made to resolve the matter, as amicably as possible, by the parties concerned. Where this is not possible, it may be necessary to refer the disagreement to a form of alternative dispute resolution, adjudication or arbitration. The parties to a contract usually agree to take any dispute initially to adjudication rather than to litigation. This can save time, costs and adverse publicity, which may be damaging to both parties. Where the dispute cannot be resolved, it is taken to court to establish a legal opinion. Such opinions, if held, eventually become case law and can be cited should similar disputes arise in the future.

It is always preferable to use one of the standard forms available, rather than to devise one's own personal form. Several private forms have been established by the large industrial corporations for their own use. The imposition of conditions of contract which are biased in favour of the employer are not to be recommended. Contractors will tend to overprice the work, even in times of shortage, to cover the additional risks involved. Unless there are very good reasons to the contrary, the architect should attempt to persuade the building owner to use one of the standard forms available. The modification of some of the standard clauses, or the addition of special clauses, should only be made in exceptional cases.

The majority of the standard forms of contract comprise, in one way or another, the following three sections.

Articles of agreement

The articles of agreement are the part of the contract which the parties sign. Note that the contract is between the employer (building owner) and the contractor (building contractor). The blank spaces in the articles are filled in with:

- The names of the employer, contractor, architect and quantity surveyor
- The date of signing of the contract
- The location and nature of the work
- The list of the contract drawings
- The amount of the contract sum

If the parties make any amendments to these articles of agreement or to any other part of the contract, then the alterations should be initialled by both parties.

In some circumstances it may be necessary or desirable to execute the contract under seal. This is often the case with local authorities and other public bodies. The spaces for the signatures are then left blank and the seals are affixed in the appropriate spaces indicated. After sealing, the contract must be taken to the Department of Customs and Excise, where upon payment of stamp duty, a stamp will be impressed on the document. Without this the contract will be unenforceable.

Conditions of contract

The conditions of contract in any of the forms in general have a large degree of comparability, but are different in their details. They include, for example, the contractor's obligations to carry out and complete the work shown on the drawings and described in the bills to the satisfaction of the architect (or supervising officer). They cover matters dealing with the quality of the work, cost, time, nominated suppliers' and subcontractors' insurances, fluctuations and VAT. Their purpose is to attempt to clarify the rights and responsibilities of the various parties in the event of a dispute arising. The JCT 98 standard form of building contract contains 45 clauses, the last three of which deal with fluctuations and are published separately from the other clauses. The clauses from JCT 98 are discussed in detail in Chapters 17 to 25.

Appendix

The appendix to the conditions of contract needs to be completed at the time of signing of the contract. The completed appendix includes that part of the contract which is peculiar to the project in question. It includes information on the start and completion dates, the periods of interim payment and the length of the defects liability period for which the contractor is responsible. The appendix includes recommendations for some of the information.

CONTRACT DRAWINGS

The contract drawings should ideally be complete and finalized at tender stage. Unfortunately, this is seldom the case, and both clients and architects rely too heavily upon the clause in the conditions allowing for variations (JCT 98 clause 13). Occasionally the reason is due to insufficient time being made available for the pre-contract design work or, frequently, because of indecision on the part of the client and the design team. One of the intentions of the standard method of measurement (SMM7), was only to allow bills of quantities to be prepared on the basis of complete drawings. To invite contractors to price work that has yet to be designed does not appear to be a sensible course of action to follow. Tenderers should be given sufficient information to enable them to understand what is required in order that they may submit as accurate and realistic a price as possible. The contract drawings will include the general arrangement drawings showing the site location, the position of the building on the site, means of access to the site, floor plans, elevations and sections. Where these drawings are not supplied to the contractors with the other tendering information, the contractors should be informed where and when they can be inspected. The inspection of these and other drawings is highly recommended, since it may provide the opportunity for an informal discussion on the project with the designer.

The preparation of the architect's drawings should comply with the appropriate British Standard, BS 1192. Each drawing should include:

- The name and address of the consultant (architect, engineer or surveyor).
- The drawing number is for reference and recording purposes.
- The scale; if two or more scales are used, they should be sufficiently dissimilar as to be readily distinguishable by sight.
- The title will indicate the scope of the work covered on the drawing.

The contractor, upon signing the contract, will be provided with two further copies of the contract drawings. The contract drawings may include copies of the drawings sent to the contractor with the invitation to tender, together with those drawings that have been used in the preparation of either the bill of quantities or specification. In JCT 98 the contract drawings are defined in the third recital of the articles of agreement as those which have been signed by both parties to the contract. It will be necessary during the construction phase for the architect to supply the contractor with additional drawings and details. These may either explain and amplify the contract drawings, or because of variations identify and explain the changes from the original design. An information release schedule is to be provided (clause 5.4.1) and the provision of further drawings and details is the norm and is expected (clause 5.4.2).

The contractor must have an adequate filing system for the drawings. Superseded copies should be clearly marked. They should not, however, be discarded or destroyed until the final account has been agreed, since they may contain relevant information used for contractual claims. The drawings should be clear and accurate, but because paper expands and contracts, only figured

Table 10.1 Drawings supplied by the architect

Drawing title	Suggested scale
Survey plan	1 : 1,250
Site plan and draining	1 : 500
Floor plans	1 : 200
Roof plan	1 : 200
Foundation plan	1 : 200
Elevations	1 : 200
Sections	1 : 100
Construction details	1 : 10, 1 : 5
Engineering services	1 : 200

dimensions should be used. Scaling dimensions is therefore very much a last resort. Depending upon the size and type of the project, the architect may supply the building contractor with some or all of the drawings in Table 10.1.

Return of drawings

Once the contractor has received the final payment under the terms of the contract, the drawings may be requested to be returned to the architect along with details, schedules and other documents which bear the name of the architect (clause 5.6). The copyright of the design is vested in the architect. If the building owner or the building contractor wishes to repeat the work, then a further fee must be paid to the architect. None of the documents prepared especially for the project can be used for any other purpose than the contract (clause 5.7), by either the employer or the contractor.

Schedules

The preparation and use of schedules is particularly appropriate for items of work such as:

- Windows
- Doors
- Access holes
- Internal finishings

Schedules provide an improved means of communicating information between the architect and the builder. They are also invaluable to the quantity surveyor during the preparation of the bills of quantities. They have several advantages over attempting to provide the same information by correspondence or further drawings. Checking for possible errors is simplified, and the schedules can also be used to place orders for materials or components.

During the preparation of schedules the following questions must be borne in mind:

- Who will use the schedule?
- What information is required to be conveyed?
- What additional information is required?
- How can the information be best portrayed?
- Does it revise information provided elsewhere?

The architect must supply the building contractor with two copies of these schedules that have been prepared for use in carrying out the works (clause 5.3.1). This should be done soon after the signing of the contract, or as soon as possible thereafter should they not be available at this time. The architect should supply the information release schedule as described in clause 5.4.1.

CONTRACT BILLS

Some form of bill of quantities (described in JCT 98 as the contract bills) or measured schedules should be prepared for all types of building projects, other than those of only a minor nature. The bill comprises a list of items of work to be carried out, providing a brief description and the quantities of the finished work in the building. A bill of quantities allows each contractor who is tendering for a project to price the construction work on exactly the same information as other contractors using a minimum of effort. The bill may include firm or approximate quantities, depending upon the completeness of the drawings and other information from which it was prepared. The contractor's rates must not be divulged to others or used for any purpose other than the contract (clause 5.7).

Uses

Although the main use of a bill of quantities is to properly assist the contractor during the process of tendering, it can be used for many other purposes, such as:

- Preparation of interim valuations in order that the architect may issue the certificate
- Valuation of variations that have either been authorized or sanctioned by the architect
- Cost control purposes
- Ordering of materials, if care is properly exercised in the checking of quantities and the implications of future variations
- Preparation of the final account, the bill is used as a basis for agreement
- Production of a cost analysis for the building project
- Determination of the quality of materials and standard of work by reference to preamble clauses
- Useful in obtaining domestic subcontract quotations for sections of the measured work
- Use as a form of cost data (noting the provision of clause 5.7)
- Essential when preparing contractual claims

Preparation

The preparation of bills of quantities should be undertaken by the quantity surveyor, adopting best practice manual or computerized procedures. The items of work should have been measured in accordance with a recognized method of measurement. Building projects are generally measured in accordance with the Standard Method of Measurement of Building Works (SMM7). There are several other methods of measurement available to the construction industry. On mass housing projects it may be preferable to use a simplified version such as the Code of Measurement of Building Works in Small Dwellings. Separate methods also exist for civil engineering work and work on petrochemical plants. Many countries abroad have their own methods of measurement, some of which are based on UK methods. An international version has been published by the RICS. On building projects the descriptions and quantities are derived from the contract drawings and SMM7. These are on the basis of the finished quantities of work in the completed building.

Contents

Preliminaries

The preliminaries cover the employer's requirements and the contractor's obligations in carrying out the work. SMM7 provides a framework for this section of the bill. Here are some of the items it includes:

- Names of parties
- Description of the works
- Form and type of contract
- General facilities to be provided by the contractor

In practice, although the preliminaries may comprise over twenty pages of the bill of quantities, only a small number of items, approximately ten to fifteen, are priced by contractors. The remainder of the items are included for information and contractual purposes only. The value of the preliminary items may account for 8–15 per cent of the contract sum.

Preambles

The preamble clauses contain descriptions relating to:

- The quality and performance of materials
- The standard of work
- The testing of materials and work
- Samples of materials and work

This section of the bill of quantities has in some cases been replaced by a set of standard and comprehensive preamble clauses. The contents of the usual preambles bill are generally extracted from a library of standard clauses. Contractors rarely price any of this section of the bill.

Measured works

The measured works include the items of work to be undertaken by the main contractor or to be sublet to domestic subcontractors. There are several forms of presentation available for this work; here are some examples:

- *Trade format*: the items in the bill are grouped under their respective trades. The advantages of this format are that similar items are grouped together, there is a minimum of repetition, and it is useful to the contractor when subletting.
- *Elemental format*: this groups the items according to their position in the building, on the basis of a recognized elemental subdivision of the project, e.g. external walls, roofs, wall finishes, sanitary appliances. This type of bill should, in theory, help in tendering by locating the work more precisely. In practice contractors tend to dislike it since it involves a considerable amount of repetition. It is, however, useful to the quantity surveyor for cost analysis purposes and during future cost planning.

A recognized order for the inclusion of the various items is important, in order to provide quick and easy reference.

Prime cost and provisional sums

Some parts of the building project are not measured in detail but are included in the bill as a lump sum item. These sums of money are intended to cover specialist work not normally undertaken by the general contractor (prime cost sum) or for work which cannot be entirely foreseen, defined or detailed at the time that the tendering documents are issued (provisional sum). They are separately described as for defined or undefined work. Prime cost sums cover work undertaken by nominated subcontractors, nominated suppliers and statutory undertakings. They include lump sums that have been based upon quotations for items of work such as electrical installations, lifts, escalators and other similar conveyancing systems, etc.

Appendices

The appendices include the tender summary, a list of the main contractors and domestic subcontractors, a basic price list of materials and nominated subcontractor's work for which the main contractor may desire to tender.

The form of tender is the contractor's written offer *to undertake and execute the works in accordance with the contract documents for a contract sum of money*, and will also state the contract period and whether the contract is on the basis of a fixed price. The tenders are submitted to the architect who will then make a recommendation regarding the acceptance of a tender to the client. If the client decides to go ahead with the project, the successful tenderer is invited to submit a bill of quantities for checking. The form of tender may also state that the employer:

- May not accept any tender
- May not accept the lowest tender
- Has no responsibility for the costs incurred in their preparation

CONTRACT SPECIFICATION

In certain circumstances it may be more appropriate to provide documentation by way of a specification rather than a bill of quantities. The types of project where this may be appropriate include:

- Minor building projects
- Small-scale alteration projects
- Simple industrial shed-type projects

The specification is similar to a bill of quantities. The difference is that it does not include a measured works section. In place of this are the detailed descriptions of the work to be performed, to assist the contractor in preparing the tender. A specification is used during tendering to help the estimator price the work that is required to be carried out, during construction by the designer in order to determine the requirements of the contract, legally, technically and financially, and by the building contractor to determine the work to be carried out on site.

SCHEDULE OF RATES

A schedule of rates is a compromise between a specification and a bill of quantities. It is essentially a bill of quantities that has omitted to include any quantities for the work to be carried out. Its main purpose is therefore in valuing the items of work once they have been completed and measured. A schedule of rates may be used on:

- Jobbing work
- Maintenance or repair contracts
- Projects that cannot be adequately defined at the time of tender
- Urgent works
- Painting and decorating

MASTER PROGRAMME

The contractor must provide the architect with a master programme showing when the works will be carried out (clause 5.3.1.2). Unless otherwise directed by the architect, the type of programme and the details to be included are to be at the discretion of the contractor. If the architect agrees to a change in the completion date because of an extension of time (clause 25.3.1) then the contractor should provide the architect with amendments and revisions to the programme within 14 days.

INFORMATION RELEASE SCHEDULE

This schedule is described in clause 5.4.1. of JCT 98. It informs the contractor when information will be made available by the architect. The schedule is not annexed to the contract. However, the architect must ensure that the information is released to the contractor in accordance with that agreed in the information release schedule. In practice the architect would need to coordinate this with the contractor's master programme for the works.

CONTRACTUAL PROVISIONS

Copies of contract documents

Clause 5.2 requests the architect to supply the contractor with the following documents free of charge:

- One copy of the form of contract certified by the employer
- Two further copies of the contract drawings
- Two copies of the unpriced bill of quantities
- Two copies of any descriptive schedules or similar documents

Clause 5.1 states that the contract drawings and the contract bills will remain in the custody of the employer. These are to be available for inspection at all reasonable times. Where there are no bills of quantities, the building contractor will be provided with copies of a specification or schedule rates as may be appropriate. The contractor must provide two copies of the master programme, without charge to either the employer or the architect.

Before the date of practical completion of the works, the contractor must supply the employer with any drawings or other information for any performance-specified works. This would include, for example, details of any operating schedules for mechanical plant or equipment that has been installed in the building by the contractor or a subcontractor (clause 5.9).

Availability of documents on site

Clause 5.5 requires the contractor to keep one copy of the following documents on site at all times for reasonable inspection:

- Contract drawings
- Bills of quantities (unpriced)
- Descriptive schedules
- Master programme
- Additional drawings and details

Discrepancies in documents

Clause 2.3 requires the contractor to write to the architect if any discrepancies between the documents are found. Only the form of contract takes precedence over the other documents. The contractor cannot therefore assume that the drawings are more important than the bills. However, drawings drawn to a larger scale will generally take precedence over drawings that have been prepared to a smaller scale. Clause 6.1.2 expects the contractor to adopt a similar course of action should a divergence be found between statutory requirements and the contract documents.

Where discrepancies or differences result in instructions to the contractor requiring variations, these will be dealt with under clause 13 of JCT 98.

CHAPTER 11

DESIGN AND BUILD

What has become known as the traditional method of constructing building projects requires the separation of the roles of the designer from those of the constructor. They each do their work in isolation from each other and have separate legal responsibilities towards the client. During the latter half of the last century numerous attempts have been made to find alternative procurement solutions, which to many is a failure of the traditional method of project development.

Design and build is a procurement arrangement where one single entity or consortium is contractually responsible to the client for both the design and construction of a project. It is sometimes referred to as design and construct, usually to incorporate works of a civil engineering nature. However, design and build is not a modern-day concept. In centuries past it was the only procurement arrangement that was available. Its roots originate in the ancient master builder concept, where responsibility for both design and construction resided with a single individual.

Design and build can be traced to ancient Mesopotamia, where the Code of Hammurabi (1800 BC) fixed absolute accountability upon master builders for both the design and construction of projects. In classical Greece, great temples, public buildings and civil works were designed and built by master builders. Enduring structures such as the Parthenon and the Theatre of Dionysus are testimony to this master builder process.

During the Renaissance, architecture and construction evolved as separate professions and the presence of the master builders disappeared. Also project complexity, in both design capability and in construction methods, increased during this time as the need arose for specialists in design and specialists in construction. As time went by, these specialists subdivided even further, being recognized for different types or classes of building. Also different specialists evolved taking on responsibility for specific functions of the project. As statutory and case law evolved during the 1800s, courts determined the extent of liability of architects. Builders also were responsible for the adequacy of their own work.

Design and build came back into prominence during the late 1960s, although it would be a decade or so later when its real impact would be felt. Subsequently, the use of project procurement would consist of range of variable design and build type arrangements. Package deal arrangements, for example, which are sometimes believed to be another name for design and build (Chapter 8) are really a special type of arrangement usually involving a contractor's own type of prefabricated or predesigned building system. Turnkey projects are a further example, where a contractor's responsibility far exceeds the designing and initial construction alone to include fitting out and long-term maintenance agreements with the client (Chapter 8).

DESIGN AND BUILD PRACTICE

In the more common design and build arrangement, a client chooses to engage a contractor in preference to a designer. In some countries, e.g. France, an architect may complete a design to about 50 or 60 per cent and then invite design and build proposals. The primary motivation of design and build, in preference to other forms of procurement, is to eliminate the inherent conflicts that exist between the architect and the contractor. The construction industry has a poor reputation for its adversarial nature. Other reasons for the choice of design and build include the greater assurance of completion on time, at an agreed and lower cost and improvements in the quality and performance of the building. These further considerations are often contested by design-only specialists and may in practice not always be achieved.

Whilst design and build procurement has many obvious advantages, clients remain wary of its ability to achieve design excellence. This concern is largely due to the importance of the contractor within a design and build organization.

Contractors' main concerns are with aspects of buildability and profitability and the fact that a design and build project places a much greater responsibility on the design and build contractor. Inevitably, innovative design principles that may never have been properly tested in practice are excluded as far as possible from design and build solutions. Since some of the design ideas are of an innovatory nature, they may not always work in practice as expected.

However, design and build contractors are wanting to expand their businesses and must therefore aim to deliver projects that clients require. Such firms are aware of the criticisms that design and build projects lack architectural merit. This view may now be outdated since there are many examples of design and build projects that have received praise from architectural critics.

ADVANTAGES AND DISADVANTAGES

There are several reasons why design and build projects offer many advantages in respect of design, price, construction and time:

- The contractor is normally involved from the commencement of the project and is thus aware directly of the client's requirements and priorities.
- The contractor is able to offer the benefits of specialized construction knowledge and methods.

- By eliminating the traditional tendering procedure, the time from inception through to completion is reduced to a minimum.
- There is a direct contract between the contractor and the client.
- A functional building at a reasonable cost should be achieved. A past criticism of design and build schemes was their apparent limitation to design projects with aesthetic merit. There is no reason why this cannot be achieved today.
- It should be possible to minimize initial tendering and pre-tender costs.
- The costs associated with the construction works, which also include the design fees, can be agreed prior to the commencement of the project on site.
- The main contractor (design and build contractor) has control over all aspects of the project, including the design and the appointment of subcontractors.
- The need to appoint nominated subcontractors is virtually eliminated in most cases.
- There can be no claims by the contractor for delays due to lack of information, since the contractor has full responsibility for the design details and project information (other than direct delays by the client).

Besides these advantages, there remain some disadvantages in the use of design and build. Many of these are actively being addressed by design and build contractors.

- Because only a performance brief is provided to the contractor, alternative design solutions may not always be fully investigated.
- The contractor's solution may be governed to some extent by the capabilities of the firm rather than by the needs of the client.
- Only a limited number of design and build contractors are able to offer in-house design facilities and this may be restrictive to a client who initially desires to appoint a firm on the basis of some competition. However, in practice many contractors employ external consultants for this purpose.
- The aesthetic aspects of the design, both internally and externally, may be sacrificed in favour of easier and more cost-effective solutions.
- By designing and building to minimum performance requirements, the client's long-term interests may be compromised.
- The normal external supervision, typical with traditional procurement arrangements, is only available where the client retains a separate representative. This increases the costs involved, although external objectivity is desirable. A quantity surveyor may be required to authorize interim payments that become due to the contractor.
- If the contractor becomes bankrupt, the client's legal and financial interests in the project can be considerable, even where independent consultants have been retained.

CHOICE OF DESIGN AND BUILD

The recommendation to adopt design and build for a construction project will depend upon a number of different factors. These may include:

- User familiarity with design and build procurement arrangements
- Preferences of a client or consultant
- The desire for single-point responsibility for the client
- Greater certainty of outcomes, initially and in the longer term

Design and build arrangements are useful in the following circumstances:

- Relatively simple and standard forms of construction, involving little architectural design such as warehouses, small factory units and farm buildings.
- Buildings using a proprietary building system, especially those described as closed systems of construction. These arrangements are more correctly described as package deals (Chapter 8).
- Construction projects where contractors may have become specialist firms within a region or locality.
- Where the cumulative experience and skills of an experienced general contractor and their ability to utilize the attributes of a range of subcontractors can be used with advantage.
- Design and build or the use of a contractor's design is a common component on many traditional contractual arrangements, for the design and construction of specialist elements.

However, design and build arrangements far exceed these five circumstances. Clients may choose to employ a contractor in preference to a designer for a whole range of reasons and types of project. The trio of design, quality and cost will need to be considered by the client prior to appointment. Where a client appoints a contractor, often an architectural practice will be employed to look after the design and therefore no project is exempt from this consideration.

Design and build contractors are also sometimes known to make extravagant claims, which cannot be substantiated, about the benefits of using this form of procurement in preference to other alternatives. These have been considered separately in Chapter 8. However, the single-point responsibility will reduce administration, contractual claims and litigation, as conflicts between a designer and constructor are eliminated.

In many cases design and build is well suited to the task of delivering standard types of building using tested construction technology. These can include building projects of any size. Where the project involves a relatively few construction trades then it will be particularly suited to design and build. Conversely, design and build may be ill-suited to more complex solutions for building projects.

CONTRACTUAL ARRANGEMENTS

The contractual arrangements that are used for design and build can vary considerably. In some cases the contractor is invited to design and then build the project. In other circumstances the client may have already instructed an architect to provide sketch plans and outline designs which, after planning consent has been received, are then passed over to the contractor to develop them into working drawings. The client may choose to retain the services of an architect or quantity surveyor to oversee the compliance with aspects of quality and to authorize payments due to the contractor.

The payment of the contractor may be based on a lump sum with arrangements for interim or stage payments. Alternatively, detailed quantities may be requested from the contractor in order to more accurately value the works for interim payment purposes and as a basis for agreeing variations and final accounts.

Some element of competition may be required, although the evaluation of such schemes as design and build is by no means straightforward. Design and build contracts are frequently awarded on a fixed-price basis which will not be subject to adjustment except in exceptional circumstances. This aspect is viewed by clients as an advantage, but it may also be partially seen as a disadvantage. Many design and build contracts exclude the possibility of client-directed variations.

PROCUREMENT: THREE MAIN METHODS

Competition

It is not unusual at the outset of the project for clients or their advisers to discuss outline proposals and ideas with a selected group of suitable building firms. On the basis of these initial contacts, firms may be requested to submit tenders and outline design solutions together with other criteria that the client may have requested, such as time for completion, standards of work, etc. However, it is an extremely difficult task to judge such solutions, since the best solution may frequently cost the most and building firms, or even clients, may not have considered the associated implications.

Competition is a useful method where the project is of a simple design, for example, a rectangular warehouse building of known size and requirements. Competition is also appropriate where a package deal solution is to be sought. The differences between design and build and package deals are described in Chapter 8.

Negotiation

Negotiation is the common approach to procurement, although it may be preceded by some form of competition, as described above. Negotiation will be sought where a relationship already exists between the client and the contractor or in those circumstances where the particular specialist skills of the contractor are sought by the client. Negotiation is used where the firm's interests, qualifications and experience meet the needs of the project.

Design competitions

In these circumstances the client awards the contract to the firm that wins the competition held between a number of design and build contractors. The successful proposals are usually highly developed and include significant detail in their preparation. Where the project is not complex then standard details and layouts may be used, but this is more common where the project is a package deal frequently representing a contractor's own patented system. Whilst unsuccessful designs do not usually receive any reimbursement, some clients are prepared to offer some minimum level of compensation.

SELECTION FACTORS

Design and build contractual arrangements offer a range of advantages to clients; the most important ones are given in Table 11.1. Clients will of course consider the various attributes differently to each other, setting their own priorities and goals. Whilst design and build organizations are vociferous about the apparent advantages of this method of procurement, the evidence in practice remains inconclusive. Some of the claims appear to be obvious, but there is a lack of empirical data and unbiased research to support them. For example, even identical projects, carried out across the road from each other, frequently display different outcomes. The evaluation therefore depends not solely upon a range of factors that are directly related to the project, but on other factors that may have little to do with design or construction methods that have been adopted.

The question of whether design and build is just a fashionable solution needs to be considered. The traditional method is flawed but design and build may be shown to be no better when an improved method is discovered. Design and build may be the best arrangement for the current time, but may not be the chosen alternative for the future. Design and build is no doubt aptly suited to certain types of project, but its uniform application for all types of construction project is not necessarily the best advice.

Cost certainty

Some owners will choose to use design and build because a fixed price can be guaranteed. There are usually fewer opportunities for variations or changes to

Table 11.1 Selection factors for design and build procurement arrangements

Selection factor	Description
Cost certainty	Determine the overall project cost before awarding the contract
Cost reduction	Increase the project's value by reducing costs in comparison to other procurement methods
Contract period	Agree the time for completion and handover to the client
Reduced contract period	Reduce the length of time from inception to completion
Quality and standards	Enhance initial quality, reduce defects and long-term maintenance
Constructability	Introduce this knowledge into the process, during the design phase
Litigation	Reduce the possibility of contractual claims

the design, because it is recognized that such changes often have a bad effect on the project as a whole. The single organization should be better able to control the costs that are involved.

Cost reduction

Although little empirical data exists to indicate that design and build offers better value for money than other procurement methods, there is sound reasoning to believe this is so. The cost reduction stems from three factors:

- The shortening of the contract period and the overall time required to completion.
- The introduction of the contractor's knowledge and expertise into the construction process and its influence upon the design formulation.
- The *hidden* design fees that are included within the overall tender sum. Whilst design fees are charged, they are set at a more realistic level. The designer is able to work closely with the constructor to achieve a solution without the possibility of future litigation between the two roles in the one single entity.

Contract period

The design and build contractor is able to exercise much greater control over the project's time management. This is achievable because the single entity is able to control not just the contract period, but also the time available for the design phase.

Reduced contract period

Design and construction are able to be carried out in parallel, so reducing the project's overall completion date. Through integrating design and construction, methods of construction can be included that will further reduce the amount of time available on site.

Quality and standards

Design and build should result in a project that has inherent quality attributes, in terms of the performance of the building in respect of the structure, materials, services and finishes. The holistic development of the design along with the construction capability should result in fewer defects that need to be remedied at completion and there should be improved performance in the longer term.

Constructability

The early involvement of the constructor is a good practice with any type of procurement. Designers know what needs to be built and constructors know how to build. Buildability is an important concept in ensuring that a design concept can be easily developed into a building programme that can be practised efficiently and effectively.

Litigation

One of the really attractive features of design and build procurement is the potential to reduce contractual claims arising out of disputes between designers and constructors. Design errors or omissions, for example, are solely the responsibility of the design and build company. Claims for delays due to a lack of information from the designer are entirely removed. Design and build will not remove litigation entirely from the process, but it does have the potential to remove a considerable amount of conflict in construction.

USE WORLDWIDE

The use of design and build projects is common in most countries around the world, as indicated in Table 11.2. It is the most common in Europe and least

Table 11.2   Use of design and build

Country	Use is common	Frequency of use		Standard contract
		Private %	Public %	
Europe				
Belgium	yes	70	5	no
Denmark	yes	25	10	yes
Finland	yes	8	4	yes
France	no	15	0	n/a
Germany	yes	20	2	no
Great Britain	yes	30	10	yes
Greece	yes	50	70	no
Ireland	yes	10	20	no
Italy	no	5	0	no
Netherlands	no	10	25	no
Norway	yes	50	30	yes
Russia	yes	60	30	no
Spain	no	5	0	no
Sweden	yes	60	20	yes
North America				
Canada	yes	15	5	no
Mexico	no	10	0	no
USA	yes	20	15	yes
Pacific				
Australia	yes	70	70	yes
Japan	yes	50	0	no
New Zealand	yes	30	5	no

Source: Hanscomb Partnership, July 1997

common in Africa and the Middle East. The increased globalization of the construction industry will help to propagate different construction methods. Debates still exist in some countries about the role of design and build on the work of the architect. In some countries design and build is seen as contrary to the role of the architect as defined in their Architects Act.

Where design and build is commonly used, the frequency of use varies between the public sector and the private sector. Typically it is much more common in the private sector. The public sector has for years been anxious about accountability, and issues relating to design and build projects appear at times to run contrary to this. For example, aspects of competition are more difficult to justify with design and build arrangements than with more traditional procurement methods. A number of countries are an exception to this rule (Table 11.2).

Table 11.3 indicates some typical projects likely to use design and build. The obvious responses included warehouses and industrial buildings, but the range of projects using design and build in Norway and Australia is interesting. Where there is a strong design and build culture in a country, contractors will permanently

Table 11.3 Typical sorts of project using design and build arrangements

Country	Types of Project
Europe	
Belgium	Industrial, some non-prestige offices
Denmark	Office, industrial, warehouses, hotels
Finland	Warehouses, projects with tight schedules
France	Any type of project
Germany	Manufacturing, residential, commercial
Great Britain	Warehouses, retail parks, educational
Greece	Any type of project
Ireland	Industrial, warehouses, decentralized public offices
Italy	Small simple projects
Netherlands	Offices, industrial, infrastructure
Norway	Most types of project
Russia	Standard offices, industrial
Spain	Industrial, warehouses
Sweden	Industrial, offices
North America	
Canada	Public sector jails, administration, industrial
Mexico	Industrial
USA	Warehouses, offices, rental
Pacific	
Australia	Schools, stadiums, universities, most large projects
Japan	Industrial, warehouses
New Zealand	Industrial, warehouses, suburban offices

Source: Hanscomb Partnership, July 1997

employ a range of design professionals such as architects, engineers and surveyors. A more common approach is for the contractor to form a joint venture with consultants. In these circumstances the contractor will frequently lead the joint venture company.

DESIGN AND MANAGE

Design and manage is the consultants' response to design and build. With the growth in subcontracting, it provides the opportunity for designer-led design and build. It is discussed more fully in Chapter 8.

PARTNERING

In some parts of the world design and build appears to be the old-fashioned word for partnering. Partnering is a very different concept and it is described in Chapter 12.

FORM OF CONTRACT

The Joint Contracts Tribunal has published a standard form of building contract with contractor's design (CD 98). This is currently based on JCT 98, the standard form of building contract. CD 98 was originally published in 1981, superseding the design and building form of contract available from the National Federation of Building Trades Employers, now known as the Construction Confederation (CC); the original form had been in use since 1970. The contractor's duties under these forms are wider than usual, since greater responsibility has to be taken for a much wider service. Many different arrangements are available nowadays.

There are 38 clauses in this form compared with 41 in JCT 98. Some of the clauses are a direct reproduction of JCT 98, whereas others incorporate minor amendments or a completely new clause. All references to 'the architect' have been deleted, and these now become 'the employer'. There is no specific provision for a clerk of works, although such a person may well come within the definition of one of the employer's agents. The intention is, however, broadly the same. Here are some of the significant differences from the standard form.

Articles of agreement

Employer's requirements

The employer in the first recital in lieu of supplying the contractor with drawings and bills of quantities has issued the contractor with basic requirements. These requirements will broadly be the same as the information that would have been provided to the architect.

Contractor's proposals

In the second recital the contractor's proposals are identified. These include details of the contract sum analysis that will be required for the execution of these proposals.

Employer's acceptance

The third recital states that the employer has accepted these proposals and the contract sum analysis and that the employer is satisfied they meet the requirements for the project. The employer's requirements, the contractor's proposals and the contract sum analysis are described in detail in Appendix 3 to the conditions. Article 3 names the employer's agent. Since there is no independent designer or supervisor for the works envisaged, provision is made for someone to act on behalf of the employer. The duties of this person may include receiving or issuing applications, consents, instructions, notices, requests or statements; acting on behalf of the employer.

The conditions

Clause 2

Clause 2 includes the provisions of the counterpart clause of the standard form, but in addition covers matters appropriate to the contractor's design and their liability for it. The contractor in clause 2.1 shall carry out and complete the works, including the selection of any specifications of materials and work standards in order to meet the employer's requirements. Clause 2.5 indicates that the contractor must carry out the design work in an equal manner to that of an independent architect or professional designer. The contractor's responsibility to the employer for any inadequacy of the design work includes the exclusion of defects or any insufficiency in the design work. Where the work includes the design and construction of dwellings, reference is made to the liability under the Defective Premises Act 1972. If the Act does not apply, the contractor' design liability for loss of use or profit or consequential loss is limited to the amount, if any, set out in Appendix 1. The contractor's design under this clause includes whatever may have been prepared by others on behalf of the contractor.

Clause 4

The contractor must comply with all instructions issued by the employer, as long as the employer has the contractual power to issue such instructions. Variations are referred to as change instructions in clause 12, and where the contractor makes a reasonable objection in writing to a change instruction there is no need to comply with it. The provisions relating to employer's instructions are generally similar to those dealing with architect's instructions under the standard form.

Clause 5

This clause requires that both the employer's requirements and the contractor's proposals are to remain in the custody of the employer. The contractor is to be allowed access to them at all reasonable times. When the contract has been signed by the parties, the employer will provide the contractor, free of charge, with:

- A copy of the certified articles of agreement, conditions and appendices
- A copy of the employer's requirements
- A copy of the contractor's proposals which includes the contract sum analysis

The contractor must then provide the employer with two copies of the drawings, specifications, details, levels and setting out dimensions which are proposed to be used on the works. All this information largely represents the equivalent of the contract documents of the standard form, but is not described as such in this form. A copy of all the above information is to be kept on site and available to the employer's agent at all reasonable times.

Prior to the commencement of the defects liability period, the employer is to be provided with copies of the 'as built' drawings and other relevant information. This information may include details of the maintenance and operation of the building works, including any installations comprised in the works. The information provided by either party is for the use of this contract only. It would appear, therefore, that copyright of the documents is vested in the party who actually prepared them.

Clause 7

The employer is required to define the boundaries of the site, and this information is to be written into the conditions of contract at this point.

Clause 12

Variations are referred to as changes in the employer's requirements. The alteration in wording from the standard form may lead one to suspect that changes are intended to be of a minor nature only, the employer's requirements in the first instance being comprehensive and complete. A change in the contractor's proposals is not envisaged, but where the contractor considers them to be necessary then the employer's permission will need to be obtained; the contractor will possibly incur extra cost where this occurs. In other respects this clause is very similar to clause 13 of the standard form, which deals with variations and provisional sums.

Clause 16

Upon practical completion of the works, the contractor will receive a written statement from the employer to that effect. The defects liability will operate from that date, and the employer must provide the contractor with a schedule of defects within 14 days. Practical completion marks the end of the contract, and no further new instructions from the employer are permissible. A notice of completion of making good defects is issued when the work has been properly rectified.

Clause 26

Clause 26.2 includes the normal grounds for claiming loss and expense. It also includes the provision for dealing with a delay in receipt of any permission or approval for the purposes of development control requirements necessary for the works to proceed. The contractor must have taken all practicable steps to avoid or reduce this delay. This can also give rise to an extension of time under clause 25.4.7 as a relevant event. It is also considered to be an important factor that could result in the determination by the contractor under clause 28.1.2.8.

Clause 27

The employer may terminate the employment of the contractor for one of the reasons suggested under clause 27.1. In the event of this occurring, the contractor must provide the employer with two copies of all the current drawings, details, schedules, etc., in order that another contractor may be engaged to complete the works. A similar condition will still apply where the contractor has terminated the contract for one of the reasons listed in clause 28.1.

Clause 30

The processes to be used for interim payments follow those of normal practice. Two alternatives are, however, specifically suggested, and these are described in Appendix 2. Alternative A is on a stage payment basis. The stages are predetermined and the appropriate amounts set against them, the total of which adds up to the contract sum. Any adjustments to the employer's requirements or for fluctuations and claims must be properly documented. Alternative B describes the periodic payment basis which is the more usual method with the standard form. Payment on this basis must be supported by the appropriate information. The other matters of retention, and when the payments are due to be paid by the employer, are in accordance with the standard form.

The final account must be presented to the employer within three months of practical completion of the works (clause 30.5.1). The employer must then agree to this within a maximum period of four months from the time of submission. Thereafter the account is conclusive evidence of the amount due. It is also conclusive evidence that the employer is fully satisfied that the project is in accordance with the terms of the contract and the employer's requirements.

CONCLUSION

The role of the building contractor in the design of any project is valid. Criticism has often been levelled at the absence of any contractor input, and the fragmentation of the design and construction process. There is no doubt, therefore, that this method of contracting plays a very vital role and function within the industry today. It does not include all the advantages over the more traditional methods, since if it did, these methods would now be moving nearer towards extinction. The more one can involve the contractor in both the constructional detailing and method, the more satisfactory buildings are likely to be. An employer, however, considering embarking upon this contractual option for a proposed project, is well advised to retain in some form professional advice of an architect and quantity surveyor. The normal building employer is unlikely to be familiar with building contracts or the processes involved. An independent adviser is therefore likely to be able to offer both constructive comments on the contractor's proposals, and also assistance during the building's erection.

The JCT have produced a number of amendments (six up to 1995) to coincide generally with revisions to JCT 98. There are also two practice notes which provide an introduction and commentary on this form.

CHAPTER 12

PROCUREMENT IN THE TWENTY-FIRST CENTURY

The procurement process is one of the components that bring construction projects into existence. The others are design and construction. Each project possesses a number of different variables that will determine the choice of procurement method in providing the most advantageous route for clients. The selection of the procurement route may have repercussions on the operation of the building throughout its life. This chapter explores the issues surrounding a client's main priorities and the methods that are available to produce the building on time, within budget and to the specified quality. Reference will be made to the interface between clients and the industry, how the industry organizes the development process and an evaluation of the different procurement methods.

The construction industry is unique. Its characteristics separate it from all other industries. These include:

- The physical nature of the product
- Manufacture normally takes place on the client's own construction site
- Projects often represent a bespoke design
- Design separate from manufacture is not mirrored in other industries
- The organization of the construction process
- The methods and manner of price determination

Traditionally clients who wished to have projects constructed would invariably commission a designer, normally an architect for building projects or a civil engineer for civil engineering projects. These would prepare drawings for the proposed scheme and, where the project was of a sufficient size, a quantity surveyor would prepare budget estimates and contract documentation on which contractors could then prepare their tenders. This was the procedure used during the early part of the century. Even up to thirty years ago there was only limited variation from this method. The term *procurement* was not used for tendering practices until the 1980s. Since the 1960s there have been several catalysts for change in the way that projects are procured. These are shown in Box 12.1.

Procedures will continue to evolve in order to meet new circumstances, situations and fashion. Procurement is similar to quality in that improvements to present procedures within current practices can always be achieved. Procurement methods

Box 12.1   Catalysts for change in procurement

- Government intervention through committees, such as the Banwell Report of the 1960s and more recently through the DETR and the Latham Report (1994).
- Pressure groups formed to encourage change for their members, most notably the British Property Federation.
- International comparisons, particularly with the USA and Japan and influence of the Single European Market in 1992.
- The apparent failure of the construction industry to satisfy the perceived needs of its clients, especially in the way in which projects are organized and executed.
- The influence of educational developments and research.
- Trends throughout society towards greater efficiency, effectiveness and economy.
- Rapid changes in information technology both in respect of office practice and manufacturing processes.
- The attitudes amongst the professions.
- The overriding wish of clients for single-point responsibility.

of a hybrid nature are being developed to help utilize best practices from the various competing alternatives. Each of the different methods has been used in the industry, and some have been used more than others, largely due to:

- User familiarity
- Ease of application
- Client insistence
- Recognition
- Reliability.

As suggested in Box 12.2, many studies concerned with the organization and management of the construction industry have also been undertaken in the twentieth century. Whilst the industry has responded in some measure to change, this has been more attributed to pragmatic courses of action and commercial pressures, rather than the advice offered by government, industry or researchers.

The construction industry also comprises many different parties, organizations and professions. It is seen as fragmented and does not speak with one voice. This is both an advantage and a disadvantage. These views represent many with vested interests and traditions that in some cases represent their own power and authority. Such bodies are clearly loath to relinquish these positions freely. Some of the recommendations have already been implemented. Other practices continue to persist, even though the structure of the industry and its clients have changed dramatically.

Clients are at the core of the process and their needs must be met by industry. Implementation after all begins with them. Clients are also dispersed and vary greatly. Government in the past used to act as a monolithic client. The privatization of many government departments and activities has changed this perception, resulting in the fragmentation of this important client base. Even existing government departments now operate different procurement strategies and practices. This became more pronounced after the demise of the Property Services Agency.

Box 12.2 Reports influencing procurement

- The Placing and Management of Building Contracts (1944), the Simon Report
- A Code of Procedure for Selective Tendering (1959), NJCC
- Survey of Problems before the Construction Industry (1962), the Emmerson Report
- The Placing and Management of Contracts for Building and Civil Engineering Works (1967), the Banwell Report
- Action on the Banwell Report (1967)
- Faster Building for Industry (1983), NEDO
- The Manual of the British Property Federation System (1983), BPF
- Construction Contract Arrangements in European Union (EU) Countries (1983)
- Thinking about Building (1985), BDP and EDC
- Faster Building for Commerce (1988), NEDO
- Building Towards 2001 (1991), Reading University
- Trust and Money (1993), Department of the Environment
- Constructing the Team (1994), the Latham Report
- A Statement on the Construction Industry (1996), the Barlow Report
- Rethinking Construction (1998) the Egan Report

RELEVANT PUBLISHED REPORTS

Several different reports, many of which have been UK government sponsored, have been issued over the past fifty years. Their overall themes have been aimed at improving the way the industry is organized and the way construction work is procured. These various reports are listed in Box 12.2.

CLIENTS' MAIN REQUIREMENTS

Clients' main requirements have already been considered in Chapter 9. The three main requirements are further examined below. They are subject to change to meet current and perceived future needs, and different client groups will each have differing priorities and requirements.

Performance

The designer prepares plans and details for the proposed project. When completed it should offer aesthetic appeal and add to the environment in which the project is to be located. The completed project must also meet the needs and requirements in terms of the spatial layout, the structure's function and the environmental controls. The specification will define the quality of the materials to be used and the standards of work to be expected.

An inadequate design concept, poor constructional detailing, an incorrect choice of materials and a wrong choice of constructor are problems that will create obstacles to achieving overall good performance. In addition such problems may create difficulties throughout the project's life, so that the project will never operate effectively.

It is also important to consider the needs of future maintenance and refurbishment. The better-informed clients will consider the project in terms of the longer-term needs, rather than solely on hoping to satisfy present-day requirements. Their evaluation will therefore represent a holistic view, taking into account the provision of an immediate design solution coupled with the elimination of possible future problems.

Cost

Before clients commit themselves to a detailed design that they cannot afford, some information on costs and prices needs to be provided by the quantity surveyor. In some circumstances, perhaps where a project has to be carried out as a matter of urgency, cost may be of less importance. But cost can never be ignored, and competing proposals must be evaluated as fully as possible. In those rare circumstances where cost does not matter, some cost advice will nevertheless prove to be beneficial.

A budget price is usually prepared on the basis of scant outline information. This may be based upon the units of accommodation or the floor space that the client is considering whether to provide. This figure will be imprecise because it is based upon imprecise design data. It cannot be too high, otherwise the client may decide not to build and a possible commission will be lost. But it cannot be too low, since as the design evolves it may become apparent that what the client desires in a design cannot be afforded. The project should be cost planned as the design develops. This should result in tenders being received that remain within budget.

Clients will not evaluate a project solely on the initial cost alone, but rather on the basis of a life cycle cost. They will want to consider the future recurring costs associated with owning or using the project. It may be possible to reduce overall costs by introducing construction that requires less maintenance or repair.

Time

Once clients decide to build, they are frequently in a hurry to see the project completed. Some research has shown that if more time could be allocated to the design stage then the project would be completed more quickly. This might also improve the project's quality. A shorter time on the construction site would be likely to reduce initial construction costs. However, prior to the design getting under way, a large amount of time is spent by clients in deciding whether to build.

The design of the project will have some influence on the time required for construction. The use of an *off-the-peg* type building will greatly reduce construction time on site. The methods adopted by the contractor on site for construction purposes will also influence the length of the contract period. The involvement of a contractor as part of the design process will help to influence the methods used for construction and may ensure buildability in the design.

One way of measuring the success of a project is whether the building is available for commissioning by the date suggested in the contract documents. Where time is considered to be very important by the client then it may be necessary to consider adopting fast-track construction methods that have the effect of reducing the length of the contract period.

MAJOR CONSIDERATIONS (SEE ALSO CHAPTER 7)

Consultants versus constructors

The arguments for engaging either a consultant or a constructor as the client's main adviser or representative are linked with tradition, fashion, loyalty and the satisfaction or disappointment with a previous project. There is also the belief (sometimes mistaken) that had an alternative approach to procurement been used, some of the difficulties would not have occurred or problems would have been more easily solved. The emphasis on a single-point responsibility for the client is an attractive proposition. However, this does not automatically mean design and build by the constructor, but a re-evaluation of existing arrangements.

Competition versus negotiation

Businesses, designers or constructors, are able to secure their work or commissions in a variety of different ways. These can include invitation, recommendation, reputation and speculation. They are usually appointed to a project by negotiation or through competition with other firms. Where some form of competition on price, quality or time exists then clients supposedly obtain a better deal. However, there are circumstances where negotiation with a single firm or organization may offer direct benefits to a client.

Measurement versus reimbursement

The alternative ways of calculating the costs of construction work are either on the basis of paying for the work against some predefined criteria or rules of measurement, or through the reimbursement of the contractor's actual costs. Measurement contracts distribute more risk and incentive to the contractor to complete the works efficiently. Reimbursement contracts result in the contractor receiving only what is spent plus an agreed amount to cover profits.

Traditional versus alternatives

Until recently the majority of the major building projects were constructed using a procurement system known as *single-stage selective tendering*. This system had evolved within the parameters of good practice and procedures. However, largely due to better understanding, the availability of other procurement procedures and an improved knowledge of practices in other countries, new methods have become available for consideration.

Various procurement procedures have been devised and developed in an attempt to address some or all of these issues. However, it is difficult to select only a single procurement method to solve all problems. There is no universal solution. None of the newer contractual procedures address all the past criticisms. In fact, some of the newer procedures have been abandoned in favour of existing practices that have been shown to perform better against a range of different criteria. Procurement practice is in reality trading off the client's clear objectives, if known, against the different methods that are available. The client's needs and wants have to be differentiated in an attempt to achieve a best possible overall solution.

PROCUREMENT SELECTION

There are a wide variety of procurement options that are aimed at addressing criticisms of poor quality, lengthy construction periods and high costs. The methods vary from traditional single-stage selective tendering, where a client uses a designer to prepare drawings and documentation on which contractors are invited to submit competitive prices, to schemes where a single construction firm will provide the truly all-in service, the turnkey project. There are methods that have been devised to get the contractor on site as quickly as possible, such as two-stage tendering and fast tracking, with the anticipation that the project will be available sooner than by using other arrangements. Other methods have recognized the contractor's improved management skills or have evolved to meet changes which have occurred in the industry, such as the proliferation of subcontracting, the increase in litigation and the need for single-point responsibility. The Latham Report identified partnering between clients, consultants and contractors as a useful arrangement for the procuring of buildings, emphasizing teamwork.

CURRENT CONSIDERATIONS

Fair construction contracts

It is recognized that the existing arrangements used in the construction industry mitigate against cooperation and teamwork and against the client's own requirements. These also contribute towards helping to perpetuate the poor image of the industry. A 1995 report from the Construction Sponsorship Directorate summarized the fundamental principles of a modern construction contract:

- Dealing fairly with each other and an atmosphere of mutual cooperation
- Firm duties of teamwork, with shared financial motivation to pursue those objectives

- An interrelated package of documents, clearly identifying roles and duties
- Comprehensible language with guidance notes
- Separation of the roles of contract administrator, project or lead manager and adjudicator
- Allocating risks to the party best able to control them
- Avoiding the need, wherever possible, of changes to pre-tender information
- Assessing interim payments through milestones or activity schedules
- Clearly setting out the periods for interim payments and automatically adding interest where this is not complied with
- Provision of secure trust funds
- Provision of speedy dispute resolution
- Provision of incentives for exceptional performance
- Provision for advance payment to contractors and subcontractors for prefabricated off-site materials and components

Trust funds

It is fundamental to trust within the construction industry that those involved should be paid the correct amounts at the right time for the work they have carried out. It may be argued that a problem does not exist and that:

- Clients only award work to firms with integrity
- Contractors are at liberty to decline work from dubious clients
- Subcontractors can adopt similar business practices
- Bonds and indemnities are already available
- Bad debts are not singularly a problem in the construction industry

However diligently clients, contractors and subcontractors verify each other, the realities of the construction industry and its markets continue to exist. In circumstances such as a recession, contractors and subcontractors are often prepared to undertake work for any client. This is frequently done at a minimal profit margin. Bad debt insurance is available, but this adds extra costs at times when firms are seeking to reduce overheads. In times of prosperity, clients are prepared to undertake work with almost any firm who is available in order to get an important project constructed.

The contractor's goods and services become part of the land ownership once incorporated within the project. Any *retention of title* clause that might be incorporated by suppliers or contractors in their trading agreements does not protect them once the materials are incorporated within the works. The building contractor is also likely to be sufficiently far down the queue, where an employer is unable to make payment within the terms of the contract. In some countries around the world, legislation has been provided to deal with the potential injustice that might be suffered. The most comprehensive is the Ontario Construction Lien Act 1993.

An effective way of dealing with this problem is to set up a trust fund for interim payments and retention monies. A client, for example, could be requested to pay into such a fund at the start of the payment period, e.g. at the beginning of the month. The authorized payment would then be paid to the contractor at the appropriate time. Where a form of stage payments is used, then the amount of the particular programme stage would be deposited in the trust fund at the commencement of the work in this stage. The amounts authorized should correspond with the contractor's approved contract programme. The main contractor and the subcontractors would be informed of the amounts deposited. If a party considered that the sums were inadequate then the adjudicator would be consulted. There may also be some rationale of making payments to the subcontractors directly from this fund rather than through the main contractor's account. Any monetary interest accrued in the trust fund belongs to the client. Where the fund necessitates bank charges, then these charges would need to be determined at the time of tender.

Trust funds are not really required for public works projects, since it is unlikely they will become insolvent. However, trust funds will be a source of reassurance for subcontractors, where a main contractor becomes insolvent during the course of the work.

Compulsory competitive tendering

The philosophy behind compulsory competitive tendering (CCT) is that if market forces are allowed to operate then services can be provided with greater efficiency and at a lower cost. Government accounting and purchasing policies have made it clear that value for money and not the lowest price should be the aim. CCT was intended to lead towards better managed, more innovative and more responsive services. However, some argue that if the Transfer of Undertakings Regulations (TUPE) apply then much of the opportunities for cost savings could be lost. The provision of publicly funded services through CCT has been growing in recent years around the world. It remains highly controversial in Australia and in the industrialized areas of Europe and the USA. It raises fundamental questions about competition and ownership in the provision of such services. Some of the more problematic issues with policy implementation are:

- Fair and effective competition
- Incentive compatibility
- Performance monitoring
- Whether CCT provides the best value for money

The preliminary assessments of contracting suggest generally successful outcomes. Empirical evidence at the present time suggests that efficiency gains have been made, and effectiveness and quality of service have been maintained, if not enhanced. Few professional consultants, who come within this directive, are likely to admit openly that they have reduced their services because of professional fees. However, a survey by the Association of Consulting Engineers found that less time, resources and consideration was given to projects where fee competition was used.

Reverse auction tendering

Under reverse auction tendering, aptly abbreviated as RAT, contractors bid against each other in a live telephone auction to offer the lowest possible price or the best value for money. Bidders remain anonymous, with their bids relayed through an auction assistant to the auctioneer, who acts for the client. The aim is to do away with the often unfair practice of one-off, sealed bids and to offer contractors a chance to lower their bids against their competitors. Whilst a number of different client groups are considering piloting this idea, contractors are understandably much less enthusiastic. Contractors are raising issues of confidentiality, cartels, inequitable pricing and intellectual property rights. It is important that clients obtain the best possible price. However, if such a system resulted in contractors bidding too low to obtain work, this might have repercussions in terms of disputes arising or even more liquidation amongst construction firms and their suppliers. This would benefit no one.

Project partnering

Most people involved in the construction industry recognize the need to move away from the confrontational relationships that cause disputes, problems, delays and ultimately expense. The definition of partnering is one of a long-term commitment between two or more organizations for the purpose of achieving specific business objectives by maximizing the effectiveness of each participant's resources.

Banwell (1967) recognized there was scope for awarding contracts without the use of competitive tendering under certain circumstances. For example, where a contractor had established a good working relationship with a client over a period of time, completing projects within the time allowed, at the quality expected and for a reasonable cost. These circumstances may have arisen through serial contracting and particularly in those situations of continuation contracts where projects had been awarded on a phase basis.

There has always been some reluctance on the part of public bodies to adopt such procedures, since they may lack elements of accountability, even though it could be shown that a good deal had been obtained for the public sector body concerned. The UK government has given official approval to partnering in the public sector. This provides an environment that is similar to the private sector approach, with a number of construction firms developing long-term relationships with public clients. Effectively partnering will be possible where:

- It does not create an uncompetitive environment
- It does not create monopoly conditions
- The partnering arrangement is tested competitively
- It is established on clearly defined needs and objectives over a specified period of time
- The construction firm does not become overdependent on this arrangement

Partnering is now well established in the USA, offering a range of benefits, and in the UK it has been used successfully by several different organizations. The partnering arrangement may last for a specific length of time or for an indefinite period. The parties agree to work together in a relationship of trust, to achieve specific objectives by maximizing the effectiveness of each participant's resources and expertise. It is most effective on large construction projects or projects of a repetitive nature where the expertise developed can be retained and used again. The McDonald's chain of restaurants provides a good example of repetition. This arrangement, coupled with the use of innovative construction techniques, has reduced time and costs by 60 per cent since the start of the 1990s. Its on–site activities have been reduced from 155 days to 15 days for a typical restaurant building.

The concept of partnering extends beyond the client, contractor and consultants and includes subcontractors, suppliers and other specialist organizations who are able to add value to the project. The establishment of good supply chain management benefits those involved.

Appointment of specialist firms

The traditional arrangement of appointing specialist firms on a construction project is to use one of the nomination procedures. Whilst many specialist firms would like this procedure to be extended, it has been estimated that as few as 11 per cent of specialist engineering contractors are appointed in this way. Alternative methods are available:

- *Joint venture*: this is a particularly helpful approach where there is a large engineering services input to the project. The joint venture arrangement is between the main contractor, who may typically carry out the role of a project manager, and the specialist contractor. The companies work together as a joint company. It is therefore suitable for design and build arrangements.
- *Separate contracts*: the client in this case lets individual contracts to different firms, i.e. the main contractor and the specialist contractor. This has in the past not been easy to administer, particularly where problems have arisen.
- *Management and construction management*: this is believed to be the most effective way of dealing with such firms. The different trade and specialist firms are appointed and a contractual arrangement formed with each company. This arrangement allows for full participation by the firms in design and commercial decisions at an early date.
- *Appointing a specialist firm as the main contractor*: where the specialist work represents the largest portion of the project, then clients may choose to reverse the arrangements and appoint a specialist firm as the main contractor. The more usual construction trades would then be employed by this firm.

Construction (Design and Management) Regulations 1994 (CDM)

The construction industry is a dangerous environment. It has a poor health and safety record. Serious injury and death happen far too frequently as a result of

Table 12.1   Injuries to employees: construction and all industries

	Rates per 100,000 employees			
	1986/87	1988/89	1990/91	1993/94
Fatal and major				
Construction	293	296	291	233
All industries	101	94	92	79
All reported injuries				
Construction	1,995	1,928	1,907	1,416
All industries	862	842	818	710

Source: DETR

Table 12.2   Injuries to employees in the construction industry 1992/93

	Fatal	Major	Total
Contact with moving machinery	2	71	73
Struck by flying object	7	310	317
Struck by moving vehicle	14	90	104
Slip, trip or fall	1	366	367
Fall from height	27	808	835
Trapped by collapse	6	76	82
Exposure to harmful substance	4	41	45
Contact with electricity	8	63	71
Other	0	231	231
Total	69	2,056	2,125

Source: DETR

construction activities, especially on site. These affect not only the construction workers employed on site, but also members of the general public. Improving the management process is essential in helping to prevent accidents and ill health in the industry.

Different governments have initiated measures aimed at reducing accidents on construction sites. The most important of this legislation is the Health and Safety at Work Act 1974, which applies to all industries. However, it is recognized that some industries are more dangerous than others. The data shown in Table 12.1 and 12.2 help to illustrate this point. Falling from a height was the most common cause of fatality in the industry. Being struck by a moving vehicle was the next most common. Lifting and carrying accounted for nearly a quarter of all reported injuries in 1992/93.

The CDM Regulations, which became operative in 1995, place new duties on clients, planning supervisors, designers and contractors to plan, coordinate and manage health and safety throughout all stages of a construction project. Anyone who appoints a designer or contractor has to ensure they are competent for the work and will allocate adequate resources for health and safety. The regulations apply to construction projects and everyone associated with their design and construction. The regulations are about the management of health and safety. Two documents must be created:

- *The health and safety plan* is prepared in two stages, before and after the the contractor is appointed:
  - It provides the health and safety focus for the construction phase of the project.
  - It shows that adequate resources have been allocated to the project.
- *The health and safety file* holds information about health and safety matters which will assist those carrying out construction, maintenance, repair or demolition work at any time before or after completion:
  - It is a record for the client and user, identifying those responsible for the structure and the risks that have to be managed during maintenance, repair or renovation.
  - It describes the services that are installed in the building, the materials used and the building construction.
  - It is given to the client when the project is complete.
  - It is provided by the client to anyone carrying out work on the structure in the future.

The health and safety file is a document that is updated during the design process as more information becomes available. The health and safety plan must be sufficiently developed to form part of the tender documentation. It must do three things:

- Clarify the health and safety issues specific to the project.
- Identify where the principal risks are likely to occur and alert tenderers to any possible unexpected hazards.
- Clarify the parameters against which to judge the selection of competent and properly resourced contractors.

There are five key parties, firms or individuals who are involved; each has specific duties to perform.

- *Clients* must ensure:
  - They are able to allocate sufficient resources, including time, to enable the project to be carried out safely.
  - Only competent and adequately resourced people are employed.
  - Reasonable enquiries are made about the land or premises being developed and the outcomes passed to the planning supervisor.
  - Construction work does not start until there is an adequate health and safety plan.
  - A health and safety file is available for inspection.

- *Designers* produce drawings and written documents, e.g. specifications. Where risks cannot be avoided, adequate information must be provided. Designers must ensure:
  - Structures are designed and specified to minimize any possible risks to health and safety during construction and subsequently their maintenance.
  - Adequate information is provided on possible risks.
  - Cooperation with planning supervisors.
- *Planning supervisors* have the overall responsibility for:
  - Coordinating the health and safety aspects of the design and planning phase.
  - The early stages of the health and safety plan.
  - Advising the client on competence and adequate resourcing of the principal contractor.
  - Delivering the client a health and safety file for each structure on its completion.
- *Principal contractors* carry out these duties and have these responsibilities:
  - Take account of health and safety issues when preparing tenders or estimates.
  - Exclude unauthorized persons from the site.
  - Cooperate with the planning supervisor.
  - Develop the health and safety plan.
  - Coordinate activities of all contractors so they comply with the health and safety plan.
  - Provide information and training on health and safety.
- *Subcontractors and the self-employed* should:
  - Cooperate with the principal contractor on health and safety matters.
  - Explain how they will control the health and safety risks in their work.

The subcontractors also have duties for the provision of other information to the principal contractor and to employees. The self-employed have similar duties to contractors. Employees on construction sites should be better informed and have the opportunity to become more involved in health and safety matters.

Whilst these new procedures add extra costs to the industry, and hence its clients, because of its implementation and monitoring, over the longer period of time they will help to reduce costs through developing better construction practices. They should help to save lives and reduce accidents and the disruption they sometimes cause to work on site. They will reinforce the need for coordinating and managing health and safety from inception to completion and during the use of the completed project.

Quality assurance

Every client in the construction industry has the right to assume a standard of quality that has been specified for the project. The Building Research Establishment has reported that defects or failures in design and construction cost the industry and its clients more than £1 billion per year. This represents 2 per cent of total turnover. The construction industry is an industry in which:

- There has never been a requirement for the workforce to be formally qualified and skills are generally developed through time serving.
- Much of the work is carried out by subcontractors in a climate in which some fifty firms come into existence every day and a similar number go into liquidation or bankruptcy.
- There is a paucity of research and development involving new materials, designs and techniques.
- There is often poor management and supervision.

Studies in the UK have indicated that about:

- 50 per cent of faults originate in the design office
- 30 per cent on site
- 20 per cent in the manufacture of materials and components

An investment in quality assurance methods can therefore reap substantial long-term benefits by helping to reduce such faults, the inevitable delays, the costs of repairs and the all too frequent legal costs that often follow.

ISO 9001 certification has been increasingly taken up within the construction industry by consultants and contractors. Quality assurance is therefore seen as a good thing for the industry. Procurement methods that fail to address this issue adequately are not doing the industry or its clients any favours. The use of quality-certified firms, who have been independently assessed and registered, therefore offers some protection within the context of getting it right first time. Work that has to be rectified is rarely as good as work that was right in the first place.

ISO 9000 is seen by some as an additional expense and an unnecessary overhead. Also, whilst it should ensure that quality standards are achieved, it does not ensure that the appropriate quality has already been set in the first place. However, some clients are now refusing to employ consultants, contractors or suppliers who do not have the relevant Kitemark. Within the total quality management scenario, quality remains an ongoing process of *continuous quality improvement*. Quality must be appropriate to the work being performed. It should only be insisted upon where it adds value to the finished construction project.

The system being employed by the main contractor for the management of quality must incorporate the integrated quality management activities of the various members of the supply chain. This is especially important in respect of the quality assurance of construction projects, as frequently the main contractor often outsources work to the various members of the supply chain.

Latent defects liability and BUILD insurance

The Construction Sponsorship Directorate of the Department of the Environment, Transport and the Regions (DETR) has suggested compulsory latent defects insurance or BUILD (Building Users' Insurance against Latent Defects). Clients frequently look for some degree of protection against:

- The risk of latent defects
- The costs of remedying any defects
- The cost of any damage caused by defect, including
  - loss of rents
  - loss of profit
  - other consequential losses

The physical complexity of the construction process and the integration of its constituent parts, the number of (temporary) teams involved, the sums of money involved and the complex tangle of potential legal and financial responsibilities of the different parties involved all require greater clarity. The issues arising from the consideration on latent defects liability appertain to:

- Joint and several liability
- Limitation periods and prescription of actions
- Transfer of clients' rights

Difficulties exist where more than one party is involved and hence the desire by some clients for a single-point responsibility for the project. The proposal in the consultative document is that the period of liability should commence from the date of practical completion of the works. This might appear unfair to those specialist firms who have completed their portion of the project, perhaps several years earlier on a large project.

The provisions of the Limitation Act 1980 and the Latent Damage Act 1986 should be brought into line with the provisions of the Consumer Protection Act 1987, which provides for a ten-year limitation period. This should apply for all future new commercial, retail and industrial building works in both the public and private sectors.

The doctrine of privity of contract means that, as a general rule, a contract cannot confer rights or impose obligations arising under it on any person except the parties to it. The tenants and the subsequent owners have no privity of contract with contractors or consultants. In July 1991 a House of Lords judgement (*Murphy* v. *Brentwood*) closed down the law of negligence as a route for recovery of economic loss, except in exceptional cases. Since then owners or tenants without privity of contract have used other contractual techniques such as collateral warranties to create contractual rights where none would otherwise have existed.

The BUILD insurance would be financed by the client at currently about 1 per cent of the contract cost. Such policies normally cover the structure, foundations and the weather shield envelope. They often exclude engineering services, which is seen as a weakness. The possibility of mechanical or electrical failure could be provided for an additional premium. Policies could also cover loss of rental or extra rental expenditure.

Added value

The Latham Report, published in 1994, recommended that initial construction costs should be able to be reduced by 30 per cent by the early part of the twenty-first

Box 12.3   Adding value in construction

- Reduce the amount of changes to the design
- Optimize specifications
- Improve design cost-effectiveness
- Apportion risk efficiently
- Improve productivity
- Reduce waste
- Examine cost-efficient procurement arrangements
- Improve the use of high technology for both design and construction
- Reduce government stop-go policies
- Develop more off-site activities
- Standardize components
- Consider construction as a manufacturing process
- Get it right first time, i.e. avoid defects
- Make better use of mechanization
- Improve the education and training of operatives
- Reduction of staff, noting their high costs and their reduction in manufacturing industry

century. Such cost reductions should not reduce quality but at least maintain it and preferably improve the overall quality. The implication is one of adding value, a principle of doing more with less. It is essential that such reductions in cost do not refocus the industry backwards fifty years, towards the emphasis on initial costs alone. The importance of ensuring that life cycle costs are given their rightful importance in the overall building process must be maintained. The principle is one of changing cultures and attitudes and benefiting from changes that have already taken place in other industries. Box 12.3 shows some of the areas of possible investigation in attempting to meet this aim.

The benchmarking of good practices in the procurement of construction projects is one way of achieving added value. The use of benchmarking techniques has gathered a wide interest in the UK (Construction Task Force), USA (Construction Industry Institute) and in Australia. Evidence for the use of benchmarking has arisen from its use in the manufacture of motor cars. Toyota were identified as the best after several years of benchmarking car production activity. This was done by the International Motor Vehicle Programme that is based at the Massachusetts Institue of Technology.

Private Finance Initiative

The purpose of the Private Finance Initiative (PFI) was to encourage partnerships between the public and private sectors in the provision of public services. The scheme is outlined in a DETR report that was published in 1993. In 1992 the then Chancellor of the Exchequer announced a new initiative to find ways of mobilizing the private sector to meet needs that had traditionally been met by the public sector. Achieving an increase in private sector investment would mean that more projects could be undertaken. This also takes into account that public spending should decline in the medium term. The broad aims of such a partnership will be to:

- Achieve objectives and deliver outputs effectively
- Use public money to best effect
- Respond positively to private sector ideas

In exploring the possibilities for the use of private finance, including proposals from the private sector, consider these questions:

- Can the project be financially free-standing?
- Is it suitable for a joint venture?
- Is there potential for leasing agreements?
- Is there potential for government to buy a service from the private sector?
- Can two or more of these elements be brought together in combination in any particular instance to form innovative solutions?

Concessionary contracting falls neatly into such an arrangement, whereby the private sector is encouraged to construct public projects such as roads and then charge a levy on this provision, for a fixed period of time specified in the contract. The contractor throughout the entire period is responsible for the maintenance of the works. Upon handover to the public sector, the contract will also specify the required condition of the asset.

Competitive advanatge

The key contribution for shaping and reshaping the thinking in the current context of world economies is described by Porter (1980). In competitive advantage the rationale is not directed towards organizational structures or change but has profitability as the strategic driver. Porter argues there are five competitive forces that determine profitability:

- Potential of new entrants into the industry
- Bargaining power of customers
- Threat of substitute products
- Bargaining powers of suppliers
- Activities of existing competitors

One of the most important concepts established by Porter is the *value chain*. This is a systematic way of examining all the activities a firm performs and how they interact. Primary activities are inbound logistics, outbound logistics, marketing, sales and service. Support activities include procurement, technology development, human resource management and infrastructure. The way in which one activity in the chain interacts with another can be crucial. This can occur within the organization, or externally with suppliers. Porter argues that a firm gains competitive advantage by performing these activities, alone or linked, more economically or in a better way than its competitors.

Business process re-engineering

The idea of business process re-engineering is to learn as much as possible from other industries who have had to respond to massive cultural changes. Improvement or added value can be achieved through construction re-engineering. No other concept has recently received more interest and criticism than re-engineering. This is because it is a concept that is easy to understand but difficult to put into practice. The successful re-engineering projects are founded on six basic principles:

- Organize around outcomes not tasks.
- Have those who produce the output of the process perform the process.
- Subsume information processing into the real work that produces the information.
- Treat geographically dispersed resources as though they were centralized.
- Link parallel activities instead of integrating their tasks.
- Put the decision point where work is performed, and build control into the process.

In spite of a few well-known successes, there is much evidence to suggest that re-engineering fails or at best produces only marginal results in a majority of organizations where it is implemented. Sometimes this is because the programmes are not sufficiently radical and only tinker with the most easily accessible processes.

RETHINKING CONSTRUCTION

The Egan Report (1998) recognizes that construction in the UK at its best is excellent. It has developed engineering ingenuity and a design flair that is recognized throughout the world. Its capability to deliver the most difficult and innovative projects matches that of any other construction industry in the world. The report recognizes that the industry as a whole is underachieving. It has low profitability and invests too little in research, development and training. Also too many clients are dissatisfied with its overall performance.

Five key areas for change are identified in the report:

- *Committed leadership*: the need for vision and change at all levels.
- *Focus on the client*: the view is that the customer is always right.
- *Integration of processes and teams*: fragmentation of operations affects success.
- *Quality driven agenda*: about getting it right first time.
- *Commitment to people*: providing appropriate health and safety, wages and site conditions.

The report also stated that training and quality are inextricably linked. Real reductions in construction costs, whilst they maintain value, are unlikely to reduce significantly until the education of its workforce is improved through a culture of teamwork. The report suggested that employers, the government and the national training organizations (NTOs) should put together an agenda for urgent action.

Performance indicators

It is easy to suggest that improvements in processes and practices are being achieved based on subjective judgment and anecdotal evidence alone. There is also often a resistance to want to measure or attempt to quantify such changes. It is also all too easy to distort the data, unless clear and precise guidelines are employed. In some cases in the past there have been improvements and their effect has then been attributed to a particular cause. Upon further investigation the cause and effect are not linked. For example, the 30 per cent reduction in cost identified by the Latham Report (1984) may appear to be achieved largely due to the suppressed costs, of both labour and materials, of the recession in the middle of the 1990s. It may be difficult when looking back to the start of the twenty-first century to identify whether improved methods of working actually achieved this goal, or whether, without them, natural effects of improved technologies were the real reasons. The debates that raged throughout the 1980s about the poor time performance of UK building when compared with countries abroad were only partly remedied through productivity agreements. Other improvements in time performance were often restricted because of the different regulations and organization that was adopted in the UK.

In this context it is important that the construction industry sets itself clear and measurable objectives. These might be achieved through the use of performance indicators or quantified targets. Measures of improvement will be required in terms of cost, time and quality, relevant to the aims and objectives of the individual client. The targets must be real and composite. They must not be achieved through cutting corners in other respects such as safety and wages. In order to make such gains last, and thereby add value, continuous improvement must be implemented.

The Egan Report identified a number of measures designed for sustained improvement. These are shown in Table 12.3.

Table 12.3   Performance indicators

Indicator	Improvement per year	Definition
Capital cost	reduce by 10%	All costs excluding land and finance
Construction time	reduce by 10%	Time from client approval to practical completion
Predictability	increase by 20%	Number of projects completed within time and budget
Defects	reduce by 20%	Reduction in the number of defects at handover
Accidents	reduce by 20%	Reduction in the number of reportable accidents
Productivity	increase by 10%	Increase in value added per head
Turnover	increase by 10%	Turnover of construction firms
Profits	increase by 10%	Profits of construction firms

Source: *Rethinking construction* (1998)

Table 12.4   Industry performance: construction versus car manufacture

Wants	Modern motor car	Modern buildings		
		Domestic	Commercial	Industrial
Value for money	●●●●	●●●●●	●●●	●●●●
Pleasing to look at	●●●●	●●●●	●●●	●●●
Largely free from faults	●●●●●	●●●	●	●●
Timely delivery	●●●●	●●●●	●●●●	●●●●
Fit for purpose	●●●●●	●●●●	●●	●●●
Guarantee	●●●●●	●●●●	●	●
Reasonable running costs	●●●●	●●●●	●●	●●●
Durability	●●●●	●●●	●●	●●
Customer delight	●●●●●	●●●	●●	●●

Source: *Constructing the team* (1994)

OTHER INDUSTRY COMPARISONS

It is always relevant when examining a subject like procurement to see how it is done elsewhere. This comparison may be made against similar or competing firms, perhaps in the form of a benchmarking study. Alternatively, comparisons can be made with firms or organizations overseas, in countries that mirror UK practices and in countries where different traditions are employed. It is also important to consider other industry comparisons, as illustrated in the Latham Report, which compared the performance of the construction industry with that of the motor car industry. Table 12.4 is an adaptation of that comparison. Other comparisons have been made with the aerospace industry (Flanagan 1999). The outcome of such studies acts as a guide to good practices found elsewhere but which might have been overlooked. Current comparisons do not place the construction industry in a good light, but act as motivators to help change the culture of the construction industry.

The motor car manufacture and aerospace sectors include the following attributes that are generally absent from the construction industry:

■ The recognition of a manufacturing culture.
■ The integration of design with production.
■ The importance of the supply chain network.
■ A focus on innovation and that this will only be secured through adequate research and development.
■ An acceptance of standardization in design, components and assembly across the product range.

CHAPTER 13

LEAN CONSTRUCTION

The lean construction process is a derivative of the lean manufacturing process. This has been a concept that has been popularized since the early 1980s in the manufacturing sector. The original thinking was developed from Japan, although it is now being considered and introduced worldwide. It is concerned with the elimination of waste activities and processes that create no added value. It is about doing more for less.

Lean production is the generic version of the Toyota Production System. This system is recognized as the most efficient production system in the world today. Incidentally Toyota's activities in the construction industry are larger than those of its more well-known automobile business! It also needs at the outset to be acknowledged that construction production is different from that of making motor cars, although it is possible to learn from and adapt successful methodologies from other industries. In the automobile manufacturing industry, spectacular advances in productivity, quality and cost reduction have been achieved in the past ten to fifteen years. Construction, by comparison, has not yet made these advances. It also remains the most fragmented of all industries, but this can be seen both as a strength and a weakness.

The application of lean production techniques in motor vehicle manufacturing has been a huge success and is associated with three important factors:

- The simplification of manufacturing dies.
- The development of long-term relationships with a small number of suppliers, to allow just-in-time management.
- Changes in work practices, i.e. the culture and ethos of practices, most notably the introduction of teamworking and quality circles.

Lean thinking is aimed at delivering what clients want, on time and with zero defects. Lean construction has identified poor design information which results in a large amount of redesign work. Several organizations around the world have established themselves as centres for lean construction development. The aim, for example, of the *Lean Construction Institute* in the USA is a dedication towards eliminating waste and increasing value.

Few products or services are provided by a single organization alone. The elimination of waste therefore has to be pursued throughout the whole value stream, including all who make any contribution to the process. Removing wasted effort represents the biggest opportunity for performance improvement.

Several companies around the world are attempting to introduce lean construction methods into their core businesses. *Rethinking construction* (Egan 1998) provides two examples and these are briefly discussed later in this chapter. One of the firms is based in Colorado and another in San Francisco. One of these firms has already reduced project times and costs by 30 per cent through developments such as:

- Improving the flow of work on site
- Using dedicated design teams
- Innovation in design and assembly
- Supporting subcontractors in developing tools for improving processes

This suggests that perhaps the most useful way of achieving cost reductions, whilst still maintaining value, is to consider profitable ways of reducing the time spent on construction work on site. Design readiness is the same principle. This suggests that to fully complete the design prior to starting work on site will save both construction time and the respective costs that are involved (Figure 8.5).

Lean construction is a philosophy that is about managing and improving the construction process to profitably deliver what the customer, the construction client, requires. Engineering, in all its different kinds, has had to develop a wide range of strategies to remain at the competitive edge and to improve its products. Comparisons have been made on several occasions between the high technology engineering approach and the low technology that is adopted generally throughout the construction industry sector.

Because it is a philosophy, lean construction can be pursued through a number of different approaches. The lean principles have been identified as follows:

- The elimination of all kinds of waste; this includes not just the waste of materials on site, but all aspects, functions or activities that do not add value to the project.
- Precisely specify value from the perspective of the ultimate customer.
- Clearly identify the process that delivers what the customer values; this is sometimes referred to as the value stream.
- Eliminate all non-added-value steps or stages in the process.
- Make the remaining added-value steps flow without interruption, through managing the interfaces between the different steps.
- Let the customer pull, do not make anything until it is needed, then make it quickly. Adopt the philosophy of *just-in-time management* to reduce stockpiles and storage costs.
- Pursue perfection through continuous improvement.

LEAN MANUFACTURING

Lean manufacturing has been adopted by the large Japanese car manufacturers. It has also been implemented by a number of Japanese, American and European manufacturers with some considerable success. It has especially been adopted by the automotive and other engineering industries.

The lean process is about designing, constructing and operating the right systems to deliver the right product first time. Essential to this is the elimination of what the industry describes as *snagging* work, i.e. the remedying of defective work prior to handing over to the employer or client. There have been examples quoted in the construction industry where snagging work has taken up almost as much time as doing the work in the first place. Activities or processes that absorb resources but create no additional value must be eliminated. This waste can include:

- Mistakes
- Working out of sequence
- Redundant activities and movement
- Delayed or premature inputs that are the result of bad programming
- Products or services that do not meet a customer's needs or requirements
- Non-conformance to specified standards or quality

The primary focus of lean thinking is moving closer and closer to providing a product that customers really desire, and, by understanding the processes involved in construction manufacture, identifying and eliminating the waste that is normally generated.

PRODUCTION AND MANAGEMENT PRINCIPLES

Lean thinking is focused on value rather than on cost. However, whilst cost is not unimportant, the emphasis has been switched to adding value across the whole range of services and processes that are used in the construction of buildings. It seeks to remove all components that do not add value, especially the various processes involved, whilst improving components that do add value.

Its aim is to define value in customer terms, identifying key points in the development and production process, where that value can be added or enhanced. The goal is the seamless, integrated process or value stream, wherein products flow from one added value step to another. This is all driven by the philosophy and the pull of the customer (the employer).

The idea of getting it right first time is fundamental to the process of the lean philosophy. In this context right means making it so that it does not require any rectification at a later stage. The approach involves an extremely rigorous questioning analysis of every detail of product development and production. This approach seeks to identify the ultimate source of problems. Only by eliminating the source is it possible to prevent a fault recurring.

Production management techniques have traditionally focused on the need to schedule discrete activities in the building process. This is in contrast to adopting a philosophy of seeing it as a manufacturing process involving the management of resources across a network of firms. This perspective has been increasingly criticized and there is already a growing body of research on supply chain management techniques in the construction industry. These include buildability, just-in-time management and lean construction practices. Production management techniques have also focused too much on the process rather than the eventual product that is produced.

DESIGN AND PRODUCT DEVELOPMENT

Extensive product development work has been undertaken in the engineering industry in recent years, in order to maintain its viability and feasibility wherever this has been possible. This industry has responded positively to the demands placed upon it by adding value to its products. Lean manufacturers, regardless of the industry, have developed systems for product development which first identify the right product to be made, in terms of customer needs and expectations. The product is then designed correctly and timely in order that it can be manufactured efficiently. For example, where an analysis of architectural detailing is carried out, simpler and less costly solutions can often be found, without any detrimental effect to the design itself.

Design in manufacturing terms is concerned with the development and integration of systems and components into coherent, efficient and manufacturable products. The construction industry has had a considerable interest in buildability solutions for over twenty-five years. There is some evidence to suggest that this aspect is now considered more frequently than previously, although the separation between the design and the construction professions remains, and probably remains at an overall disadvantage. There is too little emphasis given in design education to manufacturing problems and an insufficient consideration of design methodologies on courses concerned with building production management.

Tools have been developed to capture and analyse customer perceptions and requirements for product quality and performance. These tools enable product development and manufacturing performance targets to be established. Design development targets include reductions in design changes and process iterations. The critical success factors can be summarized as follows:

- Design is informed by extensive data on the performance of products, systems and components.
- Carry-over to new models of a high proportion of systems and components from previous successful models.
- Front-loading of resources towards design to prevent problems during construction manufacture.
- Concurrent working between manufacturers and suppliers during design development.

LEAN PRODUCTION

Lean manufacturers arrange production in closely located cells, so that work flows continuously, with each step adding more value to the product. The standard time for all activities is known and the objective is to totally eliminate all stoppages or delays, throughout the entire production process. Only minimum stocks of materials are kept as buffers between processing stages. The efficient application of materials handling, through the use of information technology, was pioneered for stock control purposes in the early 1970s.

For the system to be effective, every machine and worker must be completely capable of producing repeatable perfect quality output at the exact time required. Employees are responsible for checking quality as the product is assembled, and in some cases given the authority to halt production should defects arise. In such a system, quality problems are exposed and rectified as soon as they occur.

The workforce are kept informed of their progress towards production and cost targets. Information displays are provided in order that everyone can see the status of operations at all times. The work teams in lean manufacturing are highly trained and multi-skilled. In some cases the supervisory and management functions have been devolved to them. The critical success factors include:

- In-depth understanding of production processes and the resources required
- Responsibility and authority placed with the workforce
- Real-time feedback on performance
- Training and multi-skilling

This last point is especially important. The value of appropriate training cannot be overemphasized. Without the adequacy of this investment the whole philosophy has a danger of not becoming a reality. The construction industry is frequently seen as low technology and has not changed to meet aspirations or practices that are seen elsewhere.

The use of lean production techniques must be placed within the context of the construction industry. The comparison between mass production in factories with generally bespoke buildings on a construction site must not be minimized. Furthermore, lean manufacturing appears to achieve the greatest improvements in efficiency and quality when design and manufacture occur in close proximity. With traditional procurement arrangements in the construction industry this is frequently not the case.

SUPPLY CHAIN MANAGEMENT

Lean manufacturing is based upon the elimination of waste process and practices. This includes the time in waiting for others to complete their tasks, the delay caused through the lack of late deliveries, unnecessary storage and the value that is tied up in large stocks or parts awaiting assembly. Just-in-time delivery is an important concept of the process of lean construction. Lean construction firms have therefore had to develop their own reliable network of suppliers and subcontractors. Significant efforts need to be made to encourage these firms to adopt the same principles and systems. The fragmented nature of the construction industry makes this point of even greater importance and significance.

The adoption of lean manufacturing in engineering has had the effect of moving away from traditional relationships with suppliers towards partnering arrangements with a fewer number of firms based upon good communications and open-book accounting. These arrangements work for both parties, sharing the philosophy of continuous improvement. This is especially the case in the area of defect reduction and the cost and timeliness of deliveries. There is a sharing of business and development strategies sufficient for both parties to know enough about each other in order to make forward planning effective. The critical success factors include:

- The lack of reliance on formal contracts
- The use of benchmarking of suppliers' performance against each other on a range of generic criteria
- The development of close relationships with first-tier suppliers

BENCHMARKING

Benchmarking has been described as a method for organizational improvement that involves continuous and systematic evaluation of products, services and processes of organizations that are recognized as representing best practice. It is a system that uses objective comparisons of both processes and products. It may make internal comparisons within a single firm perhaps by comparing the performances of different building sites. It can be an external system, where comparisons of performance are made against similar and dissimilar firms. The enterprises that are internationally recognized as world leaders are described as having the best practices and are the most efficient. The Construction Industry Board in the UK has published a list of *key performance indicators* that are updated and can be used by any firm to measure its performance.

Benchmarking is an integral part of lean construction methods, since they seek to identify where improvements in processes or products are possible. There is considerable interest in and practice of benchmarking techniques across a wide range of industries. It has been suggested that probably all of the top firms and companies use them in one way or another. It is also worth remembering that benchmarking is not just practised by those firms that are lagging behind and need to improve their performance, but by the international world leaders, as a tool for maintaining their competitive edge. The focus of benchmarking is on the need for continuous improvement.

LEAN-THINKING CONSTRUCTION

The lean principles that have been used effectively in engineering manufacturing can only be applied effectively in construction through focusing on improving

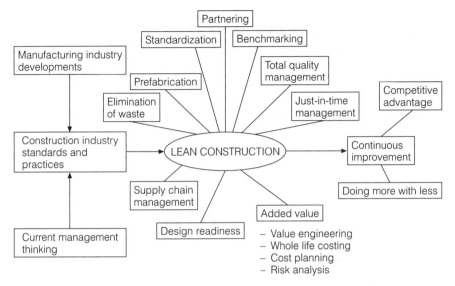

Figure 13.1 Lean construction

the whole design and construction process. This requires a commitment from all the parties involved, including the client. The obstacles that may arise through traditional contractual arrangements can then be removed. Figure 13.1 is a schematic description of lean construction. Here are some issues to consider.

Design

- Use visualization techniques such as three-dimensional CAD and virtual reality. These will help to promote product definition and provide a clearer perspective and understanding on the part of the client.
- Apply value management to achieve a greater understanding of those aspects that clients' value.
- Use integrated design and build arrangements including partnering.
- Encourage greater cooperation between designers, constructors and specialist suppliers.
- Design for greater standardization
- Use prefabrication or pre-assembly of components in order to achieve higher quality, and cost and time savings.

Procurement

- Incorporate a supply chain management process.
- Rationalize and integrate the number of suppliers who are involved.
- Provide a seamless integrated process.
- Eliminate waste in the procurement process.
- Ensure that decisions on customer value can be taken.

- Ensure confidentiality of construction costs and cash flows.
- Introduce the concept of partnering with common goals where the boundaries between companies and firms become less critical.

Production

- Introduce benchmarking practices to establish the best in the class production methods and outputs.
- Establish a stable project programme with a clear identification of a critical path.
- Apply risk management techniques to manage risk throughout the project.

Logistics

- Adopt just-in-time delivery of materials to the point of use to eliminate on-site storage and double-handling.

Construction

- Ensure clear communication of project plans to the whole team.
- Provide relevant and ongoing training.
- Encourage a teamwork approach.
- Develop multi-skilling.
- Provide daily reporting and improvement meetings.
- Develop a well-trained, highly motivated, flexible and fully engaged workforce.

CASE STUDIES

Pacific Contracting of San Francisco

This is a specialist cladding and roofing contractor. It has used the principles of lean construction to increase its annual turnover by 20 per cent within an 18 month period, using only the same numbers of staff. The key to its success was improving the design and procurement processes in order to facilitate construction on site. It did this by investing in the front end of projects to reduce costs and construction time. The firm identified two major problems in achieving a flow of the whole construction process:

- Inefficient supply of materials which prevented site operations from flowing smoothly.
- Poor design information from the prime contractor, which frequently resulted in a large amount of redesign work.

To tackle these problems, Pacific Contracting, combined more efficient use of technology with tools for improving planning of the construction processes. They used a computerized three-dimensional system to provide a better, faster method of redesign that leads to better construction information. Their design system provides a range of benefits, including isometric drawings of components and interfaces, fit coordination, planning of construction methods, motivation of work crews through visualization, first run of tests of construction sequences and virtual walk-throughs of the product. They also used a planning tool, known as last planner, to improve the flow of work on site through reducing constraints such as lack of materials and labour.

Neenan Company, Colorado

This is a design and build firm and is one of the fastest-growing construction companies in its region. The firm has worked to understand the principles of lean thinking and to look for applications in its business. It has used *study action teams* of employees to rethink the way they work. This firm has reduced project times and costs by up to 30 per cent, through developments such as:

- Improving the flow of work on site by defining units of production and using tools such as visual control of processes.
- Using dedicated design teams working exclusively on one design from beginning to end.
- Developing a tool known as *schematic design in a day* to dramatically speed up the design process.
- Innovating in design and assembly, for example, through the use of prefabricated brick infill panels manufactured off-site and pre-assembled atrium roofs lifted into position.
- Supporting subcontractors in developing tools for improving processes.

British Airports Authority

British Airports Authority (BAA) is one of the world's major airport operators. It is also a key client of the construction industry in the UK and increasingly within an international context. Typically at any one time it has more than 1,000 construction projects in operation and spends more than £1 million per day. Its core business is to operate airports. It has been highly innovative in applying the principles of lean construction to all of its activities. It has involved developing, documenting and implementing a standard construction process (now known as the BAA process), selecting partner firms of designers and constructors for its long-term relationships, implementing lean project processes and adopting information technology to support, integrate and improve the process and the product.

A first objective has been to achieve consistent best practice. This has been achieved through the BAA project process. It is a broad framework that is used to control all BAA projects. It also allows further development to take place. Lean construction cannot be achieved without enterprise integration across the supply chain. Emphasis is therefore placed on its partnering programme. A framework has been established with suppliers, construction managers and designers to meet this objective. BAA also places an emphasis on off-site fabrication, using the construction site as an assembly process. The standardization of components is also a key feature. Driving out waste means specifying correctly, i.e. not under- or overspecifying the work that needs to be carried out.

CONCLUSION

The vision of lean construction stretches across many traditional boundaries. It challenges our current practices. It has required changes in work practices and in understanding new roles and different responsibilities. It has required a change in attitudes and culture. It recognizes that technology has an important role to play today. It expects that those involved in a construction project will share a common purpose. It is setting new standards that can be measured to indicate improvement. It is a vision for the future.

The response to lean construction from the construction industry is not unanimous. Like many new ideas, when put in to practice, some of them do not always live up to the theoretical expectations. But lean construction is not just a theory since its techniques have been successfully applied in other industries. There must also be the desire for success, so a belief system is very important. Lean construction is not a phenomenon that is likely to disappear, since there is a groundswell of opinion and worldwide interest in its principles.

PART 3

PROCESS AND PARTIES

CHAPTER 14

THE CONSTRUCTION PROCESS

The statutory definition of development has been defined in the Town and Country Planning Act (1971) and is described as 'the means of carrying out of building, engineering, mining or other operations in, on, over or under land, or, making of any material change in the use of any buildings or other land'. Development can therefore be broadly classified as undertaking construction works, such as building and engineering, or making a material change of use to the land or property.

Property and construction development must first identify a need. This need may be to satisfy the placing of investment funds or to address the requirements in society for housing, health, education or other purposes. These together also require the relevant infrastructure if the project is to function and operate effectively. Factors to consider also include population and age trends, changes in patterns of lifestyles, the implications for new technology and the importance of fashion. The selection of an appropriate site, the choice of appropriate consultants and finding sources of finance are three of the early decisions that the client or developer will need to make.

DEMAND FOR DEVELOPMENT

Throughout history there has been a demand for buildings for different uses and purposes, such as housing, industry, commercial, religious, entertainment and leisure. This demand was maintained during the twentieth century in the UK, resulting from changes in the social, economic and technological aspects of society. Box 14.1 identifies some of the reasons for development during the second half of the twentieth century.

MARKETING

When the speculative development project is to be disposed of after completion then its promotion and marketing become an integral part of the development process. It is essential to have a clear understanding of what is being provided and

Box 14.1   Reasons for development during the second half of the twentieth century

- General rebuilding after damage from two world wars
- Improvements in housing standards and quality
- Schools for an increased number of pupils
- Hospitals and health buildings for the National Health Service created after the Second World War
- Universities to satisfy the increased demand for higher education by young people
- Redevelopment of commercial and retail town and city centres
- Factories and warehouses required for increased automation and new technologies
- Office buildings for the service sector and to accommodate new technologies
- Roads and motorways for the motor car
- Multi-storey car parking
- Power stations to meet increased demand from all kinds of users
- Airports to meet the needs for increased air travel
- Out-of-town shopping malls
- Change of use of existing premises, refurbishment and conservation
- Decay, obsolescence and redevelopment resulting in demolition
- Tourism and leisure developments

a knowledge and understanding of the potential market for the project. The time lag between inception and completion can be considerable. Many projects have been quickly started on site in a time of relative prosperity and could have been disposed of many times. But sometimes by the time of completion, perhaps five years later, the market has deteriorated or the needs have been switched to other types of property. It is then that those responsible for its disposal may have to work very hard to secure potential purchasers. A clear view, based on a clear analysis of the past and projections into the future, is essential at the outset and needs to be monitored during development. It may be possible to switch a design from say offices to flats, where the demand for offices declines and the demand for flats increases. However, this is not generally easy due to planning constraints in terms of a change in use. Alterations in respect of design and construction implications are also difficult unless the design is flexible, perhaps just a shell for fitting out later.

What to build

The first stage that a developer will need to consider is just what type of project should be envisaged. Where a scheme can be designed that might usefully serve a range of different client types, then this clearly has some advantages, if planning permission can be obtained. The developer will need to consider the other types of property in the location where the development is envisaged, particularly those that are in direct competition with the proposal. The decision involves an assessment of costs and income. It will be necessary through market research to establish if there is an unsatisfied demand in an area for a particular type of project. The relative shortage of types of property can generally be identified by agents.

Where to build

Developers need to construct projects in areas where there is the expected demand. Some developers work on a national basis, others confine themselves to a close geographical area with which they are familiar. These developers understand the regional planning framework and have built up close working relationships with designers and contractors. The where to build scenario depends upon sites looking for projects and projects looking for sites. A developer will have an interest in both of these options.

How to build

In essence the principle today must be adaptability and durability. The principle of long life, loose fit and low energy, sometimes referred to as the three Ls concept. A building's life expectancy is sixty years or more. During that time there are likely to be many changes in ownership or tenancy and even more technological developments. The property may also undergo a material change of use during its lifetime. The managing agents for a development project will be able to offer advice on letting, supervision of repairs, rent reviews, etc., and how these factors are likely to have an impact upon the designing of the development. Today there is an emphasis upon low costs in use, certainly where this approach is adopted as a design principle it will reduce the occupants' ongoing costs. This factor will then have the likely effect of enhancing the rental values of the property. Generally higher value locations attract a kind of occupier that will expect commensurate high standards, with consequently higher prices. For example, new office projects constructed in London are unlikely to find purchasers unless they are equipped with air-conditioning systems, computing and telecommunications networks.

When to build

It is desirable to build when construction costs are at their lowest. This implies a general scarcity of construction work and consequently a better availability of skilled craftspeople. In times of recession there is also a better chance of obtaining a better quality project at a less expensive price. Taking into account that projects may take at least two to three years to bring them to fruition and, assuming that a recession will not last forever, then this seems to be a good time to develop. Finance charges may also be lower during a recession. But all development relies on analysing forecasted trends, and it is essential that projects are available to meet these trends otherwise the opportunity is lost. In some countries the boom and slumps in construction activity are kept to a minimum by the public sector only building when prices are lower in times of a recession.

One of the secrets of effective property development is anticipating the building needs of tomorrow and in the future. The developer must be able to anticipate when projects will be required for occupation, be able to assess long-term trends and then backdate them to the development process. Towards the beginning of the 1990s there seemed to be an insatiable appetite for high quality office accommodation in and around London, with needs mirrored elsewhere throughout the UK. There seemed no end to the possible demand for such types of property, with site values and rents rising accordingly. The square mile of the City, with the Bank of England at the centre, could not accommodate new offices. Many too were old-fashioned, too small and most importantly did not have the space that was required to install the maze of cables for computer terminals. These were now the tools of the financial services industry. This was 1982. However, by the end of the 1980s, the country and world markets were heading into one of the worst recessions that had been seen this century and that was to last for more years than many analysts predicted. Later forecasters predicted that any upturn in the economy would only result in moderate levels of economic activity for the foreseeable future. Consequently, new development proposals were curtailed, but in their wake several well-known development companies ceased to trade, most notably the developers of Canary Wharf in London's Docklands. This emphasizes the importance of taking the longer-term view and accepting that slump follows boom as night follows day.

CUSTOMER SATISFACTION

Table 12.4 represents a survey of customer satisfaction and was published in *Constructing the team*. The data is based upon attitudes of clients who commission construction projects and have an interest in the ownership and use of property. It seeks to compare buildings with motor cars against a number of selected criteria. Some will not recognize the results of such a comparison, but the points made cannot easily be ignored. For example, the Building Research Establishment in 1982 commissioned a survey into housing defects. This found that on average there were 42 faults per dwelling after the dwellings were handed over to the client. This statistic compares poorly against the number of faults found in a new motor car.

Note that where clients are satisfied with their building projects, they are more likely to want to build again or to recommend the process to others. This information should help to inform the industry on the qualities expected and how they are perceived to occur in practice.

SITE IDENTIFICATION

The development of land and buildings is promoted by:

- Property developers who have identified the need for entrepreneurial projects
- Commercial and industrial corporations, developing for their own use
- Public and statutory authorities developing for their own functions
- House builders and housing associations
- Charitable organizations
- Government-sponsored development agencies
- Other private clients

The most important factor that influences development is the developer's perception of market conditions. This knowledge is important in identifying the type of development, the region in the country in which the development is likely to take place and finding a suitable site that satisfies its own criteria. The developer's ability and knowledge are important in identifying areas of potential growth for development, considering that it will take some time before such a project can become operational. Whilst there are always likely to be risks involved, the developer will use a range of market research techniques in order to identify levels of supply and demand, trends in expectations and measures to stay ahead of possible market competition. The are many different considerations of sites for different development purposes (see my book *Precontract studies*, Longman, 1996).

THE DEVELOPMENT PROCESS

There are several different ways of describing and presenting the property development process. In its simplest analysis it consists of design, construction, use and disposal. The process should be looked upon as a cycle of activities, where land once brought into development use undergoes these different stages. In prime city centre locations, where sites are more valuable, disposal and demolition are usually only a precursor to further property development. Returning land to its greenfield status is not a common occurrence in a society that continues to expand its range of activities, with a consequent need to use more of this scarce and limited resource.

The development cycle for a construction project can be best classified into five different stages and their associate phases (Table 14.1). These are not discrete

Table 14.1   RIBA work stages

Stage	Phase
Appraisal	Inception
Strategic briefing	Feasibility
	Viability
Outline proposals	Sketch design
Detailed proposals	Detail design
Final proposals	
Production information	
Tender documentation	Contract documentation
	Procurement
Tender action	
Mobilization	
Construction to practical completion	Project planning
	Installation
	Commissioning
After practical completion	
In-use	Maintenance
	Repair
	Modification
Demolition	Replacement

Table 14.2   The development process

Architect	Quantity surveyor	Engineer	Contractor
	Pre-contract		
	Client's brief		
Identifying the client's needs and wants in terms of space, function and aesthetics	Guide cost information Financial factors such as taxation and funding		Only involved if negotiation, design and build or two-stage tendering
	Investigation		
Feasibility and viability	Approximate estimate	Site survey and investigation	
Planning possibilities	Initial cost plan	Preliminary structural calculations	
Alternative sorts or types of construction Outline design approval			
	Sketch design		
Outline planning permission	Final cost plan		
Major planning problems solved	Life cycle cost plan		
	Design		
Development of sketch plans	Cost implications and cost checking	Constructional methods decided	
Constructional details determined Constructional methods decided		Preliminary schemes for services Design schemes for structure	
	Working drawings		
All construction drawings and details completed	Quotations from specialist firms Final cost checks Bills of quantities	All construction drawings and details completed	
	Tender stage		
All members of the design team check their calculations			Contractors prepare tenders
	Receipt of tenders and tender checking		Materials and subcontractor quotations Determination of market factors Contractor's method statements
	Post-contract construction		
Inspections of quality Subcontractor nomination Issue of instructions Issue of certificates	Valuations, forecasts and payments	Inspection of quality	Construction programme agreed Planning and progress control Subcontractor control
	Maintenance		
Contractually up to the end of the defects liability period in-use repair and maintenance, changes in use, conservation, demolition, alteration and reconstruction, asset valuation, etc.			

activities. The respective functions of the architect, quantity surveyor, engineer and contractor are shown in Table 14.2. Tables 14.1 and 14.2 are based on the RIBA work stages. The titles of most of the work stages have been revised, largely to take into account output expectations that are required. The different functions merge into each other as the project moves through its life cycle. Emphasis should always be on securing those developments which best satisfy the criteria that will have been identified by the developer or client at the beginning of the development. A brief will have been written that identifies the type and scale of the project, its standard of construction, funding availability, its costs and the date when the project should be available for handing over for occupation. During each of these activities, those involved with the development will have different tasks to perform, in respect of designing, costing, forecasting, planning, organizing, motivating, controlling and coordinating. These are some of the roles of the professions involved in managing property and construction, whether it be new build, refurbishment or maintenance. To these activities should be added research, innovation and improving quality and standards.

Appraisal

This first stage in the process involves identifying the employer's requirements and the possible constraints on development. The development process is initiated principally by either a project looking for a suitable site, or an available construction site looking for a project. The available site may sometimes be an existing building awaiting demolition or redevelopment. The development project may cover the whole spectrum of different building types constructed for the public or private sector, including housing, commercial, industrial, recreational, social or activities as remote as forestry and agriculture. The initiative may come from a developer, the site owner or a client seeking a site for a proposed development. The planning authority too may make recommendations or designate revised land use patterns in an area. Many projects arise from long-term programmes where clients consider the scheme as a part of the overall objectives of their own organization. Studies will be undertaken to enable a client to decide on whether to proceed and if so which procurement route should be selected.

Strategic briefing

The strategic briefing stage, which is done by or on behalf of the client, identifies the key requirements and constraints involved. It identifies the procedures, organizational structure and the type and range of consultants to be used. It is important during this early part of the process to consider a range of issues that are going to determine whether the project has any chance of coming to fruition. There is little point in expending large sums of money or time on a project that will never be constructed. The client or the developer will prepare an outline brief of the proposals and the issues that need to be settled. These will include an analysis of the market potential and the costs of the development in very broad terms. For example, it may be necessary to consider any public funding that might be available in respect of grants or loans for a project in the private sector. During this stage it will also be necessary to determine whether the proposed site, if one has been located, is suitable for the project envisaged in terms of its location, size and ground conditions. It will be important to establish as soon as possible whether the project will receive planning consent from the local authority. It would also be prudent to establish matters of ownership, rights of way and other factors that might affect the whole process of development. When these are established, it will be necessary to purchase the site, generally with the condition that outline planning permission will be granted.

The developer will need to arrange for land transfer and the finance for the development. Two sorts of finance are generally sought. The first, known as short-term finance, will cover the costs of the development until the project is disposed of to a client. The second, long-term finance, sometimes referred to as funding for development, is to cover the costs of owning the development as an investment property.

FEASIBILITY AND VIABILITY

During the feasibility and viability phases, the client's or the developer's objectives for the project become established. The clients who are involved in single projects often come to their chosen consultants with a broad outline of their aspirations, a sum of money which is often insufficient and a timescale for occupation which is often impossible. The better-informed clients, who are those involved in frequent capital development, usually have more realistic expectations of what can and cannot be achieved. The type of project will often determine whom the client or promoter appoints as designer. On building projects this has traditionally been the architect. However, whilst traditions die hard, the building surveyor is increasingly being appointed to oversee smaller works and schemes of refurbishment, on behalf of the client. As the different combinations of procurement such as design and build or management contracting are employed, clients are now often appointing the construction firm direct, choosing an alternative consultant as a main partner in the venture or appointing a project manager in overall charge of the scheme.

The feasibility phase seeks to determine whether the project is capable of execution in terms of its physical complexities, planning requirements and economics. The available site, for example, may be too prohibitive in terms of its size or shape or the ground conditions may make the proposed structure too costly. Planning authorities may refuse permission for the specific type of project or impose restrictions that limit its overall viability, perhaps in terms of the return on capital invested. Schemes may be feasible but they might not be viable.

Outline proposals

The alternative options and advantages of choosing a separate designer from the contractor, or perhaps preferring design and build, are well documented (Chapter 11). Each has its own advantages and disadvantages. In any event it is first necessary to prepare schematic outline proposals for approval, prior to a detailed design. These proposals will need to be accepted by the client or developer in terms of the requirements which have been outlined in the brief. They will also need to be accepted by the relevant planning authorities for their permission, and in terms of funding through the preparation of an initial cost budget. An early estimate of the proposed costs and a developer's budget will be required.

Detailed proposals

As the scheme evolves and receives its various approvals, a number of different specialist consultants will be employed. Some may be public relations consultants, particularly where a sensitive scheme such as a new road, building in the green belt or where a project which is out of character with the locality is being proposed. During the sketch design, the main decisions regarding the projects layout and form and the quality of materials and standards of construction will be agreed. A cost plan of the proposed project will be prepared by the quantity surveyor in order to guide the designers during the later stages of this process. An architect or planning consultant will be responsible for providing the well-thought-out scheme in order to secure planning permission. It has been estimated that over 60 per cent of all planning applications are now dealt with by professionals other than architects.

Final proposals

When the scheme has been agreed and approved by the client then further investigations will be undertaken in order to prepare the detailed design. Different solutions to spatial and other design problems will be considered, and some of these will require revisions to other aspects of the project which have already been agreed. Each alternative solution will need to be costed to ensure that the cost plan remains on target and where it significantly affects the client's proposals or developer's budget then it will require agreement before proceeding further with the design. It will be necessary during this stage to consult firms who supply or install any specialist equipment that may be required.

In the UK any change in use requires the approval of the appropriate planning authority. This is normally the local authority. Where planning permission is rejected there is recourse to the Secretary of State for the Environment, who may initiate a planning inquiry. The acquisition of planning permission can be a highly complex and technical activity, needing a detailed knowledge of legislation and government policies as well as local knowledge relating to the site of the proposed development. Obtaining planning permission may also involve the developer in additional planning agreements with the local authority, where additional conditions, described as planning gains, are sometimes required before planning permission is granted. These agreements inevitably increase the development costs for the proposed project. In some circumstances it will be necessary to obtain further approvals, such as listed building consent, that is the right to demolish or alter a protected building. There are about 500,000 structures in Britain protected because of historical or architectural importance. Listed buildings are grouped into three categories, graded 1, 2* and 2. Grade 2 offers minimal protection, but it is virtually impossible to alter or make any changes at all to grade 1 property. Information on this can be obtained from the Department for Culture, Media and Sport.

Clients and developers now need increasing certainty about the costs of the proposal in order to input realistic and reliable information into their budgets. When it is known that planning permission will be forthcoming, the plans should then quickly achieve a level of detail in order for the quantity surveyor to provide a detailed estimate of the likely costs of construction. The cost plan will already have been prepared and this will be frequently updated to take into account modifications arising from planning and changes in design.

Production information

The production information is considered in two separate parts. The first part is concerned with providing adequate information that is sufficient to obtain tenders. The second part includes the balance of information that will be required under the building contract to complete the information for construction purposes.

Tender documentation

The documentation which is required for tendering purposes will be prepared at the end of this process. This will depend to some extent upon the procurement method that has been selected. When the project is approaching the tender stage, the different firms which may be interested in constructing the project should be invited to tender. The long periods of time that elapse reflect the design and planning complexity which is required for solutions to bespoke designs for construction projects.

Tender action

Upon receipt of the documentation the contractors enter their estimating phase, since the awarding of the works of construction is most frequently done through some form of price competition. The contractors' bids will be evaluated against price and other considerations (Chapter 7) and a recommendation made to the client.

Mobilization

This is the letting of the building contract to the successful firm and appointing of the contractor. The production information will be issued to the contractor and arrangements are made in respect of handing over the site to the contractor.

Construction to practical completion

This is the stage when the contractor commences the work on site. It has often been referred to as the post-contract period, since it commences once the contract for the construction of the project has been signed and work has started on site. Where the project is on a design and build arrangement or a system of fast-track procurement, this stage may start before the design is finalized, and then run concurrently. Contractors are critical of the traditional arrangements since they are frequently required to price the works, which although assumed to be fully designed are in reality not so. Throughout this stage, formal instruction orders are given to the contractor for changes in the design and valuations are prepared and agreed for interim payment certificates. Contractual disputes all too frequently arise, all too often due to misunderstandings or incorrect information being made available to the contractor. The contractor is also sometimes overambitious and enters into legal agreements that become impossible to fulfil. These create grounds for damages on the part of the client. Project completion times can last from a few months up to ten years or more. Upon completion the formal signing over of the project to the responsibility of the client is made.

After practical completion

One of their main tasks is now to ensure that the project can be completed to the specified quality, the calculated costs and within the appropriate timescale. Commitments may have been made to future purchasers or occupiers, who will themselves have prepared their own plans for taking over the property. Anticipated problems need appropriate action to ensure that the project stays on target in respect of time and budget. Changing circumstances may mean that some variations to the scheme need to be instructed, in order to maximize the potential for the finished product. Some factors remain outside the control of any of the parties involved, but the essence is how effectively and quickly these are resolved. The satisfaction of clients and developers centres around completion on time, at the agreed price and to a quality and standard that has been specified in their original brief. Satisfied clients are likely to recommend the company to others, and thus offer a great marketing potential.

IN USE

This is the longest phase of the project's life cycle but one that the developers will keep at the forefront of their minds. The immediate aims of development are now hopefully satisfied and the project can be used for the purpose of its design and construction. However, no development is complete and the success of others remains until they have found occupiers or purchasers who are willing and able to pay the rents or purchase the property. Forecasting the future demand for development projects is difficult owing to the long time lag between inception and completion. The collapse in the need for property, due to sudden changes in the economy, can create financial disaster for a developer. The shrewd developers will attempt to make allowances for everything, even the unknown.

Routine maintenance will be necessary during this stage. The correct design, selection of materials, proper methods of construction and the correct use of components will help to reduce maintenance problems and their associated costs. A sound understanding, based upon feedback from project appraisals in practice, will help to reduce the possible future defects. Defects are often costly and inconvenient, and minor problems sometimes require a large amount of remedial work to rectify, sometimes out of all proportion to the actual problem that has arisen. Many projects have only a limited life expectancy before some form of refurbishment or modernization becomes necessary. The introduction of new technologies also makes previously worthwhile components obsolete. City centre retail outlets have a relatively short life expectancy before some form of extensive refitting becomes required. Fifteen years seems to be an optimum age. Whilst the shell of buildings may have a relatively long life of up to sixty years, and some are able to last for centuries, their respective components wear out and need frequent replacement. Obsolescence may also be a factor to consider in respect of component replacement.

DEMOLITION

The final stage in a project's life is its eventual disposal, demolition, and a possible new beginning of the life cycle on the same site. Demolition becomes necessary through decay and obsolescence and when no further use can made of the project (see my book *Cost studies of buildings*, Longman, 1999). Some buildings are destroyed by fire, vandalism and explosion, or may become dangerous structures that require demolition as the only sensible course of events, years before the end of their expected lives. Other projects may need to be demolished because they are located in the middle of a redevelopment area. There are relatively few projects that last forever and become historic monuments. Whilst some of this is attributable to decay, the style of living and the changing needs of space are constantly evolving to meet new challenges. Some projects of notoriety become listed buildings. The Secretary of State for the Environment has powers under the planning acts to compile lists of buildings which are of special historic interest. It then becomes difficult to demolish, alter or extend these buildings in any way that would affect their character. Where non-listed buildings are thought to have special historic or architectural interest, a planning authority may also serve a building preservation notice upon the owner. Whilst the planning regulations are onerous, their aim is to allow development to take place in an orderly fashion. This in the long-term must be the best policy for society. Projects that might have taken several years to plan and develop and have then been cherished for decades are finally removed from the urban landscape by demolition. In some cases this is swift, where a lifetime's project can be reduced to a heap of rubble in a matter of minutes.

ENVIRONMENTAL IMPACT ASSESSMENT

Environmental impact assessment was established in the USA as long ago as 1970. This is now a worldwide concept and a powerful environmental safeguard in the project planning process. The original EC (now EU) directive 85/337 was adopted in 1985 and since then the individual member states have implemented the directive through their own regulations. In the UK, the resulting environmental impact statements have increased more than tenfold between the early 1980s and the early 1990s. As a result of the directive, over 500 statements are now prepared annually in the UK. The required contents of a statement are given in Annex III of the directive and are as follows:

- Description of the project: physical characteristics, production processes carried out, estimates of residues and emissions.
- Appropriate details of alternative sites and their possible effects.
- Description of aspects of the environment that are likely to be affected, such as population, fauna, flora, soil, water, air, climatic factors, material assets, architectural and archaeological heritage, landscape and their interrelationship.
- Description of the likely effects on the environment of:
  - existence of the project
  - use of natural resources
  - emission of pollutants
- Description of measures envisaged to prevent or reduce any adverse effects on the environment.
- A non-technical summary of the above.
- An indication of any difficulties encountered by the developer in compiling the above information.

In the UK, the directive is implemented in regulations which are described as the Town and Country Planning (Assessment of Environmental Effects) Regulations 1988. Further guidance is included in *Environmental assessment: a guide to the procedures* (Department of the Environment, 1989).

CHAPTER 15

PARTIES INVOLVED IN THE CONSTRUCTION OF BUILDINGS

The main purpose of construction activity is to provide a completed project for the building owner. This project may include the substance of a building contract, a project constructed speculatively for a developer, or civil engineering infrastructure works such as roads, bridges or pipelines. The employers, clients or promoters of the construction industry are many and varied. They include the public sector bodies such as central and local government and private companies involved in building for domestic, commercial, industrial and retailing purposes. They also now include a range of the quasi-public companies that were formerly part of the nationalized industries. These include the industries of coal, electricity, gas, water supply, sewage treatment, etc. Many of these latter companies are virtual monopolies with limited competition from other organizations.

Construction activity is typically divided on a 60–40 per cent basis in favour of new projects compared with repair and maintenance activities. This figure remained almost at a constant level throughout the late 1990s. However, the division between public and private sector workloads has changed markedly, largely due to the privatization programme of different governments during this time. Typically the public sector used to account for about 50 per cent of the total output. By the start of the 1990s this had fallen to less than 25 per cent. Civil engineering works are worth about 20 per cent of the output of the industry. Work done by British companies overseas currently accounts for 10 per cent of the total output of the construction industry. Some of the largest clients of the construction industry, e.g. British Airports Authority and Tesco stores, each spend about £1 million per day on construction work.

Each separate organization is given a considerable amount of autonomy by the government of the day, and it is interesting to note the wide diversification in the methods used for the procurement and execution of major and minor capital works projects. Little uniformity exists either in the design procedures employed or the contract conditions that are used.

Clients in the public sector may be influenced by both social and political trends and needs, and the desire to build may be limited by these factors. They will nevertheless be restricted in their aspirations by the amount of capital they are allowed to borrow for these purposes. The private sector, which encompasses private housing ownership and the large multinational corporation, directs capital spending to the ventures that are considered to be money-making. But in both sectors there has recently been a particular emphasis upon securing added value. This has tended to be viewed on a building's life cycle rather than initial construction costs alone.

EMPLOYERS

The employer is one of the parties to the contract, the other being the contractor. Each client will have different priorities but essentially there will be a combination of:

- *Performance* in terms of quality, function and durability
- *Time* available for completion by the date agreed in the contract documents
- *Cost* as determined in the budget estimate and the contract sum

If employers are to be satisfied with the product, i.e. their construction project, then these three conditions must be critically examined. Box 15.1 includes some of the more important references to the employer in JCT 98.

Developers

Property development involves a range of different activities, and by their nature they will involve developers with different objectives. All of them are concerned with future projections and expectations. Property developers work on margins between the cost and the sale price, sometimes based on cost and sometimes on sale price. In the absence of this margin there would be no financial incentive for development. There are many different arrangements that can be provided for development. However, essentially developers can be broadly classified in two categories. Each may undertake work speculatively based upon market intelligence.

- *The investor developer* aims to retain ownership of the project. Short-term bridging finance may be required for construction and then a long-term loan, often from one of the institutional sources.

Box 15.1   Important references to the employer in JCT 98

- Appointment of the architect
- Powers regarding insurances
- Duty to give possession of the site to the contractor
- Powers in respect of damages for non-completion
- Powers to determine the employment of the contractor
- Powers to engage directly employed contractors
- Duties regarding certificates
- Procedure in respect of adjudication, arbitration and litigation

■ *The merchant developer* completes projects using short-term funding and then sells them to an owner or occupier. The tax advantages of merchant development are examined later in this chapter.

The aims and objectives of property development are wide and diverse. The different types of developer have their own particular needs and desires regarding the project.

Some property developers may choose to specialize in terms of location, whilst other companies may offer only certain types of property, such as offices, industrial premises or housing. In other cases the developer may provide a package deal arrangement to supply and construct factory-made (industrialized) units, that have the advantage of being available very quickly. Some will choose to concentrate on a particular process, such as conservation or refurbishment, in order to develop a niche market for their services. The advantages of specialization enable the company to gain an above average knowledge and expertise.

Occupiers

Occupiers require buildings that suit their particular needs of occupation. The prime objectives are to provide a building that best serves these personal needs, with benefits achieved from occupation and with less concern for its market valuation. An industrialist, for example, will require premises that allow the production process to be carried out in the most effective, efficient and economic manner. The profits from such a business far outweigh any changes in the market value of the property.

A developer who also intends to become the occupier is able to specify a building that meets personal requirements as closely as possible. In some cases the development may be so inflexible in its design that it cannot be easily utilized by others and its relative market value may therefore be small. Buildings such as schools and hospitals are essentially designed and constructed to suit the functions undertaken, with only limited concern for site values and possible resale opportunities. Where such projects are adapted for other uses, extensive conversion work is often required. It may also be necessary to build such projects on sites that might otherwise have only limited development potential. Occupiers are generally more concerned about spatial arrangements and function rather than the possible long-term investment.

Investors

The investor's view of property is similar to that of the property company – financial gain. However, investors tend to take a longer-term view, expecting both the capital and income to increase over the invested life, which may be several years. The acquisition and disposal of property investments can be costly, and it is therefore necessary to have allowed an investment at least some maturity before converting it into other assets. When investors become involved in the development itself, they expect higher returns. The risks involved in project development are greater than those of a building for a client.

Investors are generally cautious, disliking unconventional investments that because of their nature may be unpredictable and difficult to dispose of at some future date. Property that has an unusual design, has been constructed with new methods or materials, has unconventional lease terms or involves substantial management capability will not be favoured as an investment potential.

Investment companies generally manage a portfolio of investments to spread risk across a number of geographical areas as well as different industry markets and commodities.

Builders and contractors

A building company may seek to enlarge its range of activities by carrying out development work of a speculative nature. This is often done for taxation and legal reasons through a separate company charged solely with this task. In this case the company will become involved with the additional risks associated with land purchase, finance acquisition and sales or lettings. In this respect the building firm is largely acting in the same way as a development company, with the added bonus of being able to profit from the building construction operations. When the firm acts only as building company, the profits accrued largely result from the activities relating to the construction work being performed. Additional benefits of combining both development and construction are that the resources the contractor employs in terms of the workforce and expertise might be able to be retained through, for example, undertaking more development work when contracting work is not available or at too competitive prices. The building developer is therefore the reverse of the property company who employs a building contractor for a proposed development.

Public sector

The public sector has had a growing involvement with development projects usually as a client. The projects include housing, hospitals, education, roads, public utilities, etc. Restrictions on public sector borrowing and spending and the privatizing of many of the nationalized industries also accounted for some of this decline.

The public sector policies are influenced by political ideals and the government who is in power and control, both nationally and locally. There is also the need for accountability in terms of raising and spending finance and the social needs relating to the community. Many of its development projects would not be undertaken by the private sector, since they are unable to show financial profitability. However, their provision can often demonstrate benefits that accrue from the expended costs, and benefits that are of tangible worth to the country and the community. It is important that a public authority can demonstrate it has acted lawfully and that all of its dealings are free from even a hint of suspicion of corruption. Public accountability often includes some element of public participation.

Most public works building projects, undertaken either through central or local government or other government-controlled agencies, are directly related to the particular interests of the authority or government department. Traditionally much of this work would be commissioned by, for example, a county or borough council. The design might have been undertaken by its own staff or private consultants, and for projects other than the smallest, a contractor would be appointed. However, in more recent years some local authorities have undertaken development projects, such as the provision of industrial units or business parks in order to attract commerce and industry to their locality. These organizations are, however, different to the private developers in several ways. They can only undertake work within their powers, otherwise they will be acting *ultra vires*.

LANDOWNERS

Landowners include the traditional groups such as the Crown Estates, the Church Commissioners and the landed aristocracy. Although certainly motivated by economics, such as return on capital invested, these landowners are also concerned with political and social issues to do with the land. Another category of landowners are the industrial corporations, who use land because it is incidental to their production processes. This group also includes farming and agriculture, retailers, etc. The former nationalized industries also fit comfortably into this group. Their principal motives for landownership are with the use of land, rather than for any financial investment that the land might otherwise provide. A further category of landowners are the financial institutions, where land is seen solely as a means of investment and this remains their prime, if only, reason for owning land. These are the most informed group regarding land and property values.

The landowners can have a major influence over the type of development that might be undertaken. Only the state has a bigger influence through its planning procedures. It can encourage and discourage development and has the powers to prohibit development that does not fit in with the plans produced by the local planning authorities.

Crown Estates

The Crown Estate is one of the most important landed estates in the UK. It includes substantial urban, rural and marine interests. It is part of the hereditary possessions of the sovereign *in right of the Crown* that is managed by the Crown Estates Commissioners. The net surplus is paid to the Exchequer.

The origins of the estate go back to the reign of Edward the Confessor. It has over 300,000 acres of agricultural land in Great Britain, making it the largest agricultural landlord in the UK. Until the time of George III, who came to the throne in 1760, the reigning sovereign received its rent and profits.

In the mid 1990s, the estate achieved a revenue surplus of almost £80 million. Its property values are now worth in excess of £3,000 million. It has a wide-ranging and quality property portfolio. In its ownership are more than 1,000 listed buildings, with 750 of these located in London. Almost 50 per cent are grade 1 listed, compared with the national average of 2 per cent. It has property located in Central London and the West End and is one of the capital's largest landowners, with more than 8 million square feet of office space, $2\frac{1}{2}$ million square feet of retail space and over 1 million square feet of miscellaneous property, including hotels, clubs and residential accommodation.

The estate does not have borrowing powers, which creates both a constraint and a discipline on its activities. It resisted the temptation to invest in Docklands and in 1990, the start of the decline in property prices, it introduced a moratorium on development.

PROFESSIONAL ADVISERS

Because the development process is so complex, and because most employers do not have the range of skills and expertise that are required, it is necessary to employ a range of different professional advisers to advise on funding, design, costs, construction, letting, etc. These advisers will vary depending upon the type, nature and size of the project being envisaged, and might include some or all of the following.

Architects

The architect has traditionally been the leader of the design team. In the building process, where design and construction are separate entities, it is the architect who receives the commission from the client. Because projects today require a large amount of specialized knowledge to complete the design, the architect may require the assistance of consultants from other professional disciplines.

The architect's function is to provide the client with an acceptable and satisfactory building upon completion. This will involve the proper arrangement of space within the building, shape, form, type of construction and materials used, environmental controls and aesthetic considerations – all within the concept of total life cycle design.

The architect's duties and powers are described under JCT 98. A contractor who believes the architect is attempting to exercise powers beyond those assigned under the contract can insist that the architect specifies in writing the conditions that allow such powers (JCT 98 clause 4.2). The architect will generally operate under the rules of agency on the part of the employer. This means that instructions given to the contractor will be accepted and paid for by the employer.

In some forms of contract the architect is termed the supervising officer. This is the name used in the GC/Works/1 form of contract, and is one of the alternative titles suggested in the JCT group of forms. It is used in the GC/Works/1 form since the designer may be an engineer rather than an architect, and the terminology has been extended to the JCT form for a similar assumption.

The scope of the work undertaken by the architect may be broadly divided into pre-contract and post-contract duties. Although it is more common for an architect to provide a fully comprehensive design and supervision service, a design-only service may be required by the employer. It is less common to expect the architect to supervise construction work only, although it could arise in situations where prefabricated buildings are used. However, even in these circumstances the architect is more likely to be asked for advice on a particular system building during the design stage.

In the normal pre-contract stage the architect's basic duty is to prepare a design for the works. This may involve three facets: architectural design, constructional detailing and administration of the scheme. This latter aspect will entail integrating the work of the various job architects and other consultants, and ensuring that the information is available for a start on site when required. The architect during the work must exhibit reasonable skill and care in the design of the works. This duty may be established in accordance with normal trade practice. The architect will also generally be held responsible for any work delegated to another. In the JCT form of contract, for example, although the quantity surveyor is responsible for most of the financial arrangements, the architect is ultimately responsible regarding the certification of monies to be paid. If part of the design is undertaken by a nominated subcontractor, some protection may be afforded by a warranty from that firm.

During the post-contract stage the work undertaken by the architect is largely supervision and administration. Some drawings and details may still need to be prepared, particularly where such information is reasonably requested by the contractor. The purpose of supervision is to ensure that the works are carried out in accordance with the contract. The amount of supervision necessary will vary from project to project. A complex refurbishment project will require more frequent visits than the construction of a large warehouse shed. On very large contracts the architect may even be resident on site. The duties of administration are used to describe the various functions, such as issuing instructions to the contractor, that must be carried out during the progress of the works. The post-contract stage involves those duties described in JCT 98. Boxes 15.2, 15.3 and 15.4 refer to responsibilities in JCT 98.

Surveyors

There are several different types of surveyor which include general practice surveyors, building surveyors and quantity surveyors. Many laypersons confine the term 'surveyor' to a land surveyor. Land surveyors are involved in mapping and surveying the land in terms of its location, line and level.

General practice surveyors are employed in four main areas of work: agency, valuations, management and investment. Their knowledge and understanding of the local property market, land and property values are the particular attributes of this profession. Valuation is one of the main skill bases being vital to investment

Box 15.2   Architect's instructions under JCT 98

- Discrepancies and divergencies between documents
- Justification of instructions
- Instructions to be in writing
- Confirmation of verbal instructions
- Divergence between statutory regulations and project documents
- Opening up of work for inspection
- Removal from site of work, materials or goods which are not in accordance with the contract
- Exclusion from the works of any person
- Instructions given to person in charge
- Variations requirements
- Expenditure of prime cost and provisional sums
- Sanction in writing of variations created by the contractor
- Defects in the contractor's work
- Postponement of any work
- Execution of protective work after an outbreak of hostilities
- War damage
- Antiquities
- Nominated subcontractors
- Nominated suppliers

Box 15.3   Architect's responsibilities in respect of certificates under JCT 98

- Practical completion of works
- Completion of making good defects
- Estimate of the approximate total value of partial possession
- Completion of making good defects after partial possession
- Failure to complete the works by the completion date
- Determination
- Interim certificates
- Final certificate

Box 15.4   Architect's other responsibilities

- Provision of documents, schedules and drawings
- Stating levels and setting out the works
- Access to site and workshops
- Limitation of assignment and subletting
- Granting an extension of time
- Reimbursement of loss and expense to the contractor
- Arbitration

work. These may be required for a variety of purposes such as sale, lease, insurance, investment or loans, and for a range of different clients such as developers, purchasers and property owners. General practice surveyors may be involved at the outset of a new development project and are sometimes the client's first point of contact on a proposed development. They also advise the financial institutions on investment in order to yield the best result for their shareholders or members.

Traditionally the building surveyor's role was in assisting other colleagues and clients with the maintenance and repair of buildings and preparing survey reports for the prospective purchasers and users of real estate. Building surveying is today a rapidly expanding profession. Some of this is due to the growing popularity of building conversion and renovation and the poor and deteriorating nature of our buildings stock. It is also due to some extent to the nature of our society with its 'make do and mend' approach and the desire for the conservation of older properties. As a profession it is somewhat unusual in being largely restricted to the UK and some of its ex-colonies at the present time. Building surveyors are, however, rarely concerned with projects of a large size, and are not specifically referred to in the forms of contract.

Quantity surveyors

The quantity surveyor has developed from the function of a measurer to a building accountant and a cost adviser. The emphasis of the quantity surveyor's work has moved from one solely associated with accounting functions, to one involved in all matters of forecasting finance and costing.

The function of the quantity surveyor in connection with construction projects is therefore threefold: (1) as a cost adviser, attempting to forecast and evaluate the design in economic terms on an initial cost basis and a life cycle cost basis; (2) preparing of much of the tendering documentation used by contractors; (3) in an accounting role during the construction period where the quantity surveyor will report on interim payments and financial progress and the preparation and control of the final expenditure for the project.

Quantity surveyors are employed on behalf of both the building owner and the building contractor, the contractor tending to specialize in post-contract functions or commercial management. Boxes 15.5 and 15.6 list the duties of the quantity surveyor in connection with a building contract.

Box 15.5   Pre-contract role of the quantity surveyor

- Initial cost advice
- Approximate estimating
- Cost planning, value engineering, life cycle costing
- Bills of quantities and tender documentation
- Specification writing (where the bills are not required)
- Procurement
- Tender evaluation

Box 15.6   Post-contract role of the quantity surveyor

- Valuations for interim certificates
- Final accounts
- Remeasurement of the whole or part of the works
- Measuring and valuing variations
- Daywork accounts
- Adjustment to prime cost sums
- Increased cost assessment
- Evaluation of contractual claims
- Cost analysis

Box 15.7   Quantity surveyor's responsibilities under JCT 98

- Confidential nature of contractor's prices
- Valuing variations
- Calculation of loss and expense
- Preparation of interim valuations
- Preparation of the final account
- Accounts of nominated subcontractors and suppliers
- Fluctuations

In addition the contractor's surveyor will be involved in the agreement of subcontractors' work, other duties of a commercial nature and possible bonus payments and ancillary functions. The quantity surveyor is seen as an essential member of the construction team. The quantity surveyor's work frequently extends beyond that described above. Loss adjusting, arbitration and auditing are other areas where quantity surveyors are employed. The roles of the quantity surveyor within the JCT form of contract are listed in Box 15.7. Whilst the ICE form allocates these duties to the engineer, in practice it is frequently the quantity surveyor who carries them out.

Engineers

A wide range of engineers are employed in the construction industry. These may range from civil and structural engineers to building services engineers. Civil engineers are responsible for the design and supervision of civil and public works engineering, and are employed in a similar way to architects employed on a building contract. In addition the engineer's counterpart working for the contractor is often a civil engineer. Their work can be very diverse, and may include projects associated with transportation, energy requirements, sewage schemes or land reclamation projects. Structural engineers are usually employed by the architect on behalf of the client. They act as consultants to design the frame and the other structural members in buildings. The building services engineers are responsible for designing the environmental conditions that are required in today's modern buildings. There has in recent years been an upsurge in their membership as greater attention is paid towards this aspect of building design.

Engineers are not mentioned by name in JCT 98, although the supervising officer that is mentioned in some of the forms may equally be an engineer, a surveyor or an architect, depending upon who is largely responsible for the design and supervision of the works.

Clerks of works

The clerk of works is employed under the direction of the architect as an inspector of the works under construction. The clerk of works may give instructions to the contractor, but these are of no effect unless they are subsequently authorized by the architect.

The contractor must give the clerk of works every reasonable facility to carry out all duties. The clerk of works is the counterpart of the person in charge that is employed on behalf of the contractor. Duties include ensuring that the contract in terms of the specification and further instructions from the architect are fully complied with. The clerk of works will attempt to make sure the materials used and the standards of work are in accordance with the contract requirements. This will involve: inspecting the materials prior to their incorporation within the works; obtaining samples where necessary for the approval of the architect; testing materials such as concrete, bricks and timber to the specified codes of practice; and generally ensuring that the construction work complies with accepted good practice.

Other professions

A wide range of other professions are associated with the construction industry. One's viewpoint will determine whether these represent an unnecessary fragmentation of the industry or a desirable specialization. The clear demarcation of activities has now become blurred, particularly as we look forward and consider the characteristics of the EU and the USA. Certainly the amount of knowledge which is now available and the skills which are required are too great for a single person to control, and some specialization even within a conglomerate of the professions is essential. Here are some of the other professions involved:

- *Building control officer*: normally employed on behalf of the local authority to ensure that the building plans and proposals comply with the building regulations and by-laws made under the public health and building laws.
- *Estimator*: responsible for calculating in advance of building the cost of the project to a particular contractor based upon the total costs of all the labour, materials and plant that will be needed.
- *Interior designer*: developing the internal shell of buildings to provide good aesthetic and working conditions to create an acceptable ambience for the owner and user.

- *Landscape architect*: helps to create the all-important context and space in which the building is set. Increasingly in modern buildings this often includes internal spaces.
- *Planner*: involved with the legislative aspects of the building's location in interpreting the structure and district plans of local authorities. Ensures that the building fits into the environment.

CONTRACTORS

Main contractors

The majority of the construction work in the UK is undertaken by a main contractor. The term 'general contractor' is now outdated since relatively few of these firms undertake the work themselves. These firms, which will be public limited companies (PLCs), will vary in size, having from just a few to many hundreds of employees. Many of the larger companies are household names and have developed only since the beginning of the twentieth century. Although there is no clear dividing line between building and civil engineering works, many firms tend to specialize in only one of these sectors. Even in the larger companies, separate divisions or companies exist, often trading and structured in entirely different ways depending upon the sector in which they are employed. Even the operatives' unions and the rules under which they are engaged are different.

The smallest building firms may specialize in one trade, and as such may act as either domestic subcontractors or jobbing builders carrying out mainly repairs and small alterations. The medium-sized firms may be a combination of trades operating as general contractors within one town or region. These firms may specialize in certain types of building projects or be speculative house builders. The largest firms may be almost autonomous units, although it is uncommon even in these companies to find them undertaking a complete range of work. On the very large projects it is generally usual to find specialist firms for piling, steelwork and high-class joinery.

It has been suggested that one-quarter to one-third of the work of the construction industry is minor in nature, being largely of repairs and maintenance. This work is often carried out by the smallest companies. A recent survey also indicated that one-third of the building firms in Britain do not employ any operatives, but the work is carried out by the partners of the firm. The larger companies may be represented by less than one hundred firms throughout the country. In more recent years there has been a trend away from the multimillion pound project, resulting in a slight reduction in the number of these firms. Overseas projects of this size have helped to keep such firms viable within the UK. The reduction in the size of projects has also meant the breaking down of some of the larger firms into smaller-sized units working on a more localized basis.

The contractors under JCT 98 and the other forms of contract agree to carry out the works in accordance with the contract documents and the instructions from the architect. They agree to do this usually within a stipulated period of time and for an agreed amount of money. The main contractor must also comply with all statutory laws and regulations during the execution of the work, and ensure that all who are employed on the site abide by these conditions. The contractor will still be responsible contractually for any defects that may occur for the period of time stipulated in the conditions of contract, which is normally six months. However, the responsibility of the contractor for the project does not end here. In common law the rights of the employer will last for six years and twelve years respectively, depending upon whether the contract was under hand or seal.

The contractor is mentioned extensively in the conditions of contract, largely because, along with the employer, they are one of the parties to the contract. Some of the more important provisions are listed in Box 15.8.

Box 15.8   Contractor's responsibilities under JCT 98

Quality

- Contractor's obligations
- Compliance with architect's instructions
- Duties in setting out
- Compliance with the standards described
- Responsibility for faulty work standards
- Duty to keep on site a person in charge
- Requirement to give the architect access to the site and workshops
- Limitations on assignments and subletting
- Right to object to nominated subcontractors
- Duty to employer's directly employed contractors

Time

- Procedure for partial possession
- Necessity to proceed diligently with the works
- Liabilities in the event of non-completion
- Duty to inform the architect of any delays
- Rights in cases of determination of the contract

Cost

- Duty to ask for any loss or expense
- Responsibility for payment to nominated subcontractors
- Procedure for certificates and payments

Others

- Liability for injuries to persons and property
- Duties regarding insurances

Person in charge

The person in charge is responsible for the effective control of the contractor's work and workpeople on site, also being responsible for organization and supervision on the contractor's behalf, and for receiving instructions from the architect. Depending upon the size and nature of the works and the type of firm, this may be a general overseer, site agent or project manager. The person in charge may have received initial training as a trade craftsperson or be a chartered builder or engineer. The responsibilities will vary with the size of the project and company policy. On the larger projects considerable assistance will be received from other site staff. The person in charge, whatever the title, is the site manager on behalf of the main contractor and is very often a member of the Chartered Institute of Building.

Suppliers

Building materials delivered to a site may be described under one of three headings: materials, components and goods.

Materials

Materials are the raw materials to be used for building purposes and include cement, bricks, timber and plaster. The items included within this description will, in total, probably represent the largest expense on the traditional building site. As more and more of the construction processes are carried out off site, so the value of this section will diminish.

Components

Components represent those items delivered to site in almost 'kit' form. They may include joinery items such as door sets or joinery fittings to be assembled on site. The industrialized building process is based to a large extent on the assumption that many items can come to site in component form to be assembled very quickly for a very small amount of money.

Goods

Goods include those items that are generally of a standard nature which can be purchased directly from a catalogue, for example, sanitary ware, ironmongery, electrical fittings.

The contractor's source of supply for these items may vary, but must in all circumstances comply with those specified in the contract documents, regarding quality and performance. The contractor will probably make extensive use of builders' merchants, because they stock a wide variety of items that can be purchased at short notice. Some of the items will need to be obtained directly from the manufacturer, and in other circumstances specialist local suppliers of timber or ready-mixed concrete will be used. Contractors are able to secure trade discounts for the items that they purchase, and such discounts will be increased either to attract trade or because of large orders. Some clients, who undertake extensive building work, are able to arrange with suppliers a bulk purchase agreement. This helps to reduce their own costs of construction because they are able to secure very reasonable rates for the items. The contractor must then obtain the appropriate items described from such suppliers.

JCT refers to two special types of suppliers, i.e. named suppliers and nominated suppliers. A named supplier is a firm specified in the contract documents from whom the contractor should obtain certain materials, components or goods. It is usual to suggest a list of alternative suppliers or sometimes to add the words 'other equal and approved'. The contractor would then need to show that the items proposed to be purchased for the work complied with those specified. In other circumstances a single supplier or manufacturer may be named, where the intention of the architect is not to diverge to any other alternative. This is often the case in respect of sanitary ware, where a particular design is selected, or for suspended ceilings and kitchen fittings.

In some situations the architect may choose to nominate a particular supplier to provide some of the various materials, components or goods. In these circumstances the architect chooses to include these items as a prime cost sum in the bills. The appropriate conditions of clause 36 then apply, and these are described in the context of that clause.

Subcontractors

It is very unusual today for a contractor to undertake all the contract work with their own workforce. Even on minor building projects the main contractor is likely to require the assistance of some trade or specialist firms. Work undertaken by firms other than the main contractor are often described as subcontractors, although in some situations it is not uncommon to find specialist firms working on the site beyond the normal jurisdiction and confines of the main contractor, and hence not a subcontractor within the generally understood description.

The employer may, for example, choose to employ such firms directly, and in this context these firms are not to be considered as subcontractors of the main contractor. Provision is made in the conditions of contract for such firms in order that they may have access to the contractor's site (clause 29). For example, a firm constructing a sculpture may come into the confines of this clause. Second, the employer may choose to nominate particular firms to undertake the specialist work that will be required. In these cases the employer may adopt this approach in order to gain a greater measure of control over those who carry out the work. These subcontractors enjoy a special relationship with the employer, as discussed in clause 35. Although after nomination they are often supposedly treated like one of the main contractor's own subcontractors, they do have some special rights, for example, in respect of their payment.

The architect may also choose to name subcontractors in the bills of quantities or specification, who will be acceptable for the execution of some of the measured work. This procedure avoids the lengthy process of nomination, but still provides a substantial measure of control on the part of the architect. Provisions are found within the conditions of contract, clause 19, and a requirement is to name at least three firms who will be acceptable to the architect. This provides for some measure of competition and allows the contractor a choice of firms. The contractor may also add to this list with the approval of the architect, and this should not unreasonably be refused.

All the remaining work is still unlikely to be carried out by the main contractor, and provision is also made for the use of the contractor's own subcontractors. These subcontractors are referred to in the conditions as the main contractor's domestic subcontractors (clause 19). Named subcontractors are employed to undertake that work (a) for which contractors make no provision within their companies, or (b) when their own employed workers are busy elsewhere. The contractor must seek the approval of the architect in this respect, but it is unusual for this approval not to be given.

A further group of subcontractors are those described as statutory undertakers, e.g. gas, water, electricity. These are separately described for the following reasons: the employer and contractor often have no choice in employing them because of their statutory rights; they sometimes require payment in advance; and they refuse to give any cash discount for prompt payment.

The intermediate form of contract identifies another type of subcontractor, known as the named subcontractor, who is akin to the nominated subcontractor. Chapter 30 covers the IFC contract and gives more information on this.

REGULATION

Planning and control of development

Government has wide powers of control over the development of construction works. It seeks to resolve the conflicting demands of industry, commerce, housing, transport, agriculture and recreation by means of a comprehensive statutory system of land use planning and development control. The government's aim is for the maximum use of urban land, sometimes called *brown land*, for new developments. It also aims to protect the countryside, sometimes called *greenfield sites*, by assisting urban regeneration in towns and cities.

The system of land use planning in Britain involves a centralized structure under the Secretary of State for the Environment. The strategic planning is primarily the responsibility of the county councils, while the district councils are responsible for local plans and development control, the main housing function and environmental health. The development plan system involves the structure and local plans. The structure plans are prepared by the county planning authorities and require ministerial approval. They set out the broad policies for land use and ways of improving the physical environment. Local plans provide detailed guidance, usually covering about a ten-year period. These are prepared by district planning authorities but must conform with the overall structure plan.

Before a building can be constructed, application for planning permission must first have been obtained. This is made to the local authority in the form required by the Town and Country Planning (Making of Applications) Regulations. In the first instance, outline approval is sought to avoid the expense of a detailed design which could fail to secure approval. Full planning permission must still be obtained. If a scheme fails to obtain approval then a planning appeal can be made to the Secretary of State.

The Building Act was introduced onto the statute book in 1984. The 1985 Building Regulations are framed within this Act and allow two administrative systems to be applied; one through the local authority building control department and the other via certification. Such controls are necessary to:

- Secure improved standards of design and construction
- Ensure the safety and health of the occupants
- Provide for the proper location of buildings and industry
- Make the best use of the land which is available
- Provide for the safety, health and welfare of those engaged in the construction process and those affected by it

Whilst successive Acts of Parliament have introduced more legislation, and this now also needs to comply with the wider EU legislation, there has been a desire to speed up the process. This has tended to be achieved through increasing flexibility, but at a cost of increasing the uncertainty of the outcome.

Local authorities

Local government is broadly divided into county councils and district councils, each with their own function. A major reorganization of local authorities was introduced by the Local Government Act 1972, which came into operation in 1974. Further reorganization occurred in 1996. The allocation of the various functions appropriate to construction are as follows:

- *Building regulations*: district councils
- *Highways*: county councils
  - district councils do maintenance by agreement
- *Refuse collection*: district councils
- *Refuse disposal*: county councils
- *Structure plans*: county councils
- *Local plans*: district councils
- *Water and sewage*: regional water authorities

County councils have to find tipping sites for refuse. Local councils receive structure plans from their county council then they complete the local details. Local government is restricted in the way it can exercise its powers as follows:

- *Area*: it is generally confined to a defined geographical area. It works with adjoining authorities on matters of common interest through joint committees and joint boards.
- *Central government*: the control of a local authority is exercised through the Secretary of State for the Environment. In some governments a local government minister has been appointed to carry out this control. Central government control may be exercised by supervision where a statutory inspection can take place, and because a local government relies to a large extent on the finances from central government. All the local authority's accounts must be properly audited. In some cases the minister concerned has powers to carry out a local authority's functions in cases where it defaults.
- *Judicial control*: because local authorities are created by statute, they derive their powers and functions in this way. They must therefore act within the laws afforded to them. Building owners and contractors are likely to encounter the local authority concerned initially for the approval of planning permission, where an appropriate fee is payable, and during construction where they must conform to the building regulations and other statutory documents.

The Crown

The Crown is used in this context to denote the governmental powers which are exercised through the Civil Service. This means central government offices rather than local government. Central government has many powers conferred upon it by Parliament. It does, for example, exert a certain measure of control over local government. This may be achieved by:

- *Supervision*: the Secretary of State has power through statutes to supervise the activities of local authorities. For example, some town planning decisions may need to be referred to the Secretary of State. Some local services such as fire, police and education are subject to statutory inspection.
- *Finance*: much of a local authority's finances are obtained from central government. The minister can therefore exercise considerable pressure on proposed expenditure, and also through the auditing of the local authority's accounts.
- *Supersession*: in some cases the minister concerned has power to supersede in certain functions in case of default by a local authority.

Because government is a major spender in the construction industry, it is considered to be a very important client. It can also severely restrict and regulate how the work will be carried out. Although the privatized Property Services Agency (PSA) still undertakes a large amount of central government's building work, other departments also have extensive construction programmes. Much of the central government's work is undertaken on the GC/Works/1 form of contract.

DIRECT LABOUR

The public sector employer remains a large client of the construction industry. It undertakes work ranging from multimillion pound engineering projects to the minor repair and maintenance of local authority dwellings. There have been discussions regarding a national building corporation operating on a regional basis, but this has yet to be developed. However, there remain the direct labour construction departments employed within a reducing number of local authorities. The expansion or contraction of all of these organizations is a politically sensitive issue, with differing opinions expressed on their necessity and efficiency. In some local authorities the direct labour departments may be responsible for no more than building and highway maintenance. In a few examples they may be large enough to undertake major projects. Whilst they were established to work only within the confines of their geographical area, they may now tender for work in other local authorities' areas. Where capital projects are envisaged, the direct labour departments are usually required to tender for this work against private contractors. Maintenance work may be undertaken by the direct labour department as a matter of course, although even within this sphere of work some element of competition is now required.

PROFESSIONAL BODIES

Designing, costing, forecasting, planning, organizing, motivating, controlling and coordinating are some of the roles of the professions involved in managing construction, whether it be new build, refurbishment or maintenance. These activities also include research, development, innovation and improving standards and performance.

The Royal Institute of British Architects

The Royal Institute of British Architects (RIBA) is the main professional body for architects in England and Wales. In Scotland, there is a similar body, the Royal Incorporation of Architects in Scotland (RIAS). Under the Architects Registration Act 1938 it continues to be illegal for anyone to carry out a business describing themselves as an architect unless they are registered with the Architects' Registration Council (ARCUK) established under an Act of 1931. Registration involves appropriate training and education as evidenced by the possession of qualifications set out in the Act or approved by the council. However, this does not prohibit anyone from carrying out architectural work such as the design of buildings.

The Royal Institution of Chartered Surveyors

The Royal Institution of Chartered Surveyors was formed in 1868 and incorporated by Royal Charter in 1881. It originated from the Surveyors' Institute and has grown as a result of a number of mergers with other institutes, most notably the Land Agents (1970) and the Institute of Quantity Surveyors (1982). The RICS is now administered in faculties which represent the varying interests of different chartered surveyors. The General Practice (42 per cent) and Quantity Surveying (37 per cent) divisions account for over three-quarters of the membership.

The Institution of Civil Engineers

The term 'civil engineer' appeared for the first time in the minutes of the Society of Civil Engineers, founded in 1771. It marked the recognition of a new profession in Britain as distinct from the much older profession of military engineer. The members of the Society of Civil Engineers were developing the technology of the Industrial Revolution. The Royal Charter of the institution contains the often quoted definition of civil engineering as being 'the art of directing the great sources of power in nature for use and convenience of mankind'.

The Institution of Structural Engineers

The Institution of Structural Engineers (IstructE) began its life as the Concrete Institute in 1908, was renamed in 1922 and was incorporated by Royal Charter in 1934. Its aims include promoting the science and art of structural engineering in all its forms and furthering the education, training and competencies of its members. The science of structural engineering is the technical justification in terms of strength, safety, durability and serviceability of buildings and other structures.

The Chartered Institution of Building Services Engineers

The grant of a Royal Charter in 1976 enabled the Institution of Heating and Ventilating Engineers which was founded in 1897 to amalgamate with the Illuminating Engineering Society of 1909 to create an institution embracing the whole sphere of building services engineering. The objectives of the Chartered Institution of Building Services Engineers (CIBSE) are set out in their charter as the promotion of the art, science and practice of building services engineering for the benefit of all, and the advancement of education and research in this area of work.

The Chartered Institute of Building

The Chartered Institute of Building (CIOB) was formed in 1834, incorporated under the Companies Acts in 1984 and granted a Royal Charter in 1980. It originally started as the Builders' Society, a small and exclusive club with a wide influence that was responsible for helping to produce the early forms of contract. The objectives of the institute are the promotion, for public benefit, of the science and practice of building, the advancement of education and science including

Table 15.1   Chartered professional bodies in the construction industry

	Total membership	Percentage chartered
Royal Institution of Chartered Surveyors (RICS)	90,000	32%
Institution of Civil Engineers (ICE)	75,000	27%
Chartered Institute of Building (CIOB)	35,000	11%
Royal Institute of British Architects (RIBA)	30,000	11%
Institution of Structural Engineers (IStructE)	20,000	8%
Royal Town Planning Institute (RTPI)	17,000	6%
Chartered Institute of Building Services Engineers (CIBSE)	12,500	5%

research, and the establishment and maintenance of appropriate standards of competence and conduct for those engaged in building. The institute encourages the professional manager and technologist to work together with their technician counterparts in order to achieve a ladder of opportunity as a main objective of the training and examination structure of the institute.

Chartered body membership

A proportion of members in the construction professions hold membership of more than one institution. Several members of the profession are also represented in other professional bodies which are more broadly based than construction alone. For example, a large proportion of the Chartered Institute of Arbitrators (CIArb) are members of the construction professions, whereas in the British Institute of Management (BIM) membership includes only a small proportion of those connected with the construction industry. Table 15.1 lists the main professional bodies in the construction industry and their relative sizes.

Non-chartered bodies

There are also a number of non–chartered professional bodies with memberships ranging from as few as 1,000 to nearly 10,000. These include a diminishing number of organizations, due to mergers and acquisitions.

Europe and the USA

There are wide cultural considerations to be taken into account in any comparison between the construction industry professions in Britain with those in other parts of the world, notably mainland Europe and the USA. Historically, the industry and practice developed differently. In much of the rest of the world, architects and engineers dominate the construction industry. The ratio between building and civil engineering works in other countries is similar, at about 80 : 20. The various professional disciplines in Britain are not mirrored elsewhere, other than in

Commonwealth and former Commonwealth countries. The role of the professional bodies also varies. In Britain a professional qualification is one by which to practice. In Europe a professional body is more of an exclusive club, of which relatively few of those engaged in practice are members. In the USA there is the emerging discipline of construction management alongside those of architect and engineer. In Britain the architect is the only registered profession and this is also the case in mainland Europe. But its practice there is more controlled. An architect, for example, must be employed to sign the plans in mainland Europe, otherwise a building cannot be constructed.

CONSTRUCTION ASSOCIATIONS

Construction Industry Council

The Construction Industry Council represents the construction industry professions. Membership includes the seven major chartered bodies (CIOB, CIBSE, ICE, IStructE, RIBA, RICS and RTPI), business organizations which represent some of these professions and a number of the non-chartered bodies. Their objectives cover the following areas:

- Membership of CIC by all professional bodies in construction
- Generate effective relationships with employer organizations, trade associations and others
- Oversee standards development and coordinate NVQs relevant to members
- Promote the integration of compatible areas of professional education and training
- Broaden the income base of CIC
- Raise the profile, and increase understanding, of the construction industry
- Promote continuing professional development
- Promote more extensive research and development
- Encourage improved industry practices and procedures and reduce duplication
- Represent members' interests to government
- Coordinate and develop members' interests in Europe
- Collect and disseminate information

Construction employers' associations

The construction industry has typically been seen as two sectors, building and civil engineering. This distinction was also seen at employer level with the separate Building Employers' Confederation (BEC) and the Federation of Civil Engineering Contractors (FCEC). These bodies merged towards the end of the 1990s to form the Construction Confederation. In addition there are a number of subsidiary groupings of firms such as the House Builders. Many of the activities which are undertaken by the different organizations are replicated and add to the costs of administration. Many of the contractors believe that the construction industry's poor image, educational framework, lobbying strength and relationships with clients and the professions will be improved only through having one united voice.

Construction Industry Training Board

The Construction Industry Training Board (CITB) was established by an Act of Parliament in 1964 to improve the quality of training, improve the facilities available for training and help to provide enough trained people for the construction industry. It is partially funded through government, but most of its income is derived from a levy system on contractors based upon the number of their employees. In addition to being appointed as the primary managing agent for the construction industry's Youth Training Scheme, the CITB also provides for a range of skills training at its national training centres. Whilst the emphasis of its activities is on practical craft skills training, it also offers courses in supervisory management. The CITB is one of the few remaining publicly funded industrial training boards. Its future is constantly under review and is by no means secure.

Construction Industry Standing Conference

The Construction Industry Standing Conference (CISC) was established in 1990 to produce a framework of qualifications for the NVQ accreditation covering technical, professional and managerial occupations within the construction industry. CISC's membership is drawn from the chartered and non-chartered professional bodies, the awarding bodies such as Edexcel, employer, business and trade organizations. There are strong links with further and higher education. An early task of CISC was to prepare an occupational map of the construction industry in order to identify the tasks and responsibilities performed by the different professional disciplines. This was published in 1991. The second objective was to develop standards by analysing the different job roles into elements of competence. These are grouped into units and will form the basis of the NVQs. It is intended to prepare instruments for assessing the standards for accreditation purposes and finally a strategy for implementing and maintaining the NVQ system. The merits of this approach are:

- Developed by the industry
- Nationally agreed
- Relevant
- Accessible
- Transferable

CHAPTER 16

SITE COMMUNICATIONS

The construction of buildings and engineering structures utilizes a range of different resources such as people, machines, money, materials, mechanical plant, management and methods. It seeks to combine them in the most effective, efficient and economic manner.

The construction industry is a unique industry. It is one of the few industries that separate design from production, although this has now changed with the proliferation of design and build projects. It also undertakes the bulk of its work on the client's premises. The industry does not build replicas, and whilst projects may appear to be similar they are different in certain respects. Even comparable housing units on an estate will differ in their location, levels, amount of work below ground level, plot size and hence the external works that are required.

Site managers do not move materials, operate plant or carry out the physical production function. Site management organizes, informs, coordinates, orders, instructs and motivates others to undertake these tasks. The effectiveness of performance will depend upon the ability to listen, read, speak and write. More importantly, effectiveness will be evident in being understood by others and to understand another person's point of view.

Site communications therefore includes gathering information to ascertain the needs of those involved with the project as a necessary tool of management to integrate the functions of departments and as a vital link between manager and subordinate in order to get the job completed.

COMMUNICATION PRINCIPLES

Communication and the social interactions involved are keys to success in this work. Communication skills and strategies are necessary to share ideas and experiences, to find out about things and to explain to others what you want. Developing ways of communicating including language and body signals is essential in expressing feelings and insights. Learning to communicate effectively may mean the difference between barely coping with life as you know it and actively shaping your world as you would like it to be. Some of the components of good communications are:

- Know what to say
- Gain the attention of the listener
- Establish and maintain relationships
- Know the listener's likes and interests
- Choose how to communicate from a range of options
- Be skilful in communicating
- Choose when and where to communicate
- Be clear, brief and coherent
- Listen actively
- Understand and clarify any messages you receive
- Try not to be easily distracted
- Know how to close conversation or communication

Effective communication starts with a purpose or objective to achieve in transferring the information. This should be done in the most suitable form to suit the receiver. It also needs to be the most effective way. Provision should also be allowed for feedback and the need for action on that feedback appreciated. The points that should be present in all communications are:

- *Clarity*, i.e. easily understood
- *Presentation*, i.e. creating a favourable impression
- *New information*, i.e. attracting interest
- *Drive*, i.e. demanding the necessary action
- *Tone*, i.e. creating a responsive attitude
- *Feedback*, i.e. ensuring the information is transferred

Forms of communication

- *Written*: letters, reports, bills of quantities, specifications, site instructions, British Standards. This method is used when the subject matter is complex, important or likely to have possible legal implications and where a permanent record is required for future reference.
- *Visual*: films, slides, posters, graphs, charts. It includes the project drawings and works programmes. This sort of communication often has a greater impact upon the receiver. Where messages are less complex, this form of communication is the most effective.
- *Oral*: this method is instant and often generates an immediate response. Attitudes and behaviour can also be observed. The face-to-face confrontation is appropriate when the subject matter may be difficult or disagreeable. It is also used in those circumstances for simple, less important and informal messages.

Barriers to communication

An organization that is suffering from poor morale or lacks confidence in its future is likely to see an increase in communications problems. In these circumstances there is sometimes a lack of will to communicate effectively. However, nearly all barriers to good communication can be classified as follows:

- *Physical barriers* might include disability on the part of the receiver such as deafness or blindness. They might include noise on site, poor telephone lines and layers of site management. These can sometimes introduce distortion and inaccuracy into the information that is being transmitted.
- *Psychological barriers* are the biggest cause of communication breakdown. They affect the attitudes, feelings and emotions of the receiver. Sometimes individuals will only hear what they expect to hear. Preconceived ideas are used to interpret the information in ways not anticipated by the giver of the information. Feelings and emotions affect the ability to receive a true message. When worried the receiver feels threatened. When angry the information may be simply rejected.
- *Intellectual barriers* affect concepts and perceptions that may have been built up over a lifetime of work. Difficulties may arise because the information was not transmitted properly in the first place. It may use excessive technical jargon or attempt to use words that are out of context.

SITE INFORMATION

In addition to the everyday running and organizing of a contract, it is also the site overseer's responsibility to maintain accurate records of the important happenings on site. This information should be properly recorded, so that whenever necessary it can be quickly retrieved for future use. The site records that are normally kept include:

- Daily reports
- Site diaries
- Materials received sheets
- Advice of variations
- Daily labour allocation sheets
- Drawing registers
- Confirmation of verbal instructions
- Weather reports
- Daywork records
- Subcontractor files
- Site correspondence

If this information is to be valuable for future reference then it must be:

- *Precise*: clear, straightforward, as simple as possible and well thought out
- *Accurate*: true and correct in all its details
- *Definite*: future users should be left in no doubt as to what the message means
- *Relevant*: to the particular situation and the people who will use it
- *Referenced*: linked wherever possible to other documents being used by the contractor

SITE MEETINGS

Site meetings are held for various purposes. They may be held with the contractor's staff only to establish how the works should be carried out or to deal with an internal matter that has arisen in connection with the project. They frequently involve other parties such as subcontractors and suppliers. The general site meeting would usually involve the client, architect, quantity surveyor, other consultants, subcontractors, clerk of works as well as the main contractor. These meetings frequently take place on a monthly basis, depending upon the size of the project and the issues that may need to be resolved. The site meeting is a vehicle to pass information from one party to another or for consultation on how to tackle a problem that has arisen. The normal site meeting is also used for decision-making purposes and it may be used to try to persuade the parties involved of a different way of solving a particular problem. In essence it is, What do we need to know? How are we getting on? What's wrong and how can this be corrected? Here is some information on meetings preparation:

- Determine the meeting's objectives
- Produce an attendance list
- Decide on the time and place
- Determine its style
- Circulate agenda and other information
- Identify if special facilities are required (e.g. audio–visual)
- Rank and set times for each item
- Assess possible areas of conflict
- Pre-discuss items with individuals where necessary
- Determine the sorts of minutes required
- Adequately brief the participants
- Other matters: refreshments, car parking, etc.

Frequently the architect will chair the site meeting and prepare an agenda, often in consultation with the contractor. The agenda will follow the format of a business meeting, with apologies for absence, matters arising from the previous minutes, any other business and the date and place of the next meeting. In addition, items of a routine nature, such as the main contractor's report and special items that address current issues, will also be included.

The minutes of the meeting are the follow-up and as such are vital to the effectiveness of the meeting. The minutes ensure the collective views of the meeting. They should also identify the areas where responsibility for further action lies. They may include an action plan in order to minimize future problems or difficulties that might be encountered. Here are some points that should help meetings to flow more easily and become more productive:

- Prepare an agenda and establish rules of procedure.
- Start promptly and use brevity.
- Practise active listening.

- Keep replies to the point and avoid wasting time.
- Clarify issues by asking relevant questions.
- Summarize progress.
- Restate important points.
- Be prepared to change strategy if necessary.
- Be supportive: That sounds like good idea.
- Confront issues: Are we really prepared to do that?
- Question critically: What exactly do you mean?
- Provide accurate supporting information.
- Do not allow the meeting to be interrupted by telephone calls, etc.
- Avoid interrupting.
- Do not be afraid to make your feelings known.
- Refrain from distracting behaviour.
- Do not talk to others during presentations.
- Never lose your temper.
- Do not embarrass others.

SITE DIARIES

The site diary, if well maintained with the correct sort of information, is a very useful document retained by the contractor. It will contain information that can be used where disputes occur between the different parties concerned with the project. Site diaries have been used in litigation as valuable evidence to help substantiate a case put forward by the contractors. The diary should record information that generally does not warrant separate records being kept. This will vary with the size and type of project, the nature of the client and the information records kept by the contractor. Figure 16.1 is a typical form that can be used for this purpose.

Weather conditions

Weather conditions will have a major influence upon the progress of the works. They must therefore be recorded, especially where they cause a delay or suspension of the works.

Drawings received

The site manager must ensure there is an adequate system for the receipt and recording of architect's and engineer's drawings. Although there will be a drawing register, the site diary should also record when this information is received. There will also be times when information will be requested from the consultants. This should always be in writing, giving sufficient notice to avoid the possibility of delays. The late receipt of information is one of the reasons available for an extension of time under JCT 98 (clause 25.4.6) where this is considered to be a relevant event. In addition, if the delay in the receipt of such information causes

Details	Date
1. Weather conditions	Temperature
2. Drawings received	Number
3. Instructions: oral/written	Given by
4. Variation orders received	Number
5. Dayworks: reasons/descriptions	Sheet numbers
6. Delays: reasons	Labour on-site required
7. Lost time: reasons	
8. Urgent requirements	Visitors to site
9. Unusual occurrences	

Fig. 16.1   Site diary (Adapted from Davies, W H, 1982, *Construction site studies 4*, Butterworth, London)

the works to be suspended for longer than the period named in the appendix, usually three months, then the contractor may determine the contract (clause 28.3.5). The contractor in each case must have requested further instructions and drawings and the date for the period of delay will commence from that date.

Architect's instructions

During the visit to the site by the architect, and more frequently through the clerk of works, instructions will be given both orally and in writing. These need to be recorded in the site diary under the heading of site instructions. Remember that in order to be valid, all instructions must eventually be in writing (clause 4.3.1). The JCT form also lays down procedures to be followed in the case of dealing with oral instructions (clause 4.3.2); see Chapter 25. The site manager therefore needs to record precisely when the instruction was given.

Variation orders

Variation orders (VOs) are a special type of architect's instruction since they may alter or modify the design, quality or quantity works shown on the drawings and in the bills (clause 13.11). A variation may also alter the conditions under which the work is to be carried out. In many other respects VOs are similar to those of other architect's instructions (Chapter 19).

Dayworks

Clause 13.5.4 defines daywork as 'work which cannot be properly valued by measurement'. The site manager should record in the site diary reasons why some of the work should be paid for on this basis. Prior to arriving at this decision the contractor's own quantity surveyor should also be consulted for an opinion. Where dayworks are envisaged then the architect should be informed in order that the clerk of works can carry out the necessary checks on the labour and materials used on such work. The inclusion of large numbers of dayworks in the site manager's diary may lead towards the conclusion that the nature of the project has changed from what was originally tendered.

Delay

The contractor will have prepared a master programme for the project (clause 5.3) for use both by the contractor and the architect. The contractual implications of this are slightly different under the ICE conditions of contract, nevertheless the architect should have assessed its appropriateness. The programme will indicate how the contractor intends to carry out and complete the works. It also assists the architect on the dates when the contractor requires the various pieces of information. The programme provides the contractor with the best indication that the works are being constructed as planned, when progress is measured against it. Where the progress shown is different then the reasons for such variation will need to be established.

Lost time

A record of any delays will be made in the site diary at the time they actually occur. This will be the most accurate record of these events. Where the delays are due to factors that the contractor should have controlled, these will need to be remedied at the contractor's own expense. This will then avoid liquidated damages being applied for late completion.

Urgent requirements

The site diary will also be used to remind the site manager of any urgent requirements. These may include a note, in advance, of items that are expected for delivery or action on a certain date. This information can be included in the following manner:

- Sanitary fittings for delivery next Tuesday; storage provision required.
- Latest date for forwarding design details of external screens, since the manufacturer requires six weeks prior to delivery.
- Inform refuse chute subcontractor when they will be required on site. They require three weeks' notice.

Unusual occurrences

Any unusual happening on the site will need to be recorded for possible future reference by the contractor. Disputes may occur between the parties concerned and it is important to establish as soon as possible the reasons why such disputes have arisen. Here are some typical examples:

- Local stoppage on site for two hours due to a disagreement over bonus payments.
- Mr R Taylor fractured his ankle. An accident report has been completed.
- Intruders entered the site last evening. No damage or theft occurred. Police have been informed.
- Raining-in below patent glazing. Architect and subcontractor have been informed.

Labour records

The site manager will need to keep a careful record of labour that was required and labour employed on site.

Visitors to the site

Visitors should always be recorded, especially those who might have a direct impact upon the work. The diary record may show:

> The architect visited the site at 10.00 am. A tour was made with the site manager and clerk of works. Satisfaction with the quality and progress of the works was noted.

The contractor cannot refuse reasonable entry to the site for the architect or representatives (clause 11). The building owner owns the site and is therefore allowed access. This must be done to cause as little inconvenience to the contractor as possible. Building control officers, health and safety inspectors and similar officials have a legal right to entry to the site at any reasonable time.

Recording of information

If the site diary is to be of any value then it is important that the events are entered each day in a logical, careful and legible manner. The information will thus provide an accurate assessment of the site's daily progress together with a record of the labour and plant that have been used. Any matters that affect the following items must be regularly and accurately recorded:

- Completion date
- Costs
- Quality and standards of work
- Contractor's performance

The site diary must be completed daily in order that the salient points are not forgotten forever. The completion of the task will require some self-discipline on the part of the site manager, particularly when other aspects of work are requiring attention. The information should be accurate and represent a fair picture of the day's events. Exaggerated information or hearsay remarks should not be included. Where proper and adequate records are not maintained then the contractor may suffer because of:

- A loss of reputation of a well-run organization
- Having liquidated damages imposed through late completion of the works
- Being refused additional payments for losses incurred
- Having the contract unfairly terminated due to client dissatisfaction
- Being levied with further damages to redress the client's loss

PLANNING AND PROGRAMMING

A programme or schedule is developed by breaking down the work involved in a construction project into a series of operations which are then shown in an ordered stage-by-stage representation. Without a programme of work which specifies the time and resources allocation in order to undertake each stage of the project, the execution of the contract will be haphazard and disordered. There are several different methods that may be used for this purpose, such as bar charts (sometimes referred to as the Gantt chart) or network analysis (critical path analysis). A detailed description of these methods is outside the scope of this book. The bar chart is perhaps the best known of all the planning techniques. In its simplest form the sequential relationships between activities are not completely prescribed. However, they can be linked to show the relationship between an activity and the preceding and succeeding activities. Thus, dependency between the activities highlights the effect of delays. The resources required can also be calculated.

The linkages between preceding and succeeding activities, combined with a set of arrows to represent the bars of the bar chart, give rise to a simple network diagram. This forms the basis of a network analysis, which identifies the longest irreducible sequence of events. It also defines quickly those parts of the programme which could benefit from the use of increased resources and thereby benefit the project. As networks are rarely the best method for communication, the output of the analysis is often presented as a bar chart. Network analysis has been used for many large and complex projects. It is claimed that network analysis can reduce project times by up to 40 per cent.

Resources

The time taken to complete an activity in the programme depends on the resources allocated to that activity. The approaches used in assessing the required resources can be based on completing the project in a given time or completing the project with specified limited resources. Once the level of resourcing has been finalized, the overall resource demands are smoothed if necessary by rescheduling activities to ensure an acceptable overall demand for the project.

There is a measure of uncertainty in estimating the time for each activity, particularly as delays in delivery of materials and adverse weather can delay work in progress. Probabilistic distributions have been used for the generation of the most likely times for activities. An approach that was used in early projects was the project evaluation and review technique (PERT). The flexibility of the contractor's workforce has changed in recent years. Previously a contractor, using its own workforce, would operate a number of sites in an area. There were thus ample opportunities for transferring workpeople between the different sites. However, the widespread use of labour-only subcontracting has to some extent curtailed this opportunity. This has reduced the adaptability of day-to-day site management, resulting in a greater measure of pre-site planning. Also the construction industry, when compared with manufacturing, remains a labour-intensive industry. The widespread use of plant and machinery on site is perhaps less developed than it ought to be.

Monitoring and control

With a programme of work and the resource requirements for each activity having been determined, it is possible to monitor the construction work as it progresses. In practice, updating will take place and control will be exercised. This will involve the rescheduling of activities and the revision of resources. The planning model is often used to explore the overall development of the project before work on site is undertaken. This assists in investigating the influence of different construction techniques and the timing of the individual activities to better optimize the use of resources.

Ensuring that the works are constructed to the specified level of quality is essential. This extends from the initial setting out of the project to the inspection, storage, handling and incorporation of the specified materials within the finished building. All site operations must be governed by appropriate safety measures which start with a safe design and erection procedures for permanent and temporary works. The construction industry remains one of the largest single contributors to fatal accidents (Chapter 12).

Planning into practice

The construction industry is usually involved in one-off projects; these are invariably managed with a new team. As the location of each project varies widely,

Table 16.1 Delays in construction of industrial buildings

Cause of hold-up and delay	Percentage of case study projects
Subcontracting	49%
Tenant/client variations	45%
Ground problems – water, rock, etc.	37%
Bad weather	27%
Materials delivery	25%
Sewer/drains obstruction or rerouteing	20%
Information late	20%
Poor site management and supervision	18%
Steel strike	16%
Statutory undertakers	14%
Labour shortage	10%
Design complexity	6%

Source: *Faster building for industry*, NEDO, 1983

the workforce is largely new and the conditions under which the project is undertaken can differ depending on the site conditions, climate, etc. Unlike manufacturing industry, the approach to a project is rarely uniform. The size of the main contractor's site organization, which is comprised of technical and non-technical staff, often depends on the size of the works. However, as the use of subcontractors has become more widespread, the main contractor's team has reduced in size.

The work can be carried out efficiently if the site is laid out in such a way that the temporary buildings, for example offices, stores and workshops, are conveniently located with respect to the permanent works. This results in an orderly arrangement which facilitates the economy of construction and administration. The construction of a high-rise building on a confined city centre site, for example, requires the efficient storage of materials, off-peak deliveries, high-speed vertical travel with site facilities placed to minimize operative travel.

Table 16.1 indicates some of the reasons for the delays that occurred in the construction of industrial buildings together with the frequency of their occurrence. The influence of these delays on the total site time varies. For example, ground problems or bad weather occurred on sites which achieved both fast and slow site times, suggesting there was less overall effect. However, poor site management and supervision, late information and deliveries, design complexities, difficulties with statutory undertakers and labour shortages were usually associated with construction overruns of two months or more. These factors were judged to be the most damaging to the more effective organization of the construction process.

PART 4

STANDARD FORM OF BUILDING CONTRACT

CHAPTER 17

INTRODUCTION, ARTICLES OF AGREEMENT, APPENDIX

The standard form of building contract is referred to throughout the construction industry as JCT 98. This is an updated version of JCT 80, which consolidates the numerous revisions that have been issued over the past twenty years. These revisions were necessary due to changes in the law and legal principles, changes brought about through case law and changes due to the way the industry is now being organized. JCT 80 replaced the previous edition of the form that was published in 1963, colloquially known as the RIBA form, although other professional bodies were involved in its agreement. JCT 98 is more than twice as large as the 1963 edition. The standard form of building contract has 104 pages, plus a number of additional supplements amounting to approximately 800,000 words. It has sometimes been compared to the Lord's Prayer that is used in our communication with Almighty God. The Lord's Prayer achieves its aim successfully in a mere seventy words.

JCT 98 is the industry standard form of building contract. It has been developed on the general presumption of fairness in allocating risk to the party that is more likely to be able to control it. Whether this allocation of risk is fair depends upon its use and interpretation in practice. The form has been written in an attempt to cover every possible eventuality, so that should a dispute arise there are principles, processes and procedures that can be followed and applied. However, several surveys of users in the building industry indicate that they are often unaware of its full implications, consider it unnecessarily complex and describe it as ambiguous. It is also written in a language that is not easily understood. There remains continued criticism of subcontracting nomination procedures.

The content of the JCT 98 form is as follows:

- Articles of agreement.
- Conditions
  - Part 1: General
  - Part 2: Nominated subcontractors and nominated suppliers
  - Part 3: Fluctuations (clauses 38, 39 and 40 are published separately)
  - Part 4: Settlement of disputes – adjudication – arbitration – legal proceedings
  - Part 5: Performance-specified work.

- Code of Practice referred to in clause 8.4.4
- Appendix
- Annex 1 to appendix: terms of bonds
- Supplemental provisions (including the VAT agreement)
- Annex 2 to the conditions: supplemental provisions for EDI

The principle of the form is that it is an entire contract which is modified by a number of its own provisions. Separate editions are available for private clients and local authority clients. Each edition has three variants: with quantities, without quantities and with approximate quantities. Hence there are six forms in total. However, the differences between them are minimal. Financially it is a lump sum contract. The two parties to the contract are the employer (client) and the contractor. The employer is responsible for making payments to the contractor, who in turn has obligations to complete the works in accordance with the contract provisions. Neither the architect nor the quantity surveyor are parties to this contract, although they figure strongly within its clauses. The architect acts on behalf of the employer but also sometimes in an impartial way, giving benefit to the contractor where this is appropriate. The quantity surveyor is impartial in accounting for the financial transactions that take place before, during and after the project has been completed.

ARTICLES OF AGREEMENT

The articles of agreement precede the conditions of contract and they include the information that must be completed for a particular contract. The articles define the various persons referred to in the contract and they describe the procedure to be followed should a dispute arise between the two parties. The following information therefore needs to be written into the articles of agreement:

- *The date*: this is the date when the agreement is made. It is not the date of tender, which may have some significance in respect of fluctuation contracts; nor is it the date when the project will start on site. This date is defined in the appendix to the form of contract as the date of possession and included in clause 1. The date in the articles of agreement will therefore have little significance as far as the contract is concerned.
- *The parties*: the contract is between the employer and the contractor, and their names and addresses must be inserted in the articles of agreement. These names, employer and contractor, are used consistently throughout the conditions of contract. Their registered office addresses are also included.
- *The works*: the employer requires the contractor to complete the works in accordance with the drawings and bills of quantities. The works are essentially the building project, but by implication in the conditions of contract may also mean the site. A description of the project and its location (address) should be

given. The contractor is given possession of the site on the date of possession as stated in clause 23 and detailed in the appendix.

■ *The drawings and bills of quantities*: these, together with the form of contract, constitute the contract documents. They are defined in clause 1 and discussed in clause 2 of the conditions of contract. The contract drawings are formally registered by their numbers and should be signed by both parties. The contract bills, which are a priced copy of the bills of quantities, are also to be signed by the parties.

■ *The architect* (article 3): the drawings and bills of quantities are prepared under the direction of the pre-contract architect. This emphasizes that the design is the responsibility of the architect and not the contractor. On some contracts the designer may not be the person who supervises the work on site. The architect referred to in the conditions comes within this latter definition. In the event of the architect's death, or when no longer fulfilling this function, the employer must nominate a replacement architect. If the contractor has a reasonable objection then the employer must appoint an architect with whom the contractor approves. Such an appointee must not revoke decisions already acted upon by the contractors. The new architect can of course ask the contractor to rebuild part of the completed works, but if these have previously been accepted then the contractor is entitled to an appropriate payment. In some cases the supervisor of the works may not be an architect within the definition of the Architects' Registration Act 1938. In this case the term 'supervising officer' is used to refer, perhaps, to an engineer or surveyor.

■ In the local authorities version, the term 'architect' has been substituted by 'architect/contract administrator'. This reflects the view held by some, that some contractual matters are probably better dealt with by someone other than the architect.

■ *The quantity surveyor* (article 4): this refers to the client's quantity surveyor. In the event of death, or when replaced by the architect, this must be done with the approval of the contractor. Any objection by the contractor must be sufficient to invoke the appointment of the arbitrator.

■ *The contract sum* (article 2): the employer's main obligation is towards the payment of the contractor for the work executed. The total amount is described as the contract sum (clause 14) and this will be paid by instalment (clause 30). Although this sum is agreed upon by the two parties, it will be subject to adjustment within the terms of the contract (clause 30) as the project progresses. It is very unusual if the contract sum remains unaltered.

■ *The employer*: the status of the employer for the purpose of the statutory tax deduction scheme must be stated in the appendix. This refers to clause 31 and generally to labour-only subcontractors.

■ *Contractor's obligations* (article 1): the contractor's obligations under the terms of the agreement are to carry out and complete the works in accordance with the contract documents. These obligations are further reinforced in clause 2 of the conditions.

- *The planning supervisor*: this is the person, either the architect or other person, appointed as a result of regulation 6(5) of the CDM Regulations.
- *The principal contractor*: this refers to the main contractor in pursuant of regulation 6(5) of the CDM Regulations.

SETTLEMENT OF DISPUTES – ADJUDICATION (ARTICLE 5)

Where disputes or differences of opinion occur between the employer (or the architect) and the contractor, the parties agree to attempt to settle the matter first of all through adjudication in accordance with clause 41 of JCT 98. Where the entry in the appendix stating that clause 41B applies and has not been deleted then subject to article 5 it can then be referred to arbitration (article 7A). The provisions of the Arbitration Act of 1996 or any subsequent amendments then apply. Where this clause has been deleted, then if adjudication does not resolve the differences, the dispute will be referred to legal proceedings.

ATTESTATION

The attestation page (page 11 of JCT 98) is for the parties to sign the contract and for the signatures to be witnessed. The signatories must have the necessary authority to sign for this purpose to provide validity to the contract. The contract may be executed under seal or under hand, and the Limitation Act of 1963 states that actions become barred after twelve years and six years, respectively. The current ruling suggests that the cause of action begins to accrue when the damage is discovered, or when it should reasonably have been discovered. The actual time of discovery may therefore not be relevant. There is no contract between the parties until the articles of agreement have been executed, unless the facts show otherwise. The carrying out of work prior to contract may indicate an intention on both parties, and therefore, if the project does not proceed, a claim for *quantum meruit* may result.

THE APPENDIX

The appendix (Box 17.1) provides a considerable amount of information about the contract that is of particular importance to the project. The purpose is to set out in one place a schedule of these details. The completion of the appendix is necessary to give the conditions their full meaning. Each subject listed in the appendix is discussed in detail under the clause concerned. There are, however, some differences in principle between the treatment of items in the appendix. Some items will always need to be completed to make sense, whereas others can be left blank. In other circumstances the appendix makes a recommendation, and unless anything is entered to the contrary this suggestion will apply.

Box 17.1   Appendix to the form of contract

Clause etc.	*Subject*	
Fourth recital and 31	Statutory tax deduction scheme	Employer at Base Date *is a 'contractor'/is not a 'contractor' for the purposes of the Act and the Regulations
Fifth recital	CDM Regulations	*All the CDM Regulations apply/ Regulations 7 and 13 only of the CDM Regulations apply
Articles 7A and 7B 41B 41C	Dispute or difference – settlement of disputes	*Clause 41B applies *Delete if disputes are to be decided by legal proceedings and article 7B is thus to apply *See the Guidance Note to JCT 98 Amendment 18 on factors to be taken into account by the Parties considering whether disputes are to be decided by arbitration or by legal proceedings*
1.3	Base Date	_____
1.3	Date for Completion	_____
1.11	Electronic data interchange	The JCT Supplemental Provisions for EDI *apply/do not apply If applicable: the EDI Agreement to which the Supplemental Provisions refer is: *the EDI Association Standard EDI Agreement *the European Model EDI Agreement
15.2	VAT Agreement	Clause 1A of the VAT Agreement *applies/does not apply [x]

Footnotes	*Delete as applicable.	

[x] Clause 1A can only apply where the Contractor is satisfied at the date the Contract is entered into that his output tax on all supplies to the Employer under the Contract will be at either a positive or a zero rate of tax.

On and from 1 April 1989 the supply in respect of a building designed for a 'relevant charitable purpose' (as defined in the

legislation which gives statutory effect to the VAT changes operative from 1 April 1989) is only zero rated if the person to whom the supply is made has given to the Contractor a certificate in statutory form: see the VAT leaflet 708 revised 1989. Where a contract supply is zero rated by certificate only the person holding the certificate (usually the Contractor) may zero rate his supply.

This footnote repeats footnote [x] for clause 15.2.

Clause etc.	Subject	
17.2	Defects Liability Period (if none other stated is 6 months from the day named in the certificate of Practical Completion of the Works)	_____
19.1.2	Assignment by Employer of benefits after Practical Completion	Clause 19.1.2 *applies/does not apply
21.1.1	Insurance cover for any one occurrence or series of occurrences arising out of one event	£_____
21.2.1	Insurance – liability of Employer	Insurance *may be required/is not required
		Amount of indemnity for any one occurrence or series of occurrences arising out of one event
		£_____ [aaa]
22.1	Insurance of the Works – alternative clauses	*Clause 22A/Clause 22B/Clause 22C applies (See footnote [cc] to clause 22)
*22A, 22B.1, 22C.2	Percentage to cover professional fees	_____
22A.3.1	Annual renewal date of insurance as supplied by Contractor	_____
22D	Insurance for Employer's loss of liquidated damages – clause 25.4.3	Insurance *may be required/is not required
22D.2		Period of time
		_____

Footnotes *Delete as applicable.

[aaa] If the indemnity is to be for an aggregate amount and not for any one occurrence or series of occurrences the entry should make this clear.

Clause etc.	*Subject*	
22FC.1	Joint Fire Code	The Joint Fire Code *applies/does not apply
		If the Joint Fire Code is applicable, state whether the insurer under clause 22A or clause 22B or clause 22C.2 has specified that the Works are a 'Large Project': *YES/NO
		(where clause 22A applies these entries are made on information supplied by the Contractor)
23.1.1	Date of Possession	_____
23.1.2, 25.4.13, 26.1	Deferment of the Date of Possession	Clause 23.1.2 *applies/does not apply
		Period of deferment if it is to be less than 6 weeks is
		_____
24.2	Liquidated and ascertained damages	at the rate of £ _____ per _____
28.2.2	Period of suspension (if none stated is 1 month)	_____
28A.1.1.1 to 28A.1.1.3	Period of suspension (if none stated is 3 months)	_____
28A.1.1.4 to 28A.1.1.6	Period of suspension (if none stated is 1 month)	_____
30.1.1.6	Advance payment	Clause 30.1.1.6 *applies/does not apply
		If applicable: the advance payment will be
		**£ _____ / _____ % of the Contract Sum and will be paid to the Contractor on
		_____
		and will be reimbursed to the Employer in the following amount(s) and at the following time(s)
		_____ _____ _____
		An advance payment bond *is/is not required

Footnotes	*Delete as applicable.	**Insert either a money amount or a percentage figure and delete the other alternative.

Clause etc.	Subject	
30.1.3	Period of Interim Certificates (if none stated is 1 month)	_____
30.2.1.1	Gross valuation	A priced Activity Schedule *is/is not attached to this Appendix
30.3.1	Listed items – uniquely identified	*For uniquely identified listed items a bond as referred to in clause 30.3.1 in respect of payment for such items is required for £_____ *Delete if no bond is required
30.3.2	Listed items – not uniquely identified	*For listed items that are not uniquely identified a bond as referred to in clause 30.3.2 in respect of payment for such items is required for £_____ *Delete if clause 30.3.2 does not apply
30.4.1.1	Retention Percentage (if less than 5 per cent) [bbb]	_____
35.2	Work reserved for Nominated Sub-Contractors for which the Contractor desires to tender	_____
37	Fluctuations: (if alternative required is not shown clause 38 shall apply)	clause 38 [ccc] clause 39 clause 40
38.7 or 39.8	Percentage addition	_____
40.1.1.1	Formula Rules	rule 3: Base Month _____ 19_____ rules 10 and 30 (i): Part I/Part II [ddd] of Section 2 of the Formula Rules is to apply

---

Footnotes

*Delete as applicable.

[bbb] The percentage will be 5 per cent unless a lower rate is specified here.

[ccc] Delete alternatives not used.

[ddd] Strike out according to which method of formula adjustment (Part I – Work Category Method or Part II – Work Group Method) has been stated in the documents issued to tenderers.

Clause etc.	Subject	
41A.2	Adjudication – nominator of Adjudicator (if no nominator is selected the nominator shall be the President or a Vice-President of the Royal Institute of British Architects)	President or a Vice-President or Chairman or a Vice-Chairman: *Royal Institute of British Architects *Royal Institution of Chartered Surveyors *Construction Confederation *National Specialist Contractors Council *Delete all but one
41B.1	Arbitration – appointor of Arbitrator (if no appointor is selected the appointor shall be the President or a Vice-President of the Royal Institute of British Architects)	President or a Vice-President: *Royal Institute of British Architects *Royal Institution of Chartered Surveyors *Chartered Institute of Arbitrators *Delete all but one
42.1.1	Performance Specified Work	Identify below or on a separate sheet each item of Performance Specified Work to be provided by the Contractor and insert the relevant reference in the Contract Bills [zz]

| Footnotes | *Delete as applicable. [zz] See Practice Note 25 'Performance Specified Work' paragraphs 2.6 to 2.8 for a description of work which is | not to be treated as Performance Specified Work. *This footnote repeats footnote [zz] for clause 42. |

CHAPTER 18

QUALITY OF WORK DURING CONSTRUCTION

The following clauses from the conditions of contract have a particular influence in the way that the work is carried out, and hence the resultant quality and standards of the finished project.

- Clause 6   Statutory obligations, notices, fees and charges
- Clause 6A   Provisions for use where the appendix states that all the CDM regulations apply
- Clause 7   Levels and setting out of the works
- Clause 8   Work, materials and goods
- Clause 9   Royalties and patent rights
- Clause 10   Person in charge
- Clause 11   Access for architect to the works
- Clause 12   Clerk of works
- Clause 42   Performance-specified work

The combination of ensuring that the contractor complies with the statutory regulations, and the specification of the works that have been described, should result in a building that achieves the desired standards and quality that were expected. In addition there is the provision in the contract for inspection by the clerk of works regularly, and by the architect intermittently. There is also the added precaution for the inspection of goods and materials in the workshops off-site, should this be so desired. The contractor must also seek to ensure that competent tradespeople are employed to do the work that is required. A person must be constantly on the site who is able to accept instructions from the architect and who will also be responsible for the daily site management of the project and the people working on it. In addition there are clauses covering levels and setting out – making sure the project gets off to a correct start from the outset. There is also provision for uncovering work for inspection if the architect feels this is necessary. The contract specifies the outcome of this depending upon whether the work complies with the specification requirements.

STATUTORY OBLIGATIONS, NOTICES, FEES AND CHARGES (CLAUSE 6)

Statutory requirements

The contractor must comply with and give all notices required by any Act of Parliament, any instrument, rule or order made under any Act of Parliament, or any regulation or by-law of any local authority or statutory undertaker which has any jurisdiction with regard to the works. The contractor, however, will not be liable for compliance with this requirement where the works themselves do not comply with these requirements. The contractor should not deliberately and knowingly execute work that will contravene these regulations, and as such will then require future modification. On the other hand, the contractor is not a watchdog to ensure that architects carry out their duties properly and efficiently.

If the contractor finds any divergence between the statutory requirements and any of the contract documents or architect's instructions, the architect should immediately be informed in writing, pointing out the discrepancy. The architect must then, within seven days of receiving this notice, issue instructions in relation to the divergence. If the instruction requires the work to be varied then this should be treated in the same way as other variations, in accordance with clause 13.

In some circumstances it may be necessary for the contractor to comply urgently with a statutory regulation; for example, where an existing structure on the site is in danger of collapsing or where health is in danger. This may require the contractor to supply materials or execute work prior to receiving instructions from the architect. The contractor should do what is reasonably necessary to secure the immediate compliance with the statutory requirement. The contractor must then inform the architect of this action. The materials and work executed in these circumstances will then be treated as a variation under clause 13.

The contractor will not be liable to the employer for work carried out in accordance with the contract that is subsequently found not to be in compliance with any statutory requirements.

Divergence – statutory requirements and the contractor's statement

Where a divergence from statutory requirements and the contractor's statement is discovered, then the contractor must inform the architect. The contractor must then rectify the work at the contractor's own expense.

Change in statutory requirements after base date

If after the base date there are changes in the statutory requirements that require some alteration or modification to the works then these will be treated as a variation under clause 13.2.

Fees or charges

The contractor must pay and indemnify the employer against liability in respect of any fees, charges, rates, taxes, etc. These may be required under an Act of Parliament, regulation or by-law. The amount of such fees and charges will be added to the contract sum unless:

- They arise in respect of goods or work done by a local authority or statutory undertaker
- They are included in the contract sum by the contractor
- They are included as a provisional sum in the contract bills

Exclusion of provisions on domestic subcontractors and nominated subcontractors

The provisions of clause 19 (assignment and subcontracts) and clause 35 (nominated subcontractors) do not apply to the execution of part of the works by a local authority or statutory undertaker. When executing work solely in pursuance of their statutory obligations, these bodies are not considered as subcontractors within the terms of the contract.

PROVISIONS FOR USE WHERE THE APPENDIX STATES THAT ALL THE CDM REGULATIONS APPLY (CLAUSE 6A)

Employer's obligations – planning supervisor – principal contractor

The CDM Regulations are the Construction (Design and Management) Regulations 1994 and their subsequent amendments. The employer shall ensure:

- That the planning supervisor (defined as an architect or other person named in article 6.1) carries out the relevant duties under the CDM Regulations.
- Where the contractor is not the principal contractor, that contractor carries out the duties of the principal contractor.

Contractor – compliance with duties of a principal contractor

The contractor, as the principal contractor, must ensure that the health and safety plan has the features required by regulation 15(4) of the CDM Regulations. Any amendment to this plan by the contractor shall be notified to the employer, who in turn will notify the planning supervisor and the architect. The contractor should also ensure the compliance with this provision of the subcontractors working on the site.

Health and safety file

The principal contractor must maintain, within regulations 14 of the CDM Regulations, a health and safety file for the project.

LEVEL AND SETTING OUT OF THE WORKS (CLAUSE 7)

The architect is responsible for providing the contractor with all the information necessary in order that the contractor can set out the works at ground level. This information should comprise properly dimensioned drawings and levels. The actual setting out of the works from this information is entirely the responsibility of the contractor. If mistakes are made during this process, the contractor must correct them without cost or charge to the employer. Delays in the presentation of the information from the architect could result in an extension of time. One possible problem could arise from the inaccurate setting out of the works by trespassing on the land of an adjoining owner. In this situation the contractor would be liable to indemnify the employer under clause 20. With the consent of the employer, the architect may instruct that such errors will not be amended and an appropriate deduction for such errors will be made from the contract sum.

WORK, MATERIALS AND GOODS (CLAUSE 8)

Kinds and standards, etc.

This clause seeks to ensure that the quality of materials and standard of work will be as specified in the contract bills. It therefore reinforces the information contained in the bills of quantities. This includes the performance-specified work from the contractor's statement. All the work must be carried out in a proper and workmanlike manner and in accordance with the health and safety plan.

Substitution of materials or goods – performance-specified work

Where the quality specified becomes no longer possible to obtain then, before proceeding with some alternative, the contractor should seek instructions from the architect. The contractor must, whenever required to do so, prove that the materials being used are in accordance with the specified requirements (clause 2.1). This may be done by providing the architect with applicable quotations and invoices. The contractor shall not substitute any materials or goods contained in a performance specification without the architect's consent.

Vouchers – materials and goods

The architect may require documentary proof that the materials are in accordance with those that have been specified. The contractor may need to produce vouchers such as invoices for this purpose.

Executed work

Work that does not comply with the contract will be rejected by the architect. This can include an incorrect choice of materials by the contractor or standards of work that do not comply with the standards stated in the contract documents. This dissatisfaction by the architect must be expressed within a reasonable time from the execution of the unsatisfactory work.

Inspection – tests

The architect may issue instructions to the contractor regarding tests on materials and work. The tests expected will be those that have been described in the contract bills and may involve the contractor in testing the materials, goods or work. If the architect requires tests to be performed that have not been specified, then these must be carried out by the contractor, but will form an additional charge to the contract sum. Routine tests at regular intervals are generally described adequately and comprehensively in the contract bills. The contractor must therefore expect that some of these tests will be required. It may be useful, in other circumstances, to include a provisional sum for any unusual testing in case this is required.

The architect also has the powers to request the inspection or test of any part of the works up to the issue of the final certificate. This may require the contractor to open up the works for inspection or testing. The purpose of this is to ensure that the work is in accordance with that specified in the contract. The costs of opening up the works for this purpose will be borne by the contractor if the inspection or tests show that it is not in accordance with the contract. Where it is in agreement to that specified, then the costs of opening up and making good will be added to the contract sum. This may also result in other repercussions such as an extension of time (clause 25.4.5.2) or a claim for loss and expense (clause 26.2.2).

Powers of architect – work not in accordance with the contract

The architect may also issue instructions in regard to the removal from site of materials that do not conform with the contract. In other circumstances, once the materials arrive on site they cannot be removed without the architect's permission (clause 16). The materials that are removed must be replaced with materials that do conform, and at no extra expense to the contract. There is no limit in time when such defective materials should be discovered. The architect will, however, seek to inspect the works reasonably as they progress. If, however, a failure in the foundations did not occur until the building was suitably loaded, and such a failure was due to the use of defective materials, the contractor would be liable fully to rectify the problem to the satisfaction of the architect. (Note article 1 of the articles of agreement.) The architect should, however, express any dissatisfaction within a reasonable time from the execution of the unsatisfactory work (clause 8.2).

If any of the work, materials or goods is not in accordance with the contract then the architect may:

- Issue instructions for its removal.
- Allow such work to remain and confirm this in writing to the contractor. This is not to be construed as a variation, but will probably result in an appropriate deduction from the contract sum.
- Issue instructions requiring a variation but allow no addition to the contract sum or an extension of time.
- Issue instructions for opening up inspection using the following code of practice.

A code of practice has been written to assist in the operation of clause 8.4.4. The architect and contractor should agree to the amount and method of opening up or testing in accordance with the following criteria:

- The event of the non-compliance is unique and is unlikely to happen again.
- The need to discover whether any non-compliance in a primary structural element is a failure of work standards and/or materials such that rigorous testing of similar elements must take place. Where the non-compliance is less significant then it can simply be repaired.
- The significance of the non-compliance having regard to the nature of the work in which it has occurred.
- The consequence of any similar non-compliance on the safety of buildings, its effect on users, adjoining property, the public, and compliance with any statutory requirements.
- The level and standard of supervision and control of the works by the contractor.
- The relevant records of the contractor and any relevant subcontractor resulting from the supervision and control referred to above.
- Any codes of practice or similar advice issued by a responsible body which are applicable to the non-complying work, materials or goods.
- Any failure by the contractor to carry out, or to secure the carrying out of, any tests specified in the contract documents or in an instruction to the architect.
- The reason for the non-compliance when this has been established.
- Any technical advice that the contractor has obtained in respect of the non-complying work, materials or goods.
- Currently recognized testing procedures.
- The practicability of progressive testing in establishing whether any similar non-compliance is reasonably likely.
- If alternative testing methods are available, the required time and the consequential costs for such alternative testing methods.
- Any proposals of the contractor.
- Any other relevant matters.

Powers of architect – non-compliance with clause 8.1.3

Where the contractor fails to carry out the work in a proper and workmanlike manner and in accordance with the health and safety plan (clause 8.1.3), then the architect may issue instructions to ensure compliance. Such instructions will not result in any addition to the contract sum and no extension of time will be given.

Exclusion from the works of persons employed thereon

The architect may also issue instructions excluding anyone from the site. This must not be done unreasonably or to the specific annoyance of the contractor. The contractor should therefore have been previously warned of this impending situation, in order that appropriate corrective action may be taken. It is an uncommon occurrence, but it may arise in circumstances where work standards are persistently bad or where a member of the contractor's staff has consistently failed to carry out the architect's instructions as requested.

ROYALTIES AND PATENT RIGHTS (CLAUSE 9)

Treatment of royalties, etc. – indemnity to the employer

All royalties and other similar sums included for work described in the contract's bills are deemed to be included in the contract sum by the contractor. The contractor must also indemnify the employer against any actions that might be brought in connection with royalties. The contractor may, for example, have used a patented system of scaffolding, the royalties of which are entirely the contractor's responsibility.

Architect's instructions – treatment of royalties, etc.

If, however, the contractor is requested to use a patented system under an architect's instruction, the royalties that may be payable in these circumstances will be added to the contract sum. When agreeing the value of variations such royalties would then be included.

PERSON IN CHARGE (CLAUSE 10)

The contractor must keep a site manager constantly on site. This is the person who is responsible for the daily running of the site, and to whom the architect can give instructions, or to whom the clerk of works can give directions. In the construction industry several names are used from project manager to general overseer. A contracts manager, however, who is often responsible for several sites, will not easily fit within this description since such a person is not 'kept up on the works'. The instructions given by the architect to the person in charge will be deemed to have been issued to the contractor. In order to satisfy this last requirement the person in charge may delegate an assistant who can receive instructions in their absence. The term 'constantly on site' means at all reasonable times, to ensure that the work is correctly executed and to receive any instructions from the architect.

ACCESS FOR ARCHITECT TO THE WORKS (CLAUSE 11)

The architect and the architect's representatives have the right of access, at all reasonable times, to the construction site and the contractor's workshops. This clause is extended to cover the workshops of domestic and nominated subcontractors. The term 'architect and the architect's representatives' can also, presumably, be extended to quantity surveyors and engineers of different kinds. In some circumstances, for example, it may be important for the structural engineer to be allowed access to the fabrication process of the steelworker. The purpose of this clause, therefore, is largely to inspect the quality control of the firm concerned. It should be noted, however, that suppliers are not listed here, since their contract is a contract of sale of completed goods, and inspection may be inappropriate, particularly in the case of mass-produced goods.

Due to the increasing amount of specialist work included in buildings, provision is now necessary and is provided to protect the proprietary interests of the contractor or subcontractor, whilst not detracting from the right of access given to the architect. This might be achieved through confidentiality agreements or limiting the inspections of specialist manufacturing processes where some form of indemnity could be provided.

CLERK OF WORKS (CLAUSE 12)

The employer has the option of appointing a clerk of works whose duty is to act 'solely as an inspector' on behalf of the employer. The clerk of works is employed for this purpose under the directions of the architect. The contractor should provide the clerk of works with all reasonable facilities in order that the designated duties can be performed. The clerk of works is not employed in the capacity of the architect's representative, since employment is solely for inspection purposes. If the clerk of works chooses to give the contractor instructions, even in writing, they will have only little contractual effect unless they are confirmed by the architect. This must be done within two working days of the instruction, and must be in writing, which will then constitute an architect's works instruction. If the architect chooses to confirm a clerk of work's suggestion orally, then the appropriate procedures to put this in writing must then be taken up by either the contractor or the architect (clause 4). The clerk of works is generally resident on site and in constant touch with the works, in circumstances that often require a quick decision. In practice, therefore, the contractor must receive many statements from the clerk of works in good faith.

PERFORMANCE-SPECIFIED WORK (CLAUSE 42)

Meaning of performance-specified work

The term 'performance-specified work' means:

- Identified in the appendix.
- Which is to be provided by the contractor.
- For which certain requirements have been predetermined and are shown on the contract drawings.
- In respect of which the performance which the employer requires from such work, and which is included in the bills of quantities as one of these two items:
  - work that the contractor has priced
  - work included in the contract as a provisional sum

Contractor's statement

Practice Note 25 (JCT 80) describes the sort of work which is not to be treated as performance-specified work. This is still relevant under JCT 98. Before carrying out such work the contractor will provide the architect with documents referred to as the 'contractor's statement'. The work is then carried out in accordance with the terms and conditions of this approved statement. The statement should be sufficiently detailed to adequately explain the contractor's proposals. This may include drawings and schedules. The date for the receipt of the statement may be stated in the bills or provided by means of an architect's instruction. It is usual before the contractor formally submits the statement to the architect that the draft is referred to the planning supervisor.

Architect's notice to amend contractor's statement

If the contractor's statement is deficient then the architect can require this to be amended. Such a deficiency, for example, could adversely affect the performance of the work as required by the employer. The architect should immediately give notice to the contractor specifying any deficiency.

Definition of provisional sum for performance-specified work

A provisional sum for this type of work can be included in the bills of quantities if the following information is provided:

- The performance which the employer requires from such work.
- The location of such work in the building.
- Sufficient information has been provided to allow for:
  - the contractor's programme to be prepared
  - the relevant preliminaries for adequate pricing

Preparation of contract bills

If in the contract bills there are errors or omissions of information in connection with this work then they are to be corrected by a variation order issued by the architect. The architect also has powers within the terms of the contract to issue instructions that may vary this work from what was originally envisaged. Where the bills of quantities do not provide for an analysis of this type of work, the architect can request such an analysis which must then be provided within fourteen days.

The contractor should exercise reasonable skill and care in the provision of this work, maintaining the general obligations in respect of the supply of work standards, materials and goods. Performance-specified work should not be implied to assume a guarantee of fitness for purpose. This clause specifically excludes the provision of this work under a nominated subcontract.

Performance-specified work – contractor's obligations

The contractor must exercise reasonable skill and care in the provision of performance-specified work. It must not be construed to affect the obligations of the contractor under the contract in respect of materials, goods and work standards. Nothing in the contract will operate as a guarantee of fitness for purpose of the performance specified work.

CHAPTER 19

COSTS OF CONSTRUCTION

An important factor in any construction contract is the calculation and payment of the sum of money by the employer to the building contractor. The importance of the contract sum and its relationship to the final account, together with the matters relating to the timing of payments, are discussed in this chapter.

- Clause 3   Contract sum – additions or deductions – adjustment – interim certificates
- Clause 13   Variations and provisional sums
- Clause 13A   Variation instruction – contractor's quotation in compliance with the instruction
- Clause 14   Contract sum
- Clause 16   Materials and goods unfixed or off site
- Clause 24   Damages for non-completion
- Clause 26   Loss and expense caused by matters materially affecting regular progress of the works
- Clause 30   Certificates and payments

The project starts with the contract sum and the integrity of this amount is clarified within the contract conditions. It is extremely unusual for this amount to remain unchanged and be paid as the final account upon completion. Changes beyond the contractor's control will be made, and these must eventually be costed to the appropriate amount. Changes to the contract sum occur in respect of the following items:

- Substructure items, which often cannot be fully determined until the contractor starts work on site
- Remeasurement of provisional sums
- Adjustments to the accounts of nominated subcontractors and suppliers
- Variations to the design instigated by the employer or architect, and occasionally by the building contractor
- Inclusion of daywork accounts
- Assessment of any contractual claims or *ex gratia* payments

Rules are laid down on how these changes are to be evaluated in terms of their costs. In those circumstances where contractors consider that they have been insufficiently reimbursed, there is provision in the contract to rectify this, should their case be proven. In other circumstances the employer may have suffered damages due to a late completion by the contractor. Adjustments to the final account for this, whilst uncommon, are also provided for within the terms of the contract. It is uncommon because the contractor is usually able to identify mitigating circumstances for late completion of the works. On the majority of contracts of any size, contractors are unable to complete the works without some form of interim payment or payments on account. Clause 30 describes the procedures involved for calculating the amounts owing to the contractor both during the period of construction and also upon final completion of the contract. Matters covering valuation periods, retention and the time allowed for payment to be made are also included in this clause.

CONTRACT SUM – ADDITIONS OR DEDUCTIONS – ADJUSTMENT – INTERIM CERTIFICATES (CLAUSE 3)

If the conditions refer to the addition to or deduction from the contract sum, then once this adjustment has been calculated it shall be paid and included in the value of the next interim certificate. Several clauses refer back to this clause in respect of extra expense caused by changes to the contract. Some of these items are as follows:

- Clause 6   Statutory obligations, notices, fees and charges
- Clause 7   Levels and setting out the works
- Clause 8   Opening up for inspection and testing of the works where found to be correct
- Clause 9   Royalties and patent rights
- Clause 17   Repair of defects, the costs of which are accepted by the architect
- Clause 21   Certain insurance matters

VARIATIONS AND PROVISIONAL SUMS (CLAUSE 13)

The vast majority of forms of contract used in the UK and in different countries around the world allow for variations arising at some stages during the project. The absence of such a clause within the contract conditions would necessitate a new contract being arranged if variations did arise. The disadvantage of such a clause is that it allows the architect, or other designers, to delay making some decisions almost until the last possible moment. This can have serious repercussions on the planning of the project and in executing the works efficiently.

The building owner will be bound by any variation given by the architect as long as the architect does not exceed the powers under the terms of the contract.

The question sometimes arises as to what constitutes a variation. Clause 13 aptly describes those circumstances where a variation can arise, but it must be presumed and implied that a limit on these change orders will apply in practice. If the nature of the works described in the contract documents can be shown to be 'substantially different' from that carried out then the main contractor will be entitled to reprice the work accordingly. The phrase 'substantially different', however, may require the courts to formulate a decision in individual cases. A variation resulting in a change in the size of a window will be admissible. The increase in the size of a proposed extension from 200 to 2,000 square metres of floor space will probably not be described as a variation, since the nature of the works has changed considerably from what was originally envisaged. The contractor may be willing to carry out the work, but on a different contractual basis.

Definition of variation

Clause 13.1 attempts to define a variation in normal circumstances. A variation from the contract occurs where the actual work to be carried out changes, or where the circumstances in which the work is to be carried out changes. The first category includes the alteration or modification of the design, quality or quantity of works:

- The addition, omission or substitution of any work
- The alteration of the kind or standard of materials or goods
- The removal from site of work, materials or goods that were formerly in accordance with the contract, but which have now been changed

A change in the circumstances in which the work is carried out also comes within this definition and includes:

- Access and use of the site
- Limitations of working space
- Limitations of working hours
- Changes made to the sequencing of work

Clause 13.1.3 places some restrictions on the way the architect can go about issuing variation instructions. Work that has been described in the contract bills as the main contractor's (or one of the domestic subcontractors should the intention have been to sublet it) cannot be omitted and then awarded to a nominated subcontractor. It can be omitted entirely and done after the contract has been completed. It can also be awarded to a firm employed directly by the building owner, assuming that the main contractor will allow access to the works, for this purpose, during the contract period. Of course, if the main contractor agrees to this nomination then this will override the condition.

Instructions requiring a variation

The architect may issue instructions requiring a variation. The contractor has the right to the reasonable objection to such as set out in clause 4.1.1. The valuation of these variations will adopt the normal rules of 13.4.1.1 unless the contractor's price statement method is to be used as described in clause 13.4.1.2. The architect may sanction, in writing, any variation made by the contractor. This may be done for expediency. In practice contractors should not vary the works without the express permission of the architect.

All instructions from the architect requiring a variation from the contract must be in writing. This is in pursuance of clause 4.3. The quantity surveyor has no power to measure and value varied work without this written instruction (clause 13.4). Any variation made by either the architect or the contractor (clause 13.2.4) can be confirmed in writing at any time prior to the issue of the final certificate (clause 4.3.2.2).

Instructions on provisional sums

Instructions are also required from the architect in the expenditure of provisional sums in the contract bills and also in nominated subcontracts (clause 13.3).

Valuation of variations and provisional sum work and work covered by an approximate quantity

These rules apply to each of the following circumstances:

- All variations required by an instruction of the architect or subsequently sanctioned in writing.
- All work which under the contract is to be treated as a variation under clause 13.2.
- All work executed by the contractor in accordance with architect's instructions regarding the expenditure of provisional sums.
- All work executed by the contractor for which an approximate quantity has been included in the contract bills.

The valuation of variations is to be in accordance with one of these two alternatives.

Alternative A: contractor's price statement

The contractor may within 21 days from the receipt of an architect's instruction or from the commencement of work for which an approximate quantity is included in the contract bills prepare a price statement.

The price statement will state the contractor's price for this work based upon the rules for the valuation of variations (Figure 19.1). The price statement may include direct loss and expense costs (clause 26) or costs due to an adjustment where the work cannot be completed by any revised completion date. The quantity surveyor, after consultation with the architect, must notify the contractor whether the price statement has been accepted. Where the price statement is not accepted, the contractor must be given reasons for its rejection and an amended price statement prepared by the quantity surveyor. Where the two parties cannot agree then the matter is dealt with under clause 41A (settlement of disputes).

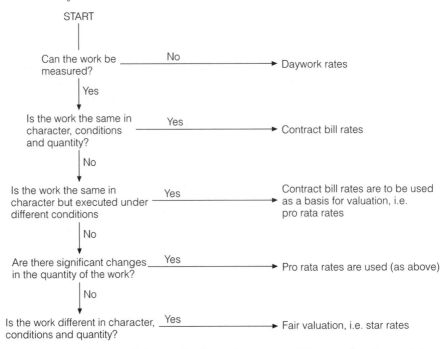

Fig. 19.1 Valuation of variations: omitted work is at contract bill rates unless the remaining work changes substantially in quantity, in which case a revaluation of these items becomes necessary

Alternative B: valuation rules

The following situations are illustrated in Figure 19.1:

- Where the additional or substituted work is the same in character, conditions and quantity to items in the contract bills, then bill rates or prices are to be used to value the variation.
- Where the additional or substituted work is the same in character, but is executed under different conditions or results in significant changes in quantity, then the bill rates or prices are to be used as a basis for valuing the variation. These are known as pro rata rates.
- If the additional or substituted work is different from those items in bills, then a fair method of valuation is to be used, often known as star rates.
- Where the approximate quantity is a reasonably accurate forecast of the quantity of work then the rate or price quoted for the approximate quantity will be used.
- Where the approximate quantity is not a reasonably accurate forecast of the quantity of work then the rate or price quoted for the approximate quantity will be used as a basis and a fair allowance for such differences will be added.
- Omissions are valued at bill rates unless the remaining quantities are substantially changed, in which case a revaluation of these items will become necessary.

The rules for measurement state that the same principles used for the preparation of the contract bills should be adopted (clause 13.5.3.1). If any percentage or lump sum adjustments have been made in the contract bills – perhaps for the correction of errors (clause 14) – then they will need to be adjusted within the variation account. Also, if it can be shown that the value of any preliminary items have changed as a result of variations, then these also will require adjustment. Preliminary items are frequently priced as either lump sums, time-related or method-related charges. Their adjustment will generally follow this pattern of recalculation.

Amendment 7 to JCT 80, issued in July 1988, extended these rules to cover approximate quantities in contract bills. If the approximate quantities are not a reasonable forecast, then the rates used in the bills will only be used as a basis for revaluation. At the extremes, i.e. where the approximate quantity does not change or where it differs by, say, 100 per cent then the above rules can easily be applied. In the grey areas, between these extremes, the test of reasonableness will need to be applied by the parties concerned. This principle only applies where the work as executed is not altered or modified other than by quantity alone. The architect does not need to give specific instructions in respect of the execution of work for which an approximate quantity is included in the contract bills. This principle is also extended to cover nominated subcontractors' work where this is appropriate.

If the varied work cannot be properly measured or valued, then the contractor or subcontractor is to be paid for the work on a daywork basis. The documents to use are the schedule *Definition of prime cost of daywork carried out under a building contract*, issued by the Royal Institution of Chartered Surveyors and the Building Employers Confederation (now the Construction Confederation), together with the percentage additions determined by the contractor in the bills. Alternatively, if the work is within the province of any specialist trade, defined as either electrical or heating and ventilating, then their appropriate definition for daywork together with the percentages inserted by the contractor in the bills should be used. Vouchers showing the daily time spent on the work, the workers' names and the plant and materials employed should be verified. This is recommended to be done weekly by the architect's representative and this is interpreted to be the clerk of works. The reason for it being done on a weekly basis is to avoid forgetfulness by either party. The verified voucher is an agreed record only; it does not mean that the method of valuation adopted will be daywork. This is a matter for the quantity surveyor to determine.

The valuation of a provisional sum follows a similar process to the valuing of the main contractor's work described in the contract bills. For example, if a provisional sum had been included in the bills for 'repairs to timber staircase', and it was agreed that daywork should apply, then the above rules for valuing daywork would be used. The term 'provisional sum' has also taken on a redefined meaning in accordance with the provisions of SMM7. This is a sum provided for work being identified as defined or undefined. General rule 10 (Box 19.1) from SMM7 should be noted in this context.

Box 19.1 Extract from SMM7

Procedure where the drawn and specification information required by these rules is not available

10.1 Where work can be described and given in items in accordance with these rules but the quantity shall be given and identified as an approximate quantity.

10.2 Where work cannot be described and given in items in accordance with these rules it shall be given as a Provisional Sum and identified as for either defined or undefined work as appropriate.

10.3 A Provisional Sum for defined work is a sum provided for work which is not completely designed but for which the following information shall be provided:

(a) The nature and construction of the work.

(b) A statement of how and where the work is fixed to the building and what other work is to be fixed thereto.

(c) A quantity or quantities which indicate the scope and extent of the work.

(d) Any specific limitations and the like identified in Section A35.

10.4 Where Provisional Sums are given for defined work the Contractor will be deemed to have made due allowance in programming, planning and pricing Preliminaries. Any such allowance will only be subject to adjustment in those circumstances where a variation in respect of other work measured in detail in accordance with the rules would give rise to adjustment.

10.5 A Provisional Sum for undefined work is a sum provided for work where the information required in accordance with rule 10.3 cannot be given.

10.6 Where Provisional Sums are given for undefined work the Contractor will be deemed not to have made any allowance in programming, planning and pricing Preliminaries.

The valuation of performance-specified works follows the similar procedure of valuing construction work generally. The valuation may need to take into account the preparation and production of drawings, schedules or other documents.

The contractor in pricing the work at tender stage will be deemed to have made due allowance in pricing for the preliminary items. If the execution of any of this work is subject to variation then an adjustment may need to be made to the value of the preliminary items. The repayment of any direct loss or expense resulting from a variation issued by the architect would be reimbursed and discussed under clause 26.

Contractor's right to be present at measurement

Variations usually necessitate the remeasurement of the work on site or from drawings. Clause 13.6 gives the contractor the right to be present at the time that such measurements are taking place. It is important that these are agreed by both of the parties prior to the work being priced in accordance with Figure 19.1.

VARIATION INSTRUCTION – CONTRACTOR'S QUOTATION IN COMPLIANCE WITH THE INSTRUCTION (CLAUSE 13A)

Contractor to submit quotation (13A quotation)

The valuation of variations will normally follow the procedures of clause 13.4.1 unless the architect's instruction states that the treatment and valuation of the variation should be dealt with under clause 13A (clause 13.2.3). This clause was added to JCT 80 through Amendment 13 (January 1994). Clause 13A can also apply to instructions that may be given to nominated subcontractors.

The application of this clause will only apply to an architect's instruction where the contractor agrees with this method of valuation. Where the contractor disagrees, then this must be notified in writing to the architect within 7 days of receiving the instruction. Valuation rules 13.4.1 will then be used as the basis for valuing the variation.

Content of contractor's 13A quotation

Clause 13A allows the contractor to provide a quotation for a variation, in advance of the work being carried out. The quotation should be based upon:

- Where relevant, the rates and prices contained in the contract bills
- Where appropriate, the adjustment of preliminary items
- Any adjustment in time required for completion of the works
- Any amount paid in lieu of direct loss and expense under clause 26.1
- A fair and reasonable amount in respect of preparing the quotation

The architect's instruction may also require the contractor to identify any additional resources that may be required and the method of carrying out the variation. The quotation should provide sufficient information to allow the work to be properly evaluated on behalf of the employer.

If a quotation has been requested and the work is then not proceeded with in this way, then the contractor is to be allowed to add to the contract sum the reasonable costs involved of preparing the quotation.

Where a clause 13A quotation has been accepted, any subsequent variations issued by the architect that may affect this work, will then be valued by the quantity surveyor on a fair and reasonable basis and not by using the rules of clause 13.5.

The effect of the valuation of variations is that such sums are either added to or deducted from the contract sum.

CONTRACT SUM (CLAUSE 14)

Quality and quantity of work included in contract sum

The contract sum is assumed to have been calculated on the basis of the quality and quantity of work as described in the bills of quantities.

Contract sum only – adjusted under the conditions – errors in computation

A part of the quantity surveying process on any project is to establish the correctness of the contract sum both arithmetically and technically. This means that the bills should have been checked to see that the item costs have been extended correctly, that page totals add up and that the various collection sheets and summaries have been correctly completed. The adequacy of the contractor's rates should also have been examined in order to avoid difficulties that might arise during the agreement of the final account. However, failure on the part of the quantity surveyor to perform these duties efficiently will not provide the contractor with any redress against the building owner. It is the contractor's responsibility to determine the sufficiency of the tender, with no alternative but to honour the pricing should any errors occur at a future date during the contract. Any errors discovered by the quantity surveyor or the contractor prior to the signing of the contract can be corrected by using one of the alternative methods described in the code of procedure for selective tendering (Chapter 8). It is prudent, therefore, on the part of the quantity surveyor to be satisfied on the nature of the prices prior to the signing of the contract.

The only errors that can be corrected are those made by the quantity surveyor during the preparation of the contract bills (clause 2.2). These may have occurred because of errors in the quantities or descriptions or because items have been omitted. If the bills depart from the method of measurement stated, and this has not been brought to the contractor's notice, then this will also be deemed an error that will need to be corrected.

MATERIALS AND GOODS UNFIXED OR OFF SITE (CLAUSE 16)

Unfixed materials and goods – on site

Materials and goods which have been delivered to the site or placed on or adjacent to works, and are intended for the works, must not be removed from the site unless written consent has been received from the architect. Once the value of these goods and materials has been included in an interim certificate and paid for by the employer, they become the property of the employer. There has been considerable case law on this point. However, the main contractor remains responsible for any loss or damage they may suffer and for the insurance of the same, as long as clauses 22B and 22C are in operation. Although the employer becomes the true owner of the materials and goods concerned, the contractor's insolvency may raise the question of whether the contractor was the reputed owner. The employer would need to ensure that the contractor's title to the goods was thus not defective.

Unfixed materials and goods – off site

The value of listed items of materials or goods intended for the works which are stored off site may be included in an interim certificate and paid for by the employer. These listed items of materials and goods then become the property of the employer. The listed items are referred to clause 30.3 and refer to the list supplied to the contractor and annexed in the contract bills. The contractor must not remove these items from the premises, usually of manufacture, except to the project for which they are intended. The contractor, however, continues to be responsible for any loss or damage that may occur, and for their insurance in accordance with clauses 22B and 22C. In these circumstances the architect has exercised an option under clause 30 to pay for materials off site. This clause has particular relevance and importance to factory-manufactured components, used with industrialized buildings. In these situations the contractor may be loath to bring to site items of manufactured joinery until their incorporation in the works is imminent. The option of paying for these materials will assist the contractor's cash flow, and should cause no problems if the provisions of clause 30 are properly complied with. The list will only normally refer to manufactured items that are specific to the project concerned and require only to be fixed or placed in position.

DAMAGES FOR NON-COMPLETION (CLAUSE 24)

Certificate of the architect

If the contractor fails to complete the works by the completion date, the architect must issue a certificate accordingly.

Payment or allowance of liquidated damages

This certificate, issued under clause 24.1, allows the employer to deduct from monies due to the contractor the liquidated and ascertained damages at the rate stated in the appendix. The amount of damages claimed is based upon the period between the completion date and the date of practical completion. If there is an insufficient amount owing to the contractor by which to offset the damages, the employer is able to recover the balance as a debt.

The amount stated in the appendix for liquidated and ascertained damages must be a realistic sum related to possible actual damages suffered by the employer. If the amount stated is shown to be a penalty then the courts will set a fair, but smaller amount, as damages payable by the contractor.

If the architect subsequently fixes a later completion date then any amounts deducted by the employer are to be adjusted and repaid to the contractor as necessary.

LOSS AND EXPENSE CAUSED BY MATTERS MATERIALLY AFFECTING REGULAR PROGRESS OF THE WORKS (CLAUSE 26)

Matters materially affecting regular progress of the works direct loss and expense

The main purpose of this clause is to reimburse the contractor in those circumstances where loss and expense have been suffered, and will not be reimbursed elsewhere under the terms of the contract. The loss and expense claim must be directly attributable to matters that have substantially affected the regular progress of the construction of the project.

In order for the contractor to claim for reimbursement a written application must first be made to the architect. In this correspondence it must be stated that direct loss and expense have been, or are very likely to be, incurred during the execution of the works, and that adequate reimbursement will not be made under any other provision of the contract. The basis of claim is that the regular progress of the works has been disturbed and this has resulted in the loss and expense. Loss in this context means that the contractor has been inadequately reimbursed for work carried out. Expense implies that additional resources were necessary to complete the works. The contractor should:

- Make the application as soon as it becomes apparent that the regular progress of the works has been or will be affected.
- Support the application with sufficient information to allow the architect to form an opinion.
- Submit to the architect or quantity surveyor details of the financial loss or expense.

List of matters

The conditions of contract list various matters that could affect the regular progress of works. The contractor must cite one or more of the following items in support of the claim for loss and expense:

- Delay in the receipt of instructions (including the expenditure of provisional sums), drawings, details or levels from the architect. The contractor must have requested this information in writing from the architect, and have allowed sufficient time to elapse prior to its use on site.
- Opening up of work for inspection or testing of materials and for their consequential making good. This is providing the items were in accordance with the work specified.
- Discrepancy or divergence between the contract drawings, contract bills and numbered document.
- Work being carried out by firms employed directly by the employer. This also includes the supply of materials and goods by the employer for the project.

- Postponement of any of the work to be executed under the provisions of the contract.
- Failure on the part of the employer to give ingress to or egress from the works by the appropriate time.
- Variations authorized by the architect under clause 13.2 or the expenditure in regard to provisional sums for performance-specified work.
- The execution of work for which an approximate quantity is included in the contract bills which is not a reasonably accurate forecast of the quantity of work required.
- Compliance or non-compliance by the employer with clause 6A (provisions for use where the appendix states that all the CDM Regulations apply).
- Suspension by the contractor of the performance of the contractor's obligations under clause 30.1.4 of the contract (interim certificates), providing the suspension was not frivolous or vexatious.

Relevance of certain extensions of completion date

Upon receipt of the correspondence from the contractor, the architect must then assess whether the claim is justified. First, it will be necessary to assess that the regular progress of the works had been disturbed by one or more of the relevant events listed above. Reference will need to be made to the contractor's master programme for this purpose. The mere fact that the project is running behind schedule may also be due in part to the contractor's inability to perform the works adequately. Some allowance on the period of delay may therefore need to be taken if this assumption is proved to be correct. Second, either the architect or the quantity surveyor will need to ascertain the amount of the loss and expense incurred by the contractor. In practice the architect may agree that the contractor has a valid case, but leave the settlement of the financial sum to the quantity surveyor.

The contractor's written application should be made as soon as it is apparent that the regular progress of the works has been affected. Relevant information should be submitted to the architect in support of the claim that will help the architect form an opinion. The contractor should also provide some indication of the amount of loss and expense that has been suffered. If any extension of time has already been agreed by the architect under the appropriate items of clause 25 then the contractor shall be informed accordingly. This may affect the calculation and assessment of the contractor's claim.

Nominated subcontractors

A nominated subcontractor can also submit a claim for the loss and expense caused by the disturbance of the regular progress of the subcontract works. Such a claim must first be submitted to the contractor. If it is appropriate then the contractor will pass it to the architect for proper consideration. This claim will then follow a similar process, as described above.

Amounts ascertained – added to the contract sum

Any amounts that the architect agrees should be paid to the contractor for a claim under this clause are to be added to the contract sum. If they can be agreed during the progress of the works, they can then be included in the next valuation.

Reservation of rights and remedies of contractor

The inclusion of this clause within the form of contract provides only one remedy for the contractor. Other courses of action are therefore open to the contractor in the event of the regular progress of the works being affected. Other actions the contractor may wish to take, if they are to succeed with the payment of damages, must not be too remote in the eyes of the law. The loss of profit is, however, a reasonable claim to be made under this heading. The provisions of clause 26 are without prejudice to any other rights and remedies which the contractor might possess.

CERTIFICATES AND PAYMENTS (CLAUSE 30)

Interim certificates and valuations

Interim certificates are issued by the architect. Although these may be required at more frequent intervals by the architect (clause 30.1.2), the contract will stipulate a maximum period for interim certificates. The appendix to the form of contract, which is completed for each project, recommends this period to be one month, although the parties do have the option of stating any period that they desire. The period of one month is common in practice. The ICE conditions of contract also stipulate that the valuation must achieve a minimum amount (to be stated in the contract) before any monthly payment will be made. There is no comparable condition in JCT 98, although if this were desirable then it could be inserted as a supplementary condition. The purpose of such a recommendation is to attempt to ensure that the contractor makes regular progress of the works. On very large contracts it may be necessary to provide approximate valuations on a weekly basis in order to satisfy the contractor's cash flow. These would be stated in the appendix to the contract. More accurate payments would continue to be made at the monthly interval. This may be required at peak construction periods, for example during the summer months on mass earth-moving contracts. In some contracts a stage payment system (clause 30.2) may be used as a preference.

The certificate will state:

- The amount due to the contractor from the employer
- To what the amount relates
- The basis on which the amount was calculated
- The final date for payment
  - 14 days from the issue date of each interim certificate

The employer must within 5 days after the issue of an interim certificate give written notice to the contractor specifying the amount of the payment, to what the payment relates and the basis on which the amount is calculated. Not later than 5 days before the final date for payment the employer may give a written notice to the contractor specifying any amounts that are to be withheld or deducted from the due amount and the grounds for deducting the amount. Where the employer does not provide any written notices to the contractor then the amount stated in the interim certificate must be paid to the contractor.

Final date for payment – interest

Where an employer fails to pay the approved amount within the stipulated period of time for honouring the certificate then interest will be added to the amount outstanding. The interest is calculated on a simple interest basis at a rate of 5 per cent above the Bank of England base rate. The rate of interest is not construed as a waiver by the contractor in lieu of the correct payment at the correct time. The contractor still has the right to suspend the performance of the contract or to determine the employment under clause 28.2.1.1 (determination by contractor or employer).

Advance payment

On some contracts it may be appropriate to make an advance payment to the contractor. This may be necessary where the project start-up costs are excessive and would not normally be recouped quickly by the contractor. The terms are stated in the appendix to the form of contract. The contractor is required to provide an advance payment bond as a surety. This will be approved by the employer and on terms agreed between the British Bankers Association and JCT.

Interim valuations

Interim valuations are usually prepared by the client's quantity surveyor in agreement with the contractor's surveyor. The valuations will be prepared on a monthly basis (or on whatever basis has been stated in the appendix) until the certificate of practical completion is issued. If the work has been accurately valued at this point, then only retention monies will be outstanding and these can therefore be released with the final certificate. There will therefore be no requirements to issue a certificate between these two certificates. The certificate of practical completion has therefore often been referred to as the penultimate certificate, although there is nothing in the contract to prohibit the issue of further certificates beyond this one. In a real contract situation it is often not possible to value accurately the works at completion, and this therefore results in further certificates being issued.

Where the formula adjustment under clause 40 applies, the effect of the timing of this on interim valuations needs to be considered. Using the traditional method of price fluctuation, i.e. increased cost sheets, these are independent from the value of the measured works. The interim valuation of the contract works was also looked upon as a means to an end, and as long as it was reasonably correct overall that was all that mattered. The contractor needed to be paid a reasonable sum for the work carried out, and the employer needed to be satisfied that certificates were reasonable. The introduction of the NEDO formulae has meant that interim valuations must be much more accurate, since the appropriate percentages used to calculate the increases (or decreases) use the valuation amounts in their computation. Although at interim valuation time the indices may only be provisional, the work valued must be actual, to closely resemble the work completed. Using the traditional method for calculating price increases, it was generally accepted that it was in the best interest of the contractor to overvalue the works wherever possible, as this helped to improve the contractor's cash flow. The adoption of the formula method will usually result in a higher overall payment should the contractor delay as long as possible the inclusion of work in a valuation.

Application by contractor – amount of gross valuation

In practice it is the contractor's surveyor who often prepares the valuation since they are more frequently on site, and therefore more familiar with the progress of the works. The quantity surveyor will visit the site at the appropriate date to examine, approve and agree the amount of the valuation. Should there be any dispute between these parties on the amount to be paid, the client's quantity surveyor's assessment will be used. The valuation is then forwarded to the architect with the provision that the work included is in accordance with the architect's requirements. The quantity surveyor will not knowingly include in a valuation work that does not conform to the contract, but the question of quality control is not really the quantity surveyor's prerogative. If, however, there are any doubts over an item, this should be brought to the notice of the architect. The architect will generally accept the valuation and use it to prepare their certificate. This is sent to the employer with a copy to the contractor, and should be paid within the 14 days stipulated in the contract.

Issue of interim certificates

Interim certificates are issued at the period of interim certificates specified in the appendix, up to and including when the certificate of practical completion is issued. The final certificate is referred to as the certificate of completion of making good defects, and is issued either at the end of the defects liability period or once the defects have all been made good, whichever date is the later. The certificate of practical completion of the works includes the release of one moiety (one-half) of the retention monies, the remainder being released with the certificate of making good defects. The architect is not required to issue certificates at this time more frequently than one calendar month (clause 30.1.3).

Right of suspension of operations by contractor

Interim payments to contractors have often been described as the lifeblood of their business. The failure by an employer to honour the agreed certificate by making payments to the contractor at the appropriate time is therefore a potentially serious situation for the contractor. JCT 98 recognizes this fact by allowing the contractor to suspend the progress of the works. This is the ultimate sanction by the contractor. However, the contractor cannot suspend the works immediately a payment becomes due, i.e. 14 days after the issue of the certificate. The contractor must then issue a written notice to the employer with a copy to the architect of the intention to suspend construction activities. The contractor must then allow time for the employer to respond by making the appropriate payment within 7 days. For clarification this suspension of the works, by the contractor, cannot be treated as the contractor's failure to proceed regularly and diligently with the works as described in clause 27.2.1.2.

Ascertainment of amounts due in interim certificates

The amount stated as due in an interim certificate is the gross valuation agreed between the parties and certified by the architect. In order to calculate the amount payable to the contractor, deductions in the form of retention, payments made under previous certificates or advance payments are deducted. The gross valuation includes the following items, which are *subject to retention* as stated in the appendix:

- The total value of work properly executed by the contractor, including variations carried out and agreed in terms of their financial consequences, but excluding any restoration, replacement or repair or loss or damage and removal and disposal of debris which in clauses 22B.3.5 and 22C.4.4 are treated as if they were a variation.
- Any adjustments to these values arising as a result of the application of the price adjustment formulae under clause 40.
- The total value of materials and goods delivered to the site. This, in practice, often only includes those items of a major financial importance. The materials on site will only include those items that are reasonably, properly and not prematurely delivered to site. The inclusion, in the valuation, of sanitary wares that are already on site at the start of the contract may be refused because they have been prematurely brought to the site. The employer, when paying for materials on site, will need to be satisfied in every respect regarding their safety from damage or theft. Materials on site for long periods are more susceptible to these occurrences. The provision for payment may also require adequate protection from the weather as a prerequisite condition. This clause also applies in respect of materials from nominated suppliers.
- The value of any materials off site as agreed with the architect as long as they conform to the requirements of the contract as described in clause 30.3.
  These include materials, goods or items prefabricated which are listed items.

- The amounts of any nominated subcontractor's work that has been properly executed. Included with the subcontractor's invoice may be materials on or off site as appropriate and agreed by the architect. It is normal practice to include these amounts only where a subcontractor supplies an invoice. In addition, in some circumstances the architect or quantity surveyor may require proof that previous payments have been made to nominated subcontractors.
- The profit of the contractor in respect of nominated subcontractors' invoices. The rates (percentages) stated in the bill of quantities are used. Should nomination be the result of a provisional sum then these rates will be used as a basis. The contractor's profit on nominated suppliers' accounts is presumably covered with the supply of the materials under the third and fourth items, and would be added as appropriate. Attendance items are properly the main contractor's work and therefore come within the definition in the first item.

In addition the following items are to be included which are *not subject to retention*:

- Items which adjust the contract sum resulting from changes in fees and charges (clause 6.2), opening up for inspection and testing of the works where found to be correct (clause 8.3), royalties or patent rights resulting from an architect's instruction (clause 9.2), certain insurance matters (clause 21.2.3) and failure of the employer to insure the works (clauses 22B.2 and 22C.2)
- Loss and expense agreed under clauses 26.1 or 34.3 or in respect of any restoration, replacement or repair or loss or damage and removal of and disposal of debris which in clauses 22B.3.5 and 22C.4.4 are treated as if they were a variation.
- Where final payment has been secured to a nominated subcontractor under clause 35.17.
- Any increased payments in respect of contributions, levies and tax fluctuations under clauses 38 and 39.
- Any amounts of a similar nature to those described above relating to nominated subcontractors' works.

The following items are not subject to retention but are to be deducted from interim certificates:

- Amounts in respect of levels and setting out of the works (clause 7); work, materials or goods which although not in accordance with the contract have been allowed to remain (clause 8.4.2); any defects, shrinkages or other faults which have been allowed to remain in the works (clauses 17.2 and 17.3) or any amounts allowable by the contractor to the employer in respect of fluctuations (clauses 38 and 39).
- Any similar allowance in respect of a reduction in nominated subcontractors' payments.

Off-site materials or goods – the listed items

The amount stated as due in an interim certificate may, at the discretion or option of the architect, include the value of such items prior to delivery to the site. The materials, goods or items of prefabrication to which this clause refers are known as the listed items. The amount stated as due in an interim certificate can include the value of any listed items before delivery or adjacent to the works provided that the following conditions have been fulfilled:

- The contractor has provided the architect with reasonable proof that the property is uniquely identified and vested in the contractor. Immediately upon payment by the employer these become the property of the employer.
- If the appendix to the form of contract requires a surety bond in favour of the employer then this must also be provided by the contractor prior to payment.
- Where the property is not uniquely identified the same principles can also apply.
- The listed items must be in accordance with the contract.
- The listed items which are at premises where they have been manufactured, assembled or stored must be set apart or clearly marked to identify the employer and their destination as the works.
- The contractor must also provide the employer with reasonable proof that the items are insured against loss or damage for their full value under a policy of insurance protecting the interests of the employer and contractor in respect of specified perils. This insurance must cover the listed items from the time of transfer of property ownership until they are delivered to the works, when they will then be covered by the insurance of the works.

In practice it is usual to interpret these conditions to mean materials or goods that are special to this contract. Common bricks or quantities of standard-sized timber which could be used on any project would not normally be considered as allowable items to be included within a certificate, although there may always be exceptions. Nothing should remain to be done to these items prior to their incorporation within the works. The typical sorts of items covered by this clause envisage some form of prefabrication for the project concerned, although other materials and goods can also be included. The listed items may therefore be at the contractor's workshops or the workshops of a subcontractor.

Retention

The contractor is not paid the full amount of the certificate, but this is subject to a deduction in the form of retention. Retention is applied to work that has not reached practical completion and to materials and goods referred to above, including nominated works. The purpose of this retention is to provide some incentive for the contractor to complete the works, and it also provides some security should the contractor default in construction.

The amount retained by the employer under this clause is recommended as 5 per cent, although a lower rate can be agreed between the parties. It would appear that higher rates of retention are not envisaged under this contract. The percentage amount of retention is shown in the appendix. Where the size of the project exceeds a contract sum of £500,000 then the recommendation of this contract is that the amount retained should be limited to 3 per cent. This implies a 3 per cent retention throughout the contract period, rather than a 5 per cent retention with a limit of 3 per cent of the contract sum. Retention is applied to all those items of work and materials described in clause 30.2.1.

The retention is released to the contractor as follows. One-half of that which is retained by the employer at the certificate of practical completion of the works, and the remainder with the issue of the certificate of making good defects.

Rules on treatment of retention

The employer has only a fiduciary interest in the retention. In this respect the retention is held on trust only, and need not therefore be invested to accrue interest. The contractor or a nominated subcontractor may require the employer to put the retention in a separate bank account and confirm this in writing to the architect. Any interest accruing from this account is for the employer's benefit.

At the issue of an interim certificate the quantity surveyor prepares a statement that specifies the amount of retention on the contractor's work and for each nominated subcontractor. This statement is issued to the architect, employer, contractor and each nominated subcontractor.

The employer may exercise the right to pay a nominated subcontractor direct, because of the contractor's default. The contractor must be informed of this action and its subsequent effect upon the retention.

Final adjustment of the contract sum – documents form the contractor

Either during the contract period or within six months after practical completion of the works, the contractor shall provide all the documents necessary for the preparation of the final account. These documents should include the appropriate details from nominated subcontractors and nominated suppliers. The quantity surveyor should within three months of receiving the information from the contractor ascertain any loss and expense under clauses 26.1, 26.4.1 and 34.3, and also prepare a statement of all contract sum adjustments as follows. The architect must send a copy of this to the contractor and to each nominated subcontractor. The contract sum is adjusted by the valuation of variations under clause 13.4.1, the amounts stated in any 13A quotations and the amounts of any price statements. The following items are to be deducted from the contract sum:

- Prime cost sums, which include the contractor's profit and attendance items where appropriate. This also includes the value of any work by a nominated subcontractor, whose employment has been determined in accordance with clause 35.24 which was not in accordance with the relevant subcontract but which has been paid or otherwise discharged by the employer.
- Provisional sums and provisional work (approximate quantities) in the contract bills.
- Amounts omitted by reason of variations caused by architect's instructions.

- Amounts deducted or deductible under clause 7 (levels and setting out of the works), clause 8.4.2 (materials and goods which although not in accordance with the contract have been allowed to remain), clauses 17.2 or 17.3 (defects, shrinkages or other faults which have been allowed to remain). It also includes amounts allowable to the employer under clause 38 (contributions, levy and tax), clause 39 (labour and materials cost and tax fluctuations) and clause 40 (use of price adjustment formulae).
- Any other amount which is required by this contract to be deducted from the contract sum, for example, liquidated damages.

The following items are to be added:

- Nominated subcontract sums as finally adjusted or ascertained under the relevant provisions of subcontract NSC/C.
- The tender sum properly adjusted against prime cost sum items which have been undertaken by the contractor.
- Nominated suppliers' accounts adjusted within the terms of the contract, including 5 per cent cash discount but excluding VAT.
- The profit and attendance items on nominated subcontractors' and nominated suppliers' items where appropriate and including cases where nomination arises through the expenditure, for example, of provisional sums.
- Additional payments authorized by the employer in respect of
  - fees and charges (clause 6.2)
  - opening up of the works for inspection and testing where the work complies with the contract (clause 8.3)
  - royalties and patent rights (clause 9.2)
  - insurance (clause 21.2.3)
- Variations authorized or approved by the architect.
- The expenditure associated with provisional sums and approximate quantities in the contract bills.
- Loss and expense ascertained under clauses 26.1 or 34.3.
- Amounts under clauses 22B and 22C (insurances by the employer) which the contractor is entitled to add to the contract sum.
- Increases in contributions, levies and taxes under clauses 38, 39 or 40.
- Any other amount which is required to be added to the contract, such as the agreement of the contractor's claims.
- Any amount to be paid in lieu of any ascertainment under clause 26.1.

Interim certificate – final adjustment for ascertainment and nominated subcontract sums

If it is practicable, but at least 28 days before the issue of the final certificate, the architect should issue an interim certificate which should include the amounts of the subcontract sums for all the nominated subcontractors adjusted and ascertained under all the relevant provisions of the contract.

Issue of final certificate

The architect must issue the final certificate and inform each nominated subcontractor of its issue within two months of the last of the following events:

- End of defects liability period
- Date of issue of certificate of making good defects
- Agreement of the final account

Within 5 days after issuing the final certificate, the employer must write to the contractor stating the amount of payment that is due, to what the payment relates and the basis of its calculation. Final payment must be made to the contractor within 28 days of the issue of the final certificate. If the employer fails to make this payment, simple interest is added at the rate of 5 per cent above the base rate of the Bank of England.

The final certificate must state:

- The amounts already paid to the contractor under interim certificates
- The contract sum adjusted within the terms of the contract
- To what the amount relates and the basis on which the statement in the final certificate has been calculated
- The difference expressed as a debt to either the employer or the contractor

Effect of the final certificate

The final certificate is conclusive evidence that:

- The quality of materials and standards of work are as described on the contract drawings and in the contract bills and to the reasonable satisfaction of the architect.
- The contract sum has been properly adjusted within the terms of the contract.
- Relevant extensions of time have been given under clause 25.
- Reimbursement of direct loss and expense is in final settlement of any contractor claims.

However, if there has been any accidental inclusion or exclusion of work, or arithmetical errors in computation, these are able to be corrected. The same principle applies in the case of fraud.

If any adjudication, arbitration or other proceedings have been commenced by either party before the issue of the final certificate, then the final certificate shall have effect as conclusive evidence as provided in clause 30.9.1 after the earlier of these two outcomes:

- Such proceedings have been concluded and the final certificate is subject to the award or judgment that has been made.
- After a period of 12 months in which neither party has taken further steps to solve the problem.

If any adjudication, arbitration or other proceedings are commenced within 28 days after the final certificate has been issued, the final certificate shall still remain conclusive evidence other than in respect of the matters relating to the proceedings.

No other certificates can be interpreted as conclusive evidence that either the works, goods or materials or performance-specified works are in accordance with the contract.

CHAPTER 20

TIME FACTOR OF CONSTRUCTION

Several clauses in the conditions of contract are of relevance and importance in respect of the time factor. These have been grouped together for commentary purposes.

- Clause 17   Practical completion and defects liability
- Clause 18   Partial possession by employer
- Clause 23   Date for possession, completion and postponement
- Clause 25   Extension of time
- Clause 27   Determination by employer
- Clause 28   Determination by contractor
- Clause 28A   Determination by employer or contractor

These clauses describe the start and finish of the project and the implications associated with these dates. The difference between them is the contract period. The distinction between completion date and date for completion should be carefully noted. This is described in clause 1 under the general rules of interpretation and definitions. Should the contractor fail to complete the works by the completion date then the architect must issue a certificate of non-completion of the works as described in clause 24, and damages are then payable by the contractor. Upon the practical completion of the project the contractor is relieved of most of the contractual obligations. The contractor is, however, still responsible for defects that may arise due to poor standards of work. This responsibility continues until the end of the defects liability period, which the JCT form recommends should extend for six months after practical completion. Currently there is some debate on whether this responsibility should extend further, for some aspects of work, up to ten years (Chapter 11). In some circumstances the contractor may be able to secure an extension of the contract period for one of the reasons listed in clause 25. The advantage of this course of action to the contractor is either the reduction or elimination of liquidated damages. The date for completion is stated in the appendix and any approved extension of time which is added to this provides the later completion date. The completion date cannot occur prior to the date of completion stated in the appendix, unless the contractor agrees to such a revision. In some extreme circumstances the works may be postponed, perhaps indefinitely, and clause 23 allows for this eventuality. The works may also offer partial possession in the case of a phased project awarded as a single contract. In this case phased completion is treated almost as a separate project as far as time is concerned. Finally, circumstances can, and sometimes do, arise where either the employer or the contractor feels that the contract should be terminated. This usually results because of a breach by the other party. The procedures and position of each party are described accordingly in clauses 27, 28 and 28A.

PRACTICAL COMPLETION AND DEFECTS LIABILITY (CLAUSE 17)

Certificate of practical completion

When in the opinion of the architect the works become practically complete, the certificate of practical completion of the works must be issued. In practice this certificate may be issued even though minor items of work still need to be carried out within the terms of the contract. This does not presume that these items are either irrelevant or need not be completed. The date of practical completion often coincides with the handover date, and this latter consideration may have resulted in this certificate being issued even though the works are only 'almost' complete. Prior to the issue of this certificate the architect needs to be satisfied that:

- The work has been carried out in accordance with the contract documents and the architect's instructions.
- The building is in an appropriate state to be taken over and used by the employer.

Usually, where the employer wants to occupy the building, the certificate should be issued.

The date of practical completion is important since the following take effect automatically:

- The start of the defects liability period (clause 17)
- The beginning of the period of final measurement (clause 30)
- The release of the first moiety of the retention fund (clause 30)
- The ending of the insurance of the works (clause 21)
- The end of any liability to liquidated damages by the employer (clause 24)
- The opening of matters referred to arbitration (article 5)
- The removal of the liability to frost damage (clause 17)

Many of these items also occur to those parts of a project affected by partial completion (clause 18). Upon achieving practical completion, the contractor is no longer obliged to accept instructions from the architect in respect of additional work since the works are now complete. The contractor may choose to do this work on the basis of a fresh agreement. For example, in connection with the payment for this work, bill rates and interim certificate periods, etc., will no longer apply.

Defects, shrinkages or other faults

The contractor is responsible for the making good of defects, shrinkages or other faults which appear within the defects liability period. These defects must be due to materials and work standards which are not in accordance with the contract, or to frost damage occurring before the practical completion of the works. If defects occur because of the inadequacy of the design, contractors are not obliged to make these good at their own expense. The architect cannot also issue a variation order after the issue of the certificate of practical completion in order to remedy design faults. Items which 'appear' during the defects liability include those presumably that were unnoticed during the contract period. These items must be included on a schedule of defects not later than 14 days after the expiration of the said defects liability period. The length of the defects liability period must be inserted in the appendix to the conditions, although six months is the recommended length. The making good of the contractor's defects should be done within a reasonable time and entirely at the contractor's own expense.

Defects, etc. – architect's instructions

The architect can issue instructions at any time for the making good of any defect, shrinkage or other fault within the defects liability period. Such defects must be due to materials or work standards that are not in accordance with the contract or frost occurring before the practical completion of the works. The contractor must within a reasonable period of time comply with such instructions and at no expense to the employer. Faults not made good by the contractor will result in a deduction from the contract sum. In practice the architect should issue a schedule of defects not later than 14 days from the expiration of the defects liability period.

Certificate of completion of making good defects

When, in the opinion of the architect, the defects requested have been made good, and the defects liability period has expired, the certificate of making good defects will be issued. The issue of this certificate releases the remainder of the retention monies due to the contractor, and removes the defects obstacle that would prevent the issue of the final certificate. Contractors, however, are not totally released from the responsibility of defective work since they will still be affected by the statute of limitations.

Damage by frost

Contractors are never responsible for making good, at their own expense, damage by frost which occurs after the issue of the certificate of practical completion of the works. Should damage of this nature be found, there may sometimes be difficulty in deciding when the damage actually occurred. The architect may be able to show that such damage occurred before practical completion of the works.

PARTIAL POSSESSION BY EMPLOYER (CLAUSE 18)

Employer's wish – contractor's consent

The employer, with the agreement of the contractor, may decide to take possession of a section of the works. This is particularly appropriate where the contract concerned may be in two distinct parts. For example, alteration work and a new extension to an existing building, two extensions to an existing building or a housing scheme where numbers of units are released to the employer at various different time intervals. The architect will issue to the contractor, on behalf of the employer, a written statement identifying the parts of the works taken into possession by the employer and the date when this occurred.

Practical completion – relevant part

This certificate has the same effect on the part completed as the certificate of practical completion has on the whole of the works. The items listed under 'Practical completion' (page 275) apply equally to this part of the works that are now complete, and have received the aforementioned certificate. (See also the sectional completion supplement to the JCT form of contracts.) Partial possession by the employer clearly has contractual advantages to the contractor.

DATE FOR POSSESSION, COMPLETION AND POSTPONEMENT (CLAUSE 23)

Date of possession – progress to completion date

The date for possession of the site by the contractor is stated in the appendix to the conditions of contract. Failure on the part of the building owner to provide the site by this date will result in a breach of contract. There is, however, provision to allow the employer, without being in breach of contract, to defer giving the contractor possession of the site for a period not exceeding six weeks. The deferment of the date of possession is referred to in the appendix and the period of deferment should be stated if less than six weeks. The contractor must start the project and 'regularly and diligently' proceed with the project to the date of completion or the completion date if this has been extended. Failure on the part of the contractor to make regular progress with the works will also result in a breach of contract, this time on the part of the contractor. The employer is able to determine the contractor's employment on the works for this reason (clause 27.1.2).

The completion date is the date at the start of the defects liability period. It is also the date, should the contractor fail to complete by this time, of the starting point of the provisions relating to liquidated damages. Note that the date for completion is the date written into the contract. The completion date is this date amended to take into account any extra time allowed under an extension of time or for the repair of war damage.

Architect's instruction – postponement

The architect can postpone the work for a short period of time or indefinitely. Where postponement exceeds the period of delay stated in the appendix, the contractor can determine the contract. In circumstances where the postponement is a relatively short period of time, the usual redress for the contractor is to claim an extension of time under clause 25 and loss and expense under clause 26.

Possession by contractor – use or occupation by employer

The contractor retains possession of the site and the works up to and including the date of the issue of the certificate of practical completion. The employer, with the consent of the contractor in writing, may use or occupy the site or the works for the purpose of storage of goods or otherwise before the date of practical completion. The provisions of the insurers should be obtained before the contractor gives this consent. If an additional insurance premium is required and the employer continues to use the site then the premium will be added to the contract sum.

EXTENSION OF TIME (CLAUSE 25)

Interpretation of delay

Any reference to delay, notice or extension of time in this clause includes further delay, further notice or further extension of time.

Notice by contractor of delay to progress

Once the main contractor realizes that the progress of the works is likely to be delayed then the architect should be informed in writing, giving the reasons for the cause of the delay. Where the delay includes reference to a particular nominated subcontractor, this firm should also be given details of the delay. The effect of the successful application of this clause by the contractor is to reduce or eliminate any damages that may be suffered due to non–completion of the works by the agreed date. The initiative to be taken in seeking either an extension of time or a further extension of time must come from the contractor.

The contractor, in giving notice to the architect, should where possible provide particulars of the expected effects of the delay, and also estimate the extent in respect of completion of the works. If the answers to these questions are not known to the contractor, the architect should still be informed of a probable delay, and be provided with the detailed information at a later date.

Fixing completion date

Upon receipt of the notice of delay, the architect shall decide whether any of the events listed are relevant. If, in the architect's opinion, these are not considered to be a reasonable basis for an extension of time, the extension to the contractor should be refused.

If the reasons listed by the contractor are accepted as a relevant event by the architect, then the architect should write to the contractor giving details of:

- The events that, in the opinion of the architect, are relevant in these circumstances for an extension of time.
- A new completion date that has been estimated to be fair and reasonable. (If variations of omissions have been issued since the fixing of a previous completion date, then they can be taken into account in the fixing of a new date.)

The new completion date fixed by the architect must be at least 12 weeks from the receipt of the notice, or at the earliest the previously agreed completion date. The architect can therefore bring forward the completion date where the contract work has been significantly reduced. However, reasonable notice must be given to the contractor of this intention (12 weeks) and in any case it must be shown to be a fair and reasonable action. The architect must also notify the contractor in writing where it is not considered to be fair and reasonable to fix a later date as a completion date.

The architect must, within 12 weeks from the date of practical completion of the works, write to the contractor to fix a completion date. This date may be fixed as follows:

- The completion date as previously fixed or stated in a confirmed acceptance of a 13A quotation.
- A later completion date, in order to take into account any extension of time that in the opinion of the architect is justifiable.
- An earlier completion date, if in the opinion of the architect the construction work has been significantly reduced. This date, however, cannot be fixed earlier than the date for completion stated in the appendix to the form of contract.

The architect can revise the completion date as necessary. It is not necessary to wait until the contractor suggests that the works will be delayed. If, however, the contractor does not request an extension of time, the architect is not bound to take any action or to attempt to enforce a later completion date than is otherwise stated in the contract.

The contractor must always, using the best endeavours available, try to prevent any delay in the progress of the works. The contractor must also, to the satisfaction of the architect, proceed with the regular progress of the works.

The architect must inform every nominated subcontractor regarding each revised completion date.

Relevant events referred to in clause 25

Force majeure

The meaning of *force majeure* is imprecise, but it is generally accepted as 'exceptional circumstances beyond the control of either of the parties of the contract'.

Exceptionally adverse weather conditions

Exceptionally adverse weather conditions are weather conditions that could not normally be expected at the time of the year in the location of the building works. For example, heavy rainfall during the winter months would not fit within the definition of 'exceptional' unless it was abnormal. For example, in the winter of 1978/79 the total rainfall was 579 mm compared to the average for that period over the preceding years of 452 mm (Manchester Weather Centre). This may have been considered exceptional, but such a decision will rest with the courts of law if the parties cannot finally agree upon an interpretation.

Loss or damage occasioned by any one or more of the specified perils

The perils referred to in this clause are generally those items which are an insurable risk, such as fire, lightning, explosion, storm, tempest, etc. All the specified perils are listed under 'Definitions' in clause 1.3.

Civil commotion, etc.

Civil commotion is one of the specified perils. It includes both local and national strikes and lockouts by management of any of the trades employed upon the works. It is also extended to similar actions in the preparation, manufacture or transportation of the goods and materials.

Compliance with architect's instructions in respect of

- Discrepancies in or divergences between documents (clauses 2.3 and 2.4.1)
- An instruction requiring a variation (clause 13.2)
  - except for a confirmed acceptance of a 13A quotation
- The expenditure of provisional sums (clause 13.3)
  - except provisional sums for defined work
- A provisional sum for performance-specified work (clause 13A.4.1
- The postponement of any work (clause 23.2)
- The finding of antiquities on the site (clause 34)
- Nominated subcontractors' work (clause 35)
- Nominated suppliers (clause 36)

Opening up of work for inspection

If the main contractor is required to open up work for inspection or testing, and if the work complies with the contract, then this becomes an admissible event. If the work is not in accordance with the contract, any delays ensuing will not be grounds for an extension of time.

Failure of the architect to comply with an information release schedule where this has been provided

Described in clause 5.4.1, this may include delays in the receipt of instructions, drawings, details or levels from the architect. If this is to be accepted as a relevant event, the contractor must have requested the information beforehand from the architect. The contractor must also use some foresight in respect of when the details are required. The architect must be given reasonable notice so that the information can be prepared and provided by the requisite time. In practice the contractor, by reference to the programme of works, should be able to ask for such information well in advance. The clause does not, however, assume that the architect will only work at the *direction* of the contractor, but that the architect will be diligently preparing the information that the contractor will require anyway. In the event of this assumption being invalid, the contractor could be involved in preparing the architect's checklist, and then be penalized for doing this inadequately.

Failure of the architect to provide further drawings or details

This is described in clause 5.4.2.

Delay on the part of the nominated subcontractors or nominated suppliers

If the main contractor has taken all practicable steps to avoid or reduce these delays, they have fulfilled this condition for an extension of time.

Delay on the part of works or persons engaged by the employer

This assumes that if the employer decides to undertake some of the work direct, outside the scope of this contract, then such work should be carried out so as not to cause a delay to the main contract. The main contractor here will have minimal control. It is up to the employer, therefore, to take the necessary steps to ensure as little inconvenience as may be expected. This clause also covers the supply by the employer of materials and goods which the employer has agreed to provide for the works.

Government intervention

If, once the contract has been signed, the government restricts the availability or use of labour which is essential to the carrying out of the works, then the contractor may apply for an extension of time. In the case of war, for example, the government may call up all 'able-bodied' men to the armed services, thus removing tradespeople from the labour market. In other circumstances it may commandeer the essential building materials. On other occasions it may restrict the use of fuel or energy resulting in limited working, as occurred in 1973 with the three-day week. Two points are worth noting. First, the government intervention must restrict the essential supplies of labour, materials, fuel or energy. If the work can be easily performed in some other way, without detriment to the project, then no claim for an extension of time will be permitted. Second, the government intervention must occur after the signing of the contract. If the contractor knew or should have known before entering into the contract that the government would introduce certain restrictive measures, this cannot be claimed as a basis of defence.

Contractor's inability to secure labour and material

The foregoing comments on government intervention apply equally to this relevant event. For example, a sudden unexpected upturn of the construction industry in a certain area may starve some sites of the essential supplies of labour and materials.

Work by local authorities or statutory undertakings

Although these may be part of the contract, the contractor may have severe difficulties in attempting to control them. In some circumstances the contractor may be delayed waiting for statutory inspections by local authorities. The contractor must, however, have taken all reasonable steps to avoid difficulties occurring in these circumstances.

Failure on the part of the employer to give access

If the employer is unable to give to the contractor, by the dates stipulated in the contract, proper ingress to or egress from the works as defined in the contract documents, then the contractor will be entitled to an extension of the contract period.

Employer defers giving possession of the site

Where clause 23.1.2 is stated in the appendix to apply, the employer can defer giving possession of the site under clause 23.1.2.

Inaccurate forecast of an approximate quantity

Where such work is executed by the contractor then it must be a reasonably accurate forecast.

Delays which the contractor has taken all practical steps to avoid

This may be due to changes in statutory requirements which necessitate some alteration or modification to performance-specified work.

The use or threat of terrorism

The authorities may take action to deal with terrorist threats.

Compliance or non-compliance by the employer with clause 6A.1

(planning supervisor).

Delays arising from a suspension by the contractor of the performance of the contractor's obligations under the contract to the employer in pursuant to clause 30.1.4

(failure to honour certificates).

Note that under clause 25 the matters discussed relate only to an extension of the contract period. The relevant events listed will form a good reason on the part of the contractor to this contractual provision. It must not be assumed to be an automatic result following any one of these events. In any case the agreement for an extension of time could sometimes result in only a few days.

Another important factor connected with this clause is the aspect of any financial adjustment to the contract sum, either in connection with damages for non-completion (clause 24) or the loss and expense to the contractor resulting from the delay (clause 26). These matters are discussed more fully with those appropriate clauses.

An extension of the contract period, for any reason, may be unacceptable to the employer. The architect may therefore reasonably expect the contractor to take appropriate action to save time. The contractor may also reasonably expect to be paid for the costs involved in such action. There is, however, no obligation under the contract for the contractor to take such action.

DETERMINATION BY EMPLOYER (CLAUSE 27)

Notices under clause 27

In certain circumstances the employer is rightfully allowed to terminate the employment of the contractor. If the contractor defaults in one of several ways, the architect may issue a written notice to the contractor. This notice must be delivered by actual, special or recorded delivery specifying the default. The reason why the notice must be delivered in this manner is to avoid any claim by the contractor that it was not received. The notice will be deemed to have been received by the contractor 48 hours after the date of posting.

Default by contractor

The following defaults must occur before the date of practical completion:

- The contractor without reasonable cause suspends the carrying out of the works. This may be wholly or substantially. The architect will interpret this accordingly.
- The contractor fails to carry out the works in a regular and diligent manner. This may also put the contractor in breach of clause 23.
- The contractor fails to comply with an instruction from the architect to remedy defective work, and this results in further damage to the works. Note that the failure to comply with other instructions merely results in the architect issuing the instructions to another firm, and recouping the necessary amount from the contractor.
- The contractor has either assigned or sublet a portion of the works without permission and has thus failed to comply with the provisions of clause 19.
- The contractor fails to comply with the requirements of the CDM Regulations.

In the case of a default the contractor should be required to remedy the contract, perhaps informally at first. If the contractor continues to ignore the warning from the architect then the notice of determination should be sent to the contractor. The contractor, upon receipt of the notice, then has 14 days in which to rectify the problem. Within 10 days after this date, assuming the contractor is still in default, the employer may then issue a notice that effectively terminates the employment of the contractor. The employer must ensure that the issue of the notice is a reasonable course of action to take, and should not be given solely to the annoyance of the contractor (clause 27.2.4). The contract would describe this as vexatious or unreasonable behaviour on the part of the employer.

If the contractor ends the specified default or the employer does not determine the contract on this occasion, this can nevertheless be proceeded with later should the contractor default on some other issue.

Insolvency of the contractor

If a contractor becomes insolvent, or arranges with creditors for voluntary winding up (Companies Act 1985 and Insolvency Act 1986), then the employment of the contractor is automatically terminated; see Table 20.1. This is described in the contract as:

- Making a proposal for a voluntary arrangement for a composition of debts
- Having a provisional liquidator appointed
- Having a winding-up order made
- Passing a resolution for voluntary winding up
- Under the Insolvency Act 1986 having an administrative receiver appointed

There is, however, provision in the contract for this employment to be reinstated where both the employer and the contractor, or liquidator, agree. In practice there are immense disadvantages to the employer where bankruptcy of the contractor

Table 20.1 Example calculation of contractor's insolvency

Position as it should have been

		£
Contract sum		500,000
Value of variations		25,000
Theoretical final account		525,000
Position at determination		
Value of work executed at time of liquidation		340,000
Interim payments	300,000	
Less retention	−15,000	
	285,000	
Nominated subcontractors paid after determination	30,000	
	315,000	315,000
Amount outstanding at determination		25,000
Position at final account		
Paid as above, at determination		315,000
New completion contract		250,000
Employer's loss and expense		25,000
		590,000
Less as it should have been		525,000
Cost of determination/bankruptcy		£65,000

occurs. It is therefore in the best interests to retain the services of the contractor, where this has been previously satisfactory. The liquidator is also in a position of wishing to continue with those contracts that are likely to be profitable and quick to execute. Contracts that are nearing completion are therefore likely to be retained, whilst those in their earlier stages of construction may be dispensed with.

Corruption

The employer is also entitled to terminate the contractor's employment in cases of corruption. This can be extended to all contracts between the employer and the contractor where an irregularity of this nature has occurred on only one contract. The contractor may, for example, be already undertaking several different contracts on behalf of the employer. A bribe might be used to secure further work. Where this is proved, then all the contractor's work may be terminated. Corruption may result from the method of obtaining work or from the execution of the contract. Corruption can also occur where one of the contractor's employees offers favours that are neither known nor sanctioned by the contractor. The offences would usually come within the remit of the Prevention of Corruption Acts 1889 to 1916. Although corruption is generally well understood, it is often difficult to draw a line between the actions that may be classified as such. For example, many would regard the offer of a bottle of spirits at Christmas time as quite acceptable, and not as a bribe or offered in return for a favour. A crate of spirits would probably prove to be a different matter!

Insolvency of contractor – option to employer

The employer's first action is to give notice to determine the employment of the contractor. The employer may choose to make an agreement with the contractor on the continuation or novation of the contract, or the contract may be simply terminated. In any event the employer and contractor may make interim arrangements for work to be carried out. The employer may take reasonable measures to ensure that site materials, the site and the works are adequately protected, and also goods and materials that are adjacent to the site. The employer may deduct the reasonable costs of taking such action.

Consequence of determination under clause 27.2 to 27.4

The respective rights of the employer and contractor, upon the termination of the contractor's employment, are as follows:

- The employer may engage another firm to complete the project. This firm is to be allowed to use all the temporary buildings, plant and equipment until completion, where they are owned by the contractor. Where they are not owned by the contractor, the consent of the owner must be obtained by the employer.
- Unless the termination occurs because of bankruptcy, the contractor must, without payment, assign to the employer any benefits for the supply of goods and materials or for the execution of subcontracts. In practice the contractor could encourage bad relations to develop in order to deprive the employer of these benefits.
- Where insolvency has not occurred the employer may pay any supplier for materials or goods, or a subcontractor for work executed, who have not previously received payment. The employer may also use discretion for paying nominated subcontractors, who should have been paid previously under certificates. The employer will then try to recover these sums from the contractor. In practice those who should have been paid, but have not been because of a default on the part of the contract, may be unwilling to carry on their work unless they receive a payment. The employer may therefore be forced into this situation, in order to see progress in the works, and then try to recover the amounts from the contractor.
- The contractor must remove temporary works and other items within a reasonable time after receiving an instruction from the architect. Where the contractor fails to do this, the employer shall remove and sell the items, and after deducting costs, pass on the proceeds for the benefit of the contractor.
- If the employer suffers any loss or expense because of this termination then this will become a debt of the contractor.
- The employer is not bound to make any further payments to the contractor until the project has been properly completed and the final accounts agreed for the project as a whole.

Employer decides not to complete the works

If after the determination of the contractor, the employer chooses not to proceed with the works, then the contractor must be notified. The contractor must be notified within six months of the date of such determination. The total value of work that has been properly executed is calculated, less any reasonable costs and expenses borne by the employer. The difference is then expressed as a debt or credit to the contractor.

Other rights and remedies

The above provisions are without any prejudice to other rights and remedies which the employer may possess.

DETERMINATION BY CONTRACTOR (CLAUSE 28)

Notices under clause 28

A number of situations can arise which can give the contractor the legal right to terminate the contract. If the contractor decides this is an appropriate course of action to follow then a notice in writing must be sent to the employer specifying the default that has caused this to take place. This notice must be delivered by actual, special or recorded delivery specifying the default. The reason why the notice must be delivered in this manner is to avoid any claim by the employer that it was not received. The notice will be deemed to have been received by the employer 48 hours after the date of posting. This does not imply that a contractor should take into account the peculiar state of the employer in respect of, for example, the employer's temporary shortage of funds or finance due to a cash flow crisis.

However, in a majority of cases this course of action suits no one. It is therefore introduced as a last resort. The contractor will already have attempted to solve the dispute that caused the determination in other ways. Also, whilst a contractor may be completely innocent, damage may nevertheless be suffered in terms of reputation or financial loss. Determination usually results from a series of actions rather than a single misdemeanour on the part of the employer. This might, for example, result from a repetition of one of the events listed below.

Default by employer – suspension of uncompleted works

Financial reasons

Perhaps the employer does not discharge in accordance with the contract the amount properly due in respect of a certificate or VAT in pursuant of the VAT agreement. Or perhaps the employer interferes or obstructs the issue of any certificate. If the employer fails to pay a certificate within 14 days (clause 30) and continues to default for a further 7 days, then the contractor can issue a notice of intention of termination. The first time this occurs, the contractor is more likely to approach the architect regarding the absence of the payment. The certification of monies due is not a matter of debate by the employer. Where an employer has misgivings about the performance of the contractor, they should be brought to the attention of the architect. Also the employer cannot deduct sums of money from a certificate that has already been authorized by the architect. This is the most usual circumstance that results in determination by the contractor because the employer fails to honour the conditions of an interim certificate.

Assignment and subcontracts

If the employer fails to comply with the provisions of clause 19.1.1 and attempts to assign a part of the contract to another, without the contractor's permission, then this can give rise to determination by the contractor.

CDM Regulations

The employer does not comply with the requirements of the CDM Regulations.

Suspension of the works

The project has been suspended by the architect, employer or someone else such as a local authority. The period for such a delay is entered in the appendix to the form of contract, and is normally one month. Suspension of the works may arise because of:

- The contractor not having received in due time the necessary instructions, drawings, details or levels from the architect, for which the contractor has made a specific request in writing. This would normally be done through the information release schedule. Such requests must allow the architect sufficient time to provide this information.
- Failure of the architect to comply with clause 5.4.2 on the provision of further drawings and details.
- The compliance with an architect's instructions issued under clauses 2.3 (discrepancies between documents), 13.2 (variations) or 23.2 (postponement), unless caused by the contractor.
- Delay in the execution of work caused by persons employed directly by the employer.
- Failure on the part of the employer to give in due time ingress to or egress from the site of the works through or over any land or buildings.

The employer is given sufficient time to rectify any default described above. The employer must be allowed 14 days from the receipt of the notice to take appropriate action. The contractor must then within a further 10 day period issue a further and final notice of determining employment under the contract. These notices must not be given vexatiously or unreasonably by the contractor.

Insolvency of the employer

The employer can make a composition or arrangement with creditors, become bankrupt or cease being a company in the following ways:

- Make a proposal for a voluntary arrangement for a composition of debts
- Have a provisional liquidator appointed
- Have a winding-up order made
- Pass a resolution for voluntary winding up
- Have an administrative receiver appointed under the Insolvency Act 1986

The employer must inform the contractor in writing where any of the above situations occurs. This is in accordance with the Companies Act 1985 or the Insolvency Act 1986. Determination will take effect upon the date of receipt of such notice.

Consequences of determination under clause 28.2 or 28.3

The effects of determination by the contractor result in the following financial arrangements with the relevant sums being paid by the employer:

- The reasonable costs of the removal of any temporary buildings, plant, tools, equipment and goods, including subcontractors. This must be done within a reasonable period of time, taking precautions to prevent possible injury, death or damage.
- The payment of any retention monies within 28 days.

The contractor shall within a reasonable time prepare an account setting out the following:

- The value of work already completed and the value of work under construction, values that have been properly authorized by the architect.
- Any sum resulting from loss and expenses under clause 26.
- The reasonable cost of removal in pursuing this course of action.
- The direct loss and expense resulting directly from this determination.
- The costs of other materials and goods already ordered and provided for the works.

The respective procedure, therefore, following determination by the contractor is as follows. The contractor will clear the site; the employer will pay for the work completed to date, including materials and goods for the works and retention amounts. In addition the contractor will be paid for loss and expense to cover this premature action by the employer.

Other rights and remedies

The above provisions are without any prejudice to other rights and remedies which the employer may possess.

DETERMINATION BY EMPLOYER OR CONTRACTOR (CLAUSE 28A)

Grounds for determination of the employment of the contractor

This clause results from the tribunal's decision that those grounds in clause 28.1.3, under which the contractor could give notice to determine the contract and which were not acts or defaults of the employer, should be withdrawn from this clause. They are in effect neutral events caused by neither the employer nor the contractor. They should thus be given as grounds for determination by either party. The events listed relate to the suspension of the works:

- Force majeure
- Loss or damage to the works occasioned by any one or more of the specified perils
- Civil commotion
- Architect's instructions issued under
  - clause 2.3 (discrepancies or divergencies between documents)
  - clause 13.2 (instructions requiring a variation)
  - clause 23.2 (architect's instructions – postponement)
- Hostilities in the UK, whether or not war has been declared
- Terrorist activity

Either the employer or contractor may give notice in writing to the other party by registered post or recorded delivery. The notice will deemed to have been received within 48 hours of posting. If after a further 7 days the suspension of the works has not ended, then the contract will be determined. In this situation presumably neither party wants the contract to end, but this has occurred due to events beyond the control of either party.

The contractor will not be allowed to give notice under clause 28A.1.1.2, where the loss or damage was due to the contractor's own negligence or default.

Consequences of determination under clause 28A.1.1 – clauses 28A.3 to 28A.6

The consequences of this determination are as follows. The removal from the site of the contractor's temporary buildings, plant, tools, equipment, materials and goods, including site materials. This must be done within a reasonable length of time and with precautions to prevent injury, death or damage. This applies also to subcontractors working on the site. The employer will release half of the retention monies within 28 days of the notice.

Within a further two months the contractor must provide the necessary documents to prepare the final account for the project. This is to be calculated as follows:

- Total value of works at determination
- Sum ascertained in respect of direct loss and expense under
  - clause 26 (loss and expense generally)
  - clause 34.3 (antiquities)

- Reasonable cost of removal under clause 28A.3
- Cost of all materials and goods ordered for the works
  - which then become the property of the employer
- Any direct loss or damage caused by the determination

This amount together with any outstanding retention sums should be paid to the contractor within 28 days. This could include loss due to one of the specified perils that might have been caused by the negligence of the employer.

Amounts attributable to nominated subcontractors

The employer must inform the contractor in writing of the amounts to be paid to nominated subcontractors. They are also expected to be issued with a copy of this notice.

CHAPTER 21

WORKS BY OTHER PERSONS

The following group of clauses relate to those parts of the works that are done by firms other than the main contractor. In some examples the main contractor has no choice in the matter, other than a reasonable objection to such firms or persons on the works.

- Clause 19   Assignment and subcontracts
- Clause 29   Works by employer or persons employed or engaged by employer
- Clause 35   Nominated subcontractors
- Clause 36   Nominated suppliers

The first of these clauses allows the main contractor to appoint domestic subcontractors. The number of subcontractors employed will depend on the overall trades and facilities provided by the main contractor. On some projects the contractor may work almost solely in a coordinating and organizing role. In this case the contractor chooses to subcontract the whole of the construction work. The architect must approve all these subcontractors, and the approval should not unreasonably be withheld. However, this does not in any way diminish the main contractor's responsibilities under the terms of the contract. The assignment of these responsibilities to another firm will not generally be acceptable to the architect or building owner.

Clause 29 allows the employer's own directly employed contractors or workers the right to be engaged upon the works during the normal contract period. These firms may be employees of the employer or of a very specialized nature, such as sculptors. The employer may wish to retain a more direct control over their work. These firms should not interfere or inconvenience the main contractor in the execution of the work.

Clause 35 is one of the longer clauses of JCT 98, and deals with the provision for nominated subcontractors. The clause is subdivided into 26 subclauses, as follows:

- 35.1–2   General
- 35.3–35.12   Procedure for nomination
- 35.13   Payment of nominated subcontractors

- 35.14   Extension of period for completion of nominated subcontract works
- 35.15   Failure to complete nominated subcontract works
- 35.16   Practical completion of nominated subcontract works
- 35.17–35.19 Early final payment of nominated subcontractors
- 35.20   Position of employer in relation to nominated subcontractors
- 35.21   Position of main contractor
- 35.22–35.23 Limitation of liability of nominated subcontractors
- 35.24   Circumstances where renomination is necessary
- 35.25–35.26 Determination of employment of nominated subcontractors

Nominated subcontractors include statutory undertakings; whilst these are a special type of subcontractor, sometimes neither the client, nor the architect nor the main contractor has any choice in their appointment and involvement in the project.

The works to be executed covered by clauses 35 and 36 are written into the bill of quantities as prime cost sums, and they are subject to adjustments under the terms of the contract.

ASSIGNMENT AND SUBCONTRACTS (CLAUSE 19)

Assignment

Without the written consent of the other party, neither the employer nor the contractor should assign the contract. Assignment is described as transferring the interests of one party to another.

Subletting occurs where the contractor chooses to enter into a subcontract with other firms for some part or even the whole of the contract, whilst still maintaining the existing relationship to the employer in all respects. Assignment occurs where another firm takes over the contractual rights of the contractor, and for all purposes the contractor to whom the project was contracted ceases to exist. Although subletting occurs to some degree on all construction projects, assignment is extremely unusual.

Subletting – domestic subcontractors – architect's consent

The contractor must also, prior to subletting any portion of the works, obtain the written consent of the architect. The consent must not be unreasonably withheld. The terms of the contract accept that the main contractor will probably not undertake all the trades, but that some of the work will be done by subcontracting firms. The subcontractors whom the main contractor chooses to employ are known as domestic subcontractors. (Nominated subcontractors are by distinction those firms appointed by the architect.) In order that the architect can have some influence over domestic subcontractors, it is necessary to know who they will be in order that they can be approved. This approval would only be withheld where the architect, from past experience or knowledge, considered them to be unreliable or where they previously performed a poor standard of work. The architect may have previously worked with them on other contracts or be dissatisfied with their standards of sample work for the current project.

Subletting – list in contract bills

Where the main contractor intends to sublet some of the work, the names of the firms together with work to be sublet, should be annexed to the contract bills. The contractor must list at least three suitable firms for each section that may be sublet. The architect and the contractor are also entitled, with the consent of each other, to add additional firms to these lists at any time prior to the execution of a binding subcontract agreement. These firms are known as domestic subcontractors.

Subletting – conditions of any subletting

If for any reason the main contractor's employment is determined under the contract, then the employment of these domestic subcontractors will also cease, although they may be re-engaged under a new agreement.

Where subletting occurs then all the conditions in the contract between the main contractor and the employer must apply equally to the subcontract. The main contractor will of course have to ensure these arrangements are written into the subcontract agreement. Otherwise conditions will apply to the main contractor that will not be able to be enforced on the subcontractor. These conditions also apply to materials and goods and, for example, access for the architect to the subcontractor's premises such as workshops.

The contractor's tender will be based upon subcontractors' prices and quotations. However, since the subcontractors are subject to the approval of the architect, this adds some risk to the contractor, if they are not subsequently approved. Other subcontractors will then have to be invited to tender for the subcontract works, and presumably at a higher price. This higher price is borne by the main contractor and is not normally passed on to the employer. The approval of domestic subcontractors by the architect should therefore be sought prior to the signing of the main contract.

The contractor must pay the subcontractors the correct amount and at the appropriate time. Where payments are not received then the subcontractor is allowed to add simple interest until such payments are received. The interest rate charged is 5 per cent above the base rate of the Bank of England. A subcontractor may determine the subcontract under these circumstances.

Nominated subcontractors

The main contractor has no responsibility to supply and fix materials or goods or to execute work that in the documents was intended to be carried out by nominated subcontractors (clause 19.5).

WORKS BY EMPLOYER OR PERSONS EMPLOYED OR ENGAGED BY EMPLOYER (CLAUSE 29)

Information in contract bills

This clause used to be described as the *artists and tradesmen clause*. It envisages that the employer may choose to employ some of the building or specialist trades directly. This work may be of a relatively minor nature or may be concerned with items which are only loosely related to the building industry. The clause is required to secure the right to the site for such persons or firms. The contractor, knowing that these will require entry to the site, can organize and allow for any possible inconvenience or disruption ahead of the work being carried out. This knowledge will also give the opportunity to the contractor of including for any expense that may thus be incurred. If there is any delay on the part of their work, the contractor has some redress by an extension of time (clause 25), loss and expense (clause 26) or determination (clause 28).

Information not in contract bills

If the execution of work under this heading is an afterthought on the part of the employer, the contractor must be prepared to give the same reasonable access to the works. Note that the employer is responsible for their proper organization and control. They will not fall within the jurisdiction of a subcontractor, domestic or nominated, nor the authority of the main contractor.

NOMINATED SUBCONTRACTORS (CLAUSE 35)

Definition of a nominated subcontractor

Works which are required to be carried out by nominated subcontractors are included in the bills of quantities as a prime cost sum. These subcontractors usually supply and fix materials or goods or execute work. SMM7 provides for the inclusion of such items. In addition, works undertaken by nominated subcontractors may arise from the following situations:

- The expenditure of provisional sums.
- An instruction from the architect requiring a variation. This is providing that such work is an extra to the contract, and that it is similar in content to other work already being carried out by nominated subcontractors.
- By agreement between the contractor and architect on behalf of the employer. However, note that the architect does not have the power under the contract to omit contractor's work and award this to a nominated subcontractor (clause 13.1.3). The creation of prime cost sums can only be formed with the agreement of the main contractor.

It is the architect's prerogative to nominate firms to undertake the above work. The firms selected should meet the approval of the contractor. The nomination, however, may still proceed in certain cases where the main contractor believes there are reasonable grounds for refusing such an approval. In these circumstances it may be unwise on the part of the architect to proceed with the nomination of such a firm. If the contractor has a reasonable objection to a proposed nominated subcontractor, the architect should be informed as soon as possible and preferably in writing. This obviously needs to be done prior to actual nomination, but also in sufficient time to enable the architect to nominate others without any delay occurring in the contract programme.

Contractor's tender for works otherwise reserved for a nominated subcontractor

The main contractor may also wish to carry out work for which a prime cost sum has been included in the contract documents. It is sometimes usual to supply, with the tender form, both a list of the main contractor's domestic subcontractors and also a list of prime cost sums for which the contractor desires to tender. Both of these forms are often required to be returned with the contractor's tender. The following conditions from clause 35 will generally apply regarding such prime cost sums:

- The contractor, during the ordinary course of business, must directly carry out this work. This does not preclude subletting if agreed by the architect.
- The contractor must give notice of this intention as soon as possible. This is in order to pre-empt a contract being awarded to another firm. The architect may at the tender stage have already received quotations for this work.
- The architect must be willing to allow the contractor to tender. Refusal to tender, or to accept the contractor's lowest tender should this be the case, will usually mean that the architect prefers a specialist firm to do the work.

If the main contractor is allowed to submit a tender for this work then the invitation to submit a price must make it absolutely clear how the contractor's discounts, profit and attendances will be dealt with. Where the price is submitted in competition then the importance of these items may influence the choice of the successful tender. An obvious way is to treat the main contractor's quotation in the same way as each of the other subcontractors. However, practical difficulties associated with these aspects may favour a composite price being provided from the main contractor.

Procedure for nomination of a subcontractor

Documents relating to nominated subcontractors

The following documents relate to nominated subcontractors. They are based upon a system of nomination that was introduced in 1991.

- The Standard Form of Nominated Subcontract Tender (NSC/T)
  - Part 1: The employer's invitation to tender
  - Part 2: Tender by a subcontractor
  - Part 3: Particular conditions (to be agreed by a contractor and subcontractor nominated under clause 35.6)
- The Standard Form of Articles of Nominated Subcontract Agreement between a Contractor and a Nominated Subcontractor (Agreement NSC/A)
- The Standard Conditions of Nominated Subcontract (NSC/C)
- The Standard Form of Employer/Nominated Subcontractor Agreement (NSC/W)
- The Standard Form of Nomination Instruction for a Subcontractor (NSC/N)

Contractor's right of reasonable objection

The architect should not nominate a subcontractor to whom the main contractor has a reasonable objection. The contractor should make a reasonable objection within 7 days of receiving the instruction from the architect. Where possible, the architect should attempt to remove the objection, so that nomination can then proceed. Alternatively a different firm should be nominated.

Architect's instruction on NSC/N – documents accompanying the instruction

The architect issues an instruction to the contractor on NSC/N, nominating the subcontractor. This is accompanied by:

- NSC/T Part 1, completed by the architect and NSC/T Part 2 completed and signed by the subcontractor and by the architect or the employer as approved. It includes a copy of the numbered tender documents listed in and enclosed with NSC/T Part 1, together with additional documents or amendments approved by the architect.
- NSC/W, completed by the employer and subcontractor.
- NSC/T Part 1, confirming any alterations to:
  - Item 7: obligations or restrictions imposed by employer
  - Item 8: order of works: employer's requirements
  - Item 9: type and location of access
- A copy of the principal contractor's health and safety plan.

Copy of instruction to subcontractor

A copy of the instructions is then sent by the architect to the subcontractor together with a copy of the completed appendix for the main contract.

Contractor's obligations on receipt of architect's instructions

- The contractor upon receipt of this instruction then completes NSC/T Part 3, signed by both the contractor and the subcontractor, and executes NSC/A with the subcontractor.

- Executes NSC/A with the subcontractor.
- Sends a copy of the completed NSC/A and the agreed and signed NSC/T Part 3 to the architect.

Non-compliance with clause 35.7 – contractor's obligation to notify architect

If the contractor after using the best endeavours has not within ten working days complied with clause 35.7 (executing NSC/A) then the architect must be informed in writing of the date when such compliance is expected to be done, or the reasons for the non-compliance, which might include discrepancy in or divergence between the numbered tender documents or a discrepancy or divergence between the numbered tender documents referred to in clauses 2.3 to 2.3.4.

Architect's duty on receipt of any notice under clause 35.8

Within a reasonable time after the receipt of a notice from the contractor under clause 35.8, the architect, after consultation with the contractor, should do one of three things:

- Fix a later date for the contractor's compliance, if appropriate.
- Disagree that the matters identified are of relevance, in which case the contractor must then comply with the nomination.
- Accept that the matters are relevant and issue a further instruction or nominate a new subcontractor.

Payment of a nominated subcontractor

Architect – direction as to interim payment for nominated subcontractor

Before any nominated subcontractor's work can be included in a valuation, invoices must be made available for inspection by the quantity surveyor. The invoice signifies that the subcontractor concerned requires a payment to be made. The quantity surveyor should check the invoice against the accepted quotation. The rates, the prices and the arithmetic should be checked, and a check should also be made to ensure that the quantity of work has been properly executed. Although it is not the quantity surveyor's duty to inspect the quality of materials and the standard of work, work should not be included that is obviously not meeting the requirements of the specification. Where there are any doubts, the matter should be brought to the attention of the architect. The approved invoices are then included in the quantity surveyor's valuation of the work as per clause 30.2, and subsequently in the architect's certificate.

Upon the issue of each interim certificate, the architect must inform the contractor of the amount that has been included for each nominated subcontractor. It must also be clearly stated whether these amounts are interim or final payments due to the subcontractors. The architect will also inform the individual subcontractors accordingly. The main contractor must then make these payments as directed. Deductions cannot be made for contra-charges unless these are agreed by the subcontractor concerned.

Direct payment to nominated subcontractor

Before the issue of each certificate, the contractor must provide some evidence that the appropriate amounts shown on previous certificates have been paid. Such evidence might include a receipt for the amount listed. The absence of this evidence will not automatically assume that the subcontractor has not been paid, but before proceeding to issue another certificate the architect must be reasonably satisfied that this is not the case.

The form of contract provides a remedy for paying a subcontractor direct, in those cases where the main contractor defaults in payment. In circumstances where adequate proof is unavailable to the satisfaction of the architect, the architect must issue a certificate accordingly, together with a copy to the nominated subcontractor. The employer is now able to pay the subcontractor direct and to deduct these amounts from future payments to the contractor. If, however, it is known that the main contractor is shortly due to enter into liquidation, such a direct payment will not be made by the employer. In no circumstances does the employer want to be placed in a position of having to pay a subcontractor twice. Unfortunately, situations do occur where this may be necessary in order to secure the services of a particular subcontractor, without which the project would not be completed.

Agreement NSC/W – prenomination payments to nominated subcontractor by employer

There are also provisions in the form of contract (clause 35.13) to allow an employer to pay a nominated subcontractor for design work, materials and goods prior to the issue of an instruction of nomination. Such direct payments are ignored in computing interim and final certificate payments.

Extension of period or periods for completion of nominated subcontract works

The main contractor is not able to give a nominated subcontractor an extension of time (this being an extension of the period in which the subcontract works are to be carried out) without the written consent of the architect. It is up to the architect to operate the relevant provisions of the subcontract, namely clause 11 in this respect.

Failure to complete nominated subcontract works

If any nominated subcontractor fails to complete the subcontract works within the period allocated on the contract programme, or any revision thereof, then the contractor should inform the architect of this event. The architect should then send a certificate to the contractor accordingly. A duplicate copy of the certificate should be forwarded to the nominated subcontractor concerned. The certificate must be sent within two months of the date of notification to the architect.

Practical completion of nominated subcontract works

When, in the opinion of the architect, a nominated subcontractor achieves practical completion of the works, a certificate should be sent to the contractor to this effect, with a duplicate copy to the nominated subcontractor.

Early final payment of nominated subcontractors

Final payment can be made to the nominated subcontractor prior to that of the main contract. Before this payment can be made, twelve months must elapse from the date of the subcontractor's certificate of practical completion. However, the subcontractor concerned must have remedied any defective work, and must have sent through the contractor to the quantity surveyor the necessary documents for the final adjustment of the contract sum.

Defects in nominated contract works after final payment of nominated subcontractor – before issue of final certificate

If upon discharge by the contractor to the nominated subcontractor, the subcontractor fails to rectify those items which are the subcontractor's responsibility, e.g. defects, shrinkages or other faults in the subcontract works, then in accordance with the contract, the architect will issue an instruction to others to rectify the work. The employer in these circumstances will take the necessary steps to recover the amounts for this remedial work. If the employer is unable to recover these sums from the subcontractor then the main contractor must provide for the necessary payments.

Final payment – saving provisions

The main contractor will, however, be responsible for any defects to the nominated subcontractor's work from the date of the subcontractor's completion up to the issue of the final certificate for the project as a whole. Before the subcontractor is released from the contract, the main contractor will require some form of indemnity against the possibility of defects occurring and the costs of their making good.

Whilst under these circumstances the insurance provisions of clause 22 continue to remain in full force and effect.

Position of employer in relation to nominated subcontractor

The existence of clause 35, or even other clauses within the conditions of contract, will not render the employer liable in any way to a nominated subcontractor. The only exception to this rule may be by way of the terms and agreement set out in

NSC/A. Contract bills often state that once the formal process of nomination has been completed, such subcontractors are not dissimilar to the domestic subcontractors in contractual terms.

Clause 2.1 of Agreement NSC/W – position of contractor

The contractor is not responsible to the employer, in respect of nominated subcontract works, for:

- Design, where items have been designed by a nominated subcontractor
- Selection of materials and goods that have been selected by a nominated subcontractor
- Satisfaction of any performance specification
- Provision of any information in respect of NSC/W

However, the contractor remains responsible for carrying out the work in order to ensure that it complies with the design and that the materials and work standards are in accordance with the specification.

Restrictions in contracts of sale – limitation of liability of nominated subcontractors

The liability of the main contractor to the employer cannot exceed the liability of the subcontractor to the main contractor in respect of subcontract works.

Circumstances where renomination is necessary

Renomination of a nominated subcontractor is necessary if the following conditions arise:

- If the architect is of the opinion that the nominated subcontractor has made a default regarding items in 7.1.1.1 to 7.1.1.4 of the subcontract. The main contractor must have informed the architect of the alleged default together with any observations. The architect must be of the opinion that the subcontractor has made the default.
- If the nominated subcontractor becomes insolvent or makes an arrangement with creditors for the winding up of the company, voluntary or otherwise. This is with the exception of amalgamation or reconstruction.
- If the nominated subcontractor has determined the contract under clause 7.7 of the subcontract.
- The contractor has been required by the employer to determine a subcontractor's employment under clause 7.3 of the subcontract.
- Work properly executed has to be re-executed as a result of compliance with an architect's instruction and the nominated subcontractor does not agree to carry out the work.

Where any of the above five events occurs, then the following are to apply:

- The architect must issue an instruction to the main contractor to give to the nominated subcontractor the notice specifying the default. The instruction may also require the contractor to obtain a further instruction prior to determining the employment of the nominated subcontractor.
- The contractor must inform the architect, following the issue of the notice, if the employment of the nominated subcontractor has been determined. If the second instruction under this clause has been given, then the contractor must confirm the determination to the architect.
- Once the architect has been finally informed of the determination of the nominated subcontractor, renomination is made as necessary. Where such determination has occurred because of the failure of the subcontractor to remedy the defects, the main contractor must be given the opportunity to agree the price to be charged by the substituted nominated subcontractor.

The architect must renominate as necessary in the case of the subcontractor's insolvency. If, however, the architect reasonably believes that the appointed receiver is prepared to continue to completion then the architect may postpone a renomination. This is as long as the relevant subcontract does not detrimentally affect any other person involved in the works.

If the subcontractor has reasonably determined the contract, the architect must renominate as necessary. Any excess price of the substituted nominated subcontractor over that of the originally nominated subcontractor may be recovered by the employer from the contractor. The reasonableness of this charge against the contractor is because, in this case, determination is a result of the contractor's default.

All amounts payable to the renominated subcontractors must be included in the interim certificates and added to the contract sum.

Determination or determination of employment of nominated subcontractor – architect's instructions

The contractor cannot determine the contract of any nominated subcontractor without an appropriate instruction from the architect. If the determination is made under the subcontract then the architect must direct the contractor with regards to any amounts due in an interim certificate.

NOMINATED SUPPLIERS (CLAUSE 36)

Definition of a nominated supplier

Goods and materials which are required to be obtained from a nominated supplier, are to be included in the contract bills as a prime cost sum. Where the architect knows the name of the supplier, this should be stated in the description. The fixing of these items will be undertaken by the main contractor or a domestic subcontractor where the subcontractor is approved. The costs associated with fixing generally include all those items from the point of delivery to the site. The right of nomination is solely that of the architect. The nomination of firms to supply materials or goods may arise in the following ways:

- By the inclusion of a prime cost sum in the contract bills for the supply of goods or materials.
- Where a provisional sum is included in the contract bills, and an architect's instruction names the supplier of any goods or materials involved.
- Where a provisional sum results in an architect's instructions to a sole source of supply of such goods or materials.
- Where a variation made by the architect specifies goods or materials for which there is only a single source of supply.

In the conditions of contract a nominated supplier must be covered by a prime cost sum. The naming of a supplier as part of the contractor's work, even where this is a sole supplier of goods and materials, will not therefore result in a prime cost sum. Work can therefore be measured in the contract bills on a supply and fixing basis, even where only one supplier is known.

Architect's instructions

The architect must issue instructions for the purpose of nominating a supplier. This is usually for work that is covered by a prime cost sum in the contract bill of quantities.

Ascertainment of costs to be set against prime cost sums

The amounts properly chargeable to the employer in respect of nominated suppliers' accounts include the total amounts in respect of materials or goods as follows:

- A cash discount to the contractor of 5 per cent, where the contractor makes the necessary payment within 30 days of the end of the month during which delivery of the materials or goods are made. Any discount offered in excess of this must be passed on for the benefit of the employer. Where no discount has been included, the main contractor is allowed to add one-nineteenth to the net cost of the invoice. Quotations should be checked with regard to the discount offered prior to their acceptance.
- Any tax or duty that is recoverable under this contract. This excludes value added tax, which is capable of being treated as an input tax by the contractor (clause 15).
- The net cost of appropriate packaging, carriage and delivery charges. The main contractor should return the empty packing cases to the supplier for any credit that may be available. The credit is for the benefit of the employer. The cost of such carriage should be identified in the quotation.

- The charges incurred for any adjustment in price, in accordance with clauses 38 to 40.
- If, in the opinion of the architect, the contractor incurs additional expense for which there is no proper reimbursement, this amount may be added to the contract sum. Such an amount may arise as a direct charge made to the contractor by the supplier.

Sale contract provisions – architect's right to nominate supplier

The architect should only nominate firms as suppliers of goods and materials where the following conditions apply. The contractor can refuse to enter into a contract of sale, where they do not.

- The quality of the materials and goods must conform to the standards specified. This is largely a matter for the architect to determine, but it may affect the overall quality of the works and the method of fixing that will be undertaken by the contractor or a domestic subcontractor working under the contractor's direction.
- The nominated supplier must make good any defects, in the materials or goods, which appear before the end of the defects liability period. The supplier should also reimburse the contractor for any costs that may be incurred, for example, in refixing or returning the defective materials to the suppliers. Any extra costs might also presumably include loss and expense caused by delays. Reimbursement of the contractor's costs will only be paid in the following circumstances:
  - Where a reasonable examination of the goods by the contractor would not have revealed their defects.
  - Where the goods have been properly stored and not misused before, during or after their incorporation in the works.
- The delivery of the materials or goods, including unloading, is done in accordance with the instructions of the supplier, or agreed between the supplier and the contractor, or undertaken in a reasonable manner by the contractor. The supplier must also be prepared to accept the contractor's programme as far as delivery is concerned. If the supplier is unable to meet this, the contractor can reasonably expect the architect to nominate an alternative supplier, because they may be crucial to the overall contract programme. The programme may only be varied on the following grounds:
  - force majeure
  - civil commotion, local combination of workers, strike or lockout
  - any instruction of the architect under
    - clause 13.2 (variations)
    - clause 13.3 (provisional sums)
  - failure on the part of the architect to supply information at the due time
  - exceptionally adverse weather conditions
- The nominated supplier must make the appropriate allowance for the cash discount of 5 per cent.

- The nominated supplier will not usually be required to deliver goods to the site after the contractor's determination, unless they have already been paid for by the employer.
- The main contractor must pay the required amount within 30 days of the end of the month during which delivery is made.
- The ownership of the materials or goods will pass to the contractor upon delivery to site by the nominated supplier, regardless of whether payment has been made. This may be a difficult clause to substantiate in practice, particularly if some of the components or materials supplied have a defective title.
- If any dispute arises between the supplier and the contractor, they must first agree to take their differences to arbitration.
- The conditions of sale, often printed on the reverse side of suppliers' quotations and invoices, will not override the conditions as described above.

Contract of sale – restriction, limitation or exclusion of liability

In practice, if the supplier can persuade the architect or the contractor to waive any of the above conditions, this can be done by agreement. Good advice, recommended elsewhere, is that the standard clauses should not be tampered with except in extreme circumstances. However, if the architect wishes to appoint a firm whose contract of sale excludes a condition listed above, then the architect must also be prepared to relieve the contractor of this same obligation towards the employer.

The architect does not have the power to nominate a supplier beyond the scope of clause 36.4.

CHAPTER 22

INJURY AND INSURANCE OF THE WORKS

The definition of *insurance* is that one party (the insurer) undertakes to make payments to or for the benefit of the other party (the assured) upon the occurrence of certain specified events. The insurance contract between these parties is generally contained in a document called an insurance policy. The consideration, which is necessary to make such a contract binding, is provided by the assured in the way of a premium.

An insurance contract is said to be *uberrimae fidei* 'based upon good faith between the parties'. The assured must therefore make a full disclosure of every material fact that is known. A material fact is information which, if disclosed, would influence the judgment of the insurer. Filling in the proposal form incorrectly can make the policy voidable by the insurer, even where an innocent mistake occurs. The insurance policies are usually printed on standard forms by the company issuing its own terms. Each policy must, however, be clear on the terms of insurance, and be accompanied by certain exclusions of liability.

If a situation occurs that may result in a claim by the assured, this must be notified to the company within a reasonable time of the event. The claim is often forwarded by the insurance company to a loss adjuster. This is the general procedure involving sums of even a relatively minor amount. The loss adjuster will then assess the amount of damage and the sum payable in respect of the insurance policy. Although the loss adjusters work on behalf of the insurance company, they will try to achieve an equitable settlement between the parties. In some circumstances the employer or contractor may employ a loss adjuster on their behalf in order to negotiate with the insurance company's own loss adjuster.

The following clauses are examined in this chapter:

- Clause 20   Injury to persons and property and indemnity to the employer
- Clause 21   Insurance against injury to persons and property
- Clause 22   Insurance of the works
- Clause 22A   Erection of new buildings. All-risks insurance of the works by the contractor
- Clause 22B   Erection of new buildings. All-risks insurance of the works by the employer

- Clause 22C   Insurance of existing structures. Insurance of the works in or extensions to existing structures
- Clause 22D   Insurance for employer's loss of liquidated damages
- Clause 22FC   Joint fire code – compliance

The construction industry has a poor performance record in respect of death and injury, especially to workers during construction operations. At the time of going to publication, pressure groups are considering introducing a new Act of Parliament described as culpable manslaughter against employers and contractors. The introduction of the CDM Regulations has assisted in attempting to make construction sites safer in respect of working practices.

INJURY TO PERSONS AND PROPERTY AND INDEMNITY TO THE EMPLOYER (CLAUSE 20)

Liability of contractor – personal injury or death – indemnity to employer

The contractor is liable for, and must indemnify the employer against, any expense, liability, loss, claim or proceedings of any nature. These may arise because of statute legislation or at common law. They are in respect of personal injury to, or the death of, anyone caused by the carrying out of the works by the contractor or others employed, such as subcontractors. The contractor's liability does not extend to any act of negligence either by the employer or persons directly under the employer's control, such as the employer's own contractors (clause 29). Indemnity in this circumstance means the protection that one party gives to another in respect of claims by a third party. Mainly because the contractor may be forced into liquidation, the contract provides for insurance under clauses 21 and 22 of these conditions. If, however, for some reason the insurance fails, this does not relieve the contractor of the indemnity.

Liability of contractor – injury or damage to property – indemnity to employer

The contractor is also liable for, and must indemnify the employer against, injury to property. The injury applies to any type of property and must be as a result of carrying out the works. The two exceptions to this liability are the same as for personal injury, namely the employer's negligence and the employer's own directly employed contractors.

Injury or damage to property – exclusion of the works and site materials

The employer does, however, have the option of accepting some of the risks where one of the alternatives of clauses 22B or 22C is selected. Also, the contractor will not be responsible should damage occur because of an inappropriate or false design, but must ensure that even in these circumstances all reasonable precautions during the execution of the works have been taken. Where loss or damage does occur, the contractor's main area of defence is to point any negligence in the direction of the employer, architect or other agents working on behalf of the employer.

In the case of partial completion and the issue of the appropriate certificate, under clause 18, then the responsibility for insurance of the property automatically becomes that of the employer.

INSURANCE AGAINST INJURY TO PERSONS AND PROPERTY (CLAUSE 21)

Contractor's insurance – personal injury or death – injury or damage to property

The principal contractor under clause 20 has a liability to indemnify the employer against injury to persons and property. The main contractor is responsible for the whole works, including that of subcontractors. The contractor should therefore seek to ensure that all subcontractors maintain insurances to cover the liability of the contractor in respect of all those items in clause 20. If, however, a subcontractor fails to provide the appropriate insurance cover, this will not remove the overall liability of the principal contractor to the employer. The contractor is only exempt in the case of negligence caused by the employer or persons for whom the employer is directly responsible.

Insurance is required against claims submitted by third parties for injury to persons and property. These third parties will include the owners and the occupiers of adjoining properties and the general public. It will also include all those who visit the site or who have lawful entry to the works. A public liability policy will protect the contractor against claims arising through negligence or mistake. There may also be occasions when a third party will be able to substantiate a claim even when negligence cannot be proved. The policy must therefore include clauses to guard against such claims. A major risk with constructional work is the danger of subsidence to adjoining property. It is equally important to ensure that the contractor's public liability policy also covers this event.

The contractor will also be required to provide an employer's liability insurance for workers employed by the contractor. Although National Insurance provides compensation of employees for injury, the employer will still be liable at common law to compensate the employee for injury due to negligence, whether through the fault of the individual concerned or that of another employee. The premiums paid are often calculated on the basis of the annual wage bill.

The contractor must indemnify the employer against all possible claims. The amount of the insurance cover required is stated in the appendix to the conditions of contract. Whilst the sum insured must not be less than the amount stated, the contractor can insure for a greater sum than is required, as explained in a footnote to the conditions of contract.

The employer can at any time request the contractor, or any subcontractor, to produce for inspection documentary evidence that the insurances are properly maintained. This will usually require the appropriate policies being produced at least some time during the early stages of the contract. The employer can, however, require them to be produced at other reasonable times, as long as this is not done simply to irritate the contractor.

If the contractor or a subcontractor fails to insure properly or continue to insure the works as requested, then the employer may insure on their behalf. The employer will pay the necessary premiums and deduct these amounts from sums due to the contractor. The amounts recoverable from the contractor are those amounts that the employer actually pays, rather than any amount inserted by the contractor in the bills. Because this problem is more likely to occur at the beginning of the contract, the employer can deduct the appropriate amounts from interim certificates.

Insurance – liability, etc., of employer

In some circumstances a provisional sum may be included in the bills of quantities for insurances. The contractor should then maintain in the joint names of the employer and contractor, insurances to the appropriate amount as stated in the contract bills. Such indemnity shall cover damage to property other than the works caused by collapse, subsidence, vibration, weakening or removal of support, or lowering of groundwater, unless the damage occurs because of:

- Contractor's liability under clause 20.2
- Errors in the design of the works
- Damage that is likely to be inevitable having regard to the nature of the works, which is almost an uninsurable item
- The employer acceptance of responsibility for the insurance of the works under clause 22c
- Parts that are subject to the issue of a certificate of practical completion
- War, invasion, act of foreign enemy, hostilities, etc.
- Direct or indirect causes by or contributions to or arising from the excepted risks
- Direct or indirect causes arising out of pollution or contamination of buildings, structures, water, land or atmosphere
- Issues in respect of damages for a breach of contract

Any insurance required for the above should be placed with insurers approved by the employer. The contractor must also provide proof of the policy and that the premiums have been paid. The inclusion of a provisional sum may mean that some of these items are difficult to insure.

Excepted risks

In no circumstances will the contractor be liable either to indemnify the employer or to insure against personal injury or death of any person, or damage, loss or injury caused to the works or site materials, work executed, the site, or any property, by the effect of an excepted risk. The excepted risks are defined in clause 1.3 and are as follows:

- Ionizing radiations
- Contamination by radioactivity from any nuclear fuel or from any nuclear waste from the combustion of nuclear fuel
- Radioactive toxic explosive
- Other hazardous properties of any explosive nuclear assembly or nuclear component
- Pressure waves caused by aircraft
- Other aerial devices travelling at sonic or supersonic speeds

The amount of insurance cover required for any one occurrence or series of occurrences arising out of one event, in respect of injury to persons and property, is to be stated in the appendix to the form of contract. The employer may need to seek expert advice in connection with this sum, but in any event the contractor may well wish to maintain third-party policies at a much higher level.

INSURANCE OF THE WORKS (CLAUSE 22)

Insurance of the works – alternative clauses

The responsibility for the cost of making good any damage to the works, during the period of construction allows for one of three choices to be made by the employer. Two of these should therefore be deleted. The conditions do not envisage, in any circumstances, the contractor being responsible for insurance beyond the period of practical completion of the works. They assume that the date of handover of the project will become the entire responsibility of the employer as far as the insurance items are concerned.

The three choices available for the insurance of the works are as follows:

- 22A   Erection of new buildings: all–risks insurance of the works by the contractor
- 22B   Erection of new buildings: all–risks insurance of the works by the employer
- 22C   Insurance of existing structures: insurance of the works in or extensions to existing structures

Where the project is comprised of two separate parts, one being new works and the other being alterations, it is possible to use both clauses 22A and 22C. Confusion might occur regarding the common stock of materials and the apportionment of the preliminaries costs. From a procedural viewpoint, the events would be governed by the alternative invoked when a particular incident occurred. Clause 22D has also been added, to cover the insurance for employer's loss of liquidated damages under clause 25.4.3.

Essentially two kinds of policy are available for property insurance. The first type is known as an indemnity policy. To indemnify means not only to protect against harm or loss and make financial compensations, but also to secure legal responsibility. The payments made by the insurance company are based upon the damage that occurs less any depreciation in the value of the property. The alternative policy has a reinstatement clause which commits the insurers to paying the full cost of replacement, no account being taken of depreciation. If property is insured for less than its true value, the policyholder then will be paid an average sum based upon a ratio of insured value to true value. For example, if the property is only insured at half the true amount then, whatever the size of the claim, only half of it will be reimbursed. Some property owners choose to carry a part of the risk themselves, and in these circumstances the insurance company will then only be responsible for a proportion of the loss. In other circumstances an excess clause may be included to avoid claims for trivial amounts of damage. In addition to the cost of reinstatement, the insured party could be put to further expense as a result of the incident, such as the provision of alternative accommodation. This is known as consequential loss and a clause to cover this should be included in the policy.

Definitions

Clause 22 is largely concerned with definitions; for example, all-risks insurance is defined as insurance which provides cover against any physical loss or damage to work executed and site materials but excluding the cost necessary to repair, replace or rectify:

- Property which is defective due to wear and tear, obsolescence, deterioration, rust or mildew.
- Work or materials lost or damaged as a result of its own defect in design, plan, specification, material or work standards, or which relied upon the support or stability or work which was defective.
- Loss or damage caused by or arising from:
  - Any consequence of war, invasion, act of foreign enemy, hostilities (whether or not war has been declared), civil war, rebellion, revolution, insurrection, military or usurped power, confiscation, commandeering, nationalization or requisition or loss or destruction of or damage to any property by or under the order of any government *de jure* or *de facto* or public, municipal or local authority.
  - Disappearance or shortage if such disappearance or shortage is only revealed when an inventory is made or is not traceable to an identifiable event.
  - An excepted risk (as defined in clause 1.3).
- If the contract is carried out in Northern Ireland then there are two extra exclusions: civil commotion and any unlawful, wanton or malicious act committed maliciously by a person or persons acting on behalf of or in connection with an unlawful association. Reference is made here to the Northern

Ireland (Emergency Provisions) Act 1973. An all–risks policy insurance in the joint names of the employer and contractor must be provided in respect of the above.

Nominated and domestic subcontractors benefit of joint names policies – specified perils

The contractor where clause 22A applies and the employer where either clause 22B or 22C applies, shall insure the works in the joint names of both the employer and the contractor. The joint names policy must provide for either recognition of each nominated subcontractor or include a waiver by the relevant insurers of any right of subrogation which they might have against any such nominated subcontractor in respect of loss or damage by the specified perils where applicable.

The specified perils are defined in clause 1.3 and include fire, lightning, explosion, storm, tempest, flood, bursting or overflowing of water tanks, apparatus or pipes, earthquake, aircraft or other aerial devices or articles dropped therefrom, riot and civil commotion, but excluding the excepted risks defined above.

ERECTION OF NEW BUILDINGS – ALL-RISKS INSURANCE OF THE WORKS BY THE CONTRACTOR (CLAUSE 22A)

New buildings – contractor to take out and maintain a joint names policy for all-risks insurance

The contractor should take out and maintain a joint names policy for all–risks insurance as defined in clause 22.2. The insurance to be provided by the contractor is to be for the full reinstatement value including professional fees, the percentage for which is as stated in the appendix. The policy shall be maintained up to the issue of the certificate of practical completion or up to any date of determination under clauses 27, 28 or 28A. This is regardless of whether the validity of determination has been contested.

Single policies – insurers approved by employer – failure by contractor to insure

The joint names policy must be taken out with insurers approved by the employer. The contractor must send to the architect, for deposit by the employer, a copy of the policy and the receipt for the premiums paid. Where the contractor fails to provide adequate insurance cover in respect of the contract, the employer may insure against the risks described. The employer cannot extend the policy to cover other risks, but can deduct the premiums from amounts owed to the contractor under interim payments.

Use of annual policy maintained by contractor – alternative to use of clause 22A.2

If, however, the contractor maintains an all-risks policy as the normal pattern of business efficiency then this may be accepted as an alternative to the joint names policy outlined above. It will be subject to the endorsement of the employer's interest on the policy, and evidence that it is appropriate and being maintained and that the necessary premiums have been paid. The architect, on occasions, may also wish to inspect the policy and the contract provides for this possibility. The annual renewal date, as supplied by the contractor, of this insurance is stated in the appendix. The provisions outlined previously in regard to any default by the contractor in taking out and maintaining insurances apply equally in this case.

Loss or damage to works – insurance claims – contractor's obligations – use of insurance monies

If the necessity for an insurance claim arises, in respect of loss or damage affecting work executed or site materials, then the contractor shall proceed as follows:

- Inform both the architect and employer in writing of the nature, extent and location of the claim.
- Allow for inspection by the insurers.
- Diligently restore any work which has been damaged, replace or repair any site materials which have been lost or damaged, remove and dispose of any debris and proceed with the carrying out and completion of the works.

The insurance monies received for this work will be paid to the employer. The contractor will be paid for the rectification of this work by instalments under certificates of the architect on a similar basis to that of interim certificates. The sums received for professional fees will be paid to parties concerned. The contractor will not be paid anything extra to cover the restoration of the damaged work other than that received from the insurers. Before proceeding 'with due diligence' the contractor should make sure that the insurers have accepted the claim. Unless it is necessary for safety reasons this part of the works should not be disturbed until the company's loss adjusters have inspected the damage. The final amount received by the contractor and the employer for professional fees may be inadequate since the nature of the damages may not be pro rata in either situation.

ERECTION OF NEW BUILDINGS – ALL-RISKS INSURANCE OF THE WORKS BY THE EMPLOYER (CLAUSE 22B)

New buildings – employer to take out and maintain a joint names policy for all-risks insurance

Under this clause the employer is responsible for insuring the works, and the procedure is similar to the one in clause 22A. However, the insured roles of the two parties are reversed and the employer is the person who takes out and maintains a joint names policy for all risks, for no less than the cover defined in clause 22.2.

Failure of employer to insure – rights of contractor

The principles of providing documentary evidence are with the employer. A subsequent failure to insure the works can be rectified by the contractor, in this case the amount of any premiums involved are added to the contract sum.

Loss or damage to works – insurance claims – contractor's obligations – payment by employer

The procedure to be followed by the contractor in the event of an insured claim also follows the pattern outlined above:

- The contractor must give notice, in writing, to both the architect and the employer if any loss or damage arises in respect of the risks covered by the policy and referred to in clause 22B.
- The occurrence of such loss or damage is disregarded in calculating any amounts due to the contractor, under the contract.
- The work is inspected by the insurers, probably a loss adjuster.
- The contractor, with due diligence, restores the damaged work by replacement or repair.
- If materials have been lost or damaged, they should be removed from the site and replaced with the specified materials.
- The contractor and any nominated subcontractor involved will authorize the insurers to pay the money to the employer.
- The restoration, replacement or repair will then be treated as a variation under clause 13.2.

The value of rectifying the damaged work will usually be on the basis of bill rates or quotations. Agreement with the insurance company's loss adjusters may also be necessary to determine the actual amount to be paid to the employer. Whether the insurer's basis of payment is the same as that of the contract basis is a matter which must remain at the employer's own risk.

INSURANCE OF EXISTING STRUCTURES – INSURANCE OF THE WORKS IN OR EXTENSIONS TO EXISTING STRUCTURES (CLAUSE 22C)

Existing structures and contents – specified perils – employer to take out and maintain joint policies

The employer is responsible for insuring the existing works under this clause in a joint names policy, with the same contractual risks occurring as with clause 22B.

Works in or extensions to existing structures – all-risks insurance – employer to take out and maintain joint names policy

The employer takes out a joint names policy following a similar procedure to clauses 22B. The amount of cover should include the full reinstatement value, including professional fees. This is to ensure that should considerable damage occur, there remain sufficient funds to reinstate and complete the project. The full reinstatement value should include VAT.

Failure of employer to insure – right of contractor

If the employer fails to insure as requested by the contract, the contractor can insure the works and the existing building. This may involve the contractor in carrying out a survey and inventory of the existing building. The costs of the premiums involved and any associated costs are then added to the contract sum.

Loss or damage to works – insurance claims – contractor's obligations – payment by employer

A similar pattern of reinstatement by the contractor and payment by the insurance company occurs as under clause 22B. One assumes that in this case the contractor will be paid the full cost of reinstatement, in accordance with the contract and valuation rules. Any shortfall between that received by the employer from the insurance company and that paid to the contractor would thus be borne entirely by the employer. The insurer may, however, pay for reinstatement on the basis of the contractor's estimate for making good. Under this section, however, the occurrence of loss or damage by a specified peril may result in determination of the contract by the employer or the contractor. Presumably the damage could be so extensive as to make reinstatement of the existing building unwise as far as the employer is concerned. The contractor may also wish to terminate because the damage could result in a project that has become too complex to contemplate.

INSURANCE FOR THE EMPLOYER'S LOSS OF LIQUIDATED DAMAGES (CLAUSE 22D)

The appendix to the form of contract identifies to the contractor whether or not this clause will be applied. Once the contract has been signed then the contractor may be requested to obtain a quotation for such insurance. This will be on an agreed basis of a genuine pre-estimate of the damages which may occur as the result of any delay and at the rate stated for liquidated and ascertained damages included in the appendix. The amounts expended by the contractor to take out and maintain this insurance are added to the contract sum. If the contractor defaults in taking out such an insurance then the employer may insure against any of the risks involved.

JOINT FIRE CODE – COMPLIANCE (CLAUSE 22FC)

Application of clause

Insurers in the construction industry have in recent years become concerned with the high levels of claims for fire loss and damage on construction sites. They have produced the fire code that imposes obligations on all those involved with construction projects, including employers, contractors, subcontractors and the professions. The main objective of the code is to ensure that adequate protection and detection systems in respect of fire safety are in place. Clause 22FC applies where it is stated in the appendix to the conditions of contract.

Compliance with joint fire code

The code ensures that the employer and anyone who is involved in the project complies with its requirements. The contractor should ensure that the code's requirements are enforced by all those brought on to the site, including subcontractors, local authorities and statutory undertakers.

Breach of joint fire code

If a breach of the joint fire code occurs, the insurers under the joint names policy should specify the measures that are required and the time when they are to be completed. The contractor must ensure the remedial measures are carried out in accordance with the contract. Where relevant these might include instructions from the architect.

Indemnity

The contractor indemnifies the employer and vice versa in respect of consequences occuring in breach of the code.

Joint fire code amendments

If, after the base date, the joint fire code is amended then the net extra cost involved in compliance with such amendments is to be added to the contract sum.

CHAPTER 23

FLUCTUATIONS IN COSTS

Contracts are often described as either fixed-price contracts or fluctuating-price contracts. The essential difference is that fixed-price contracts expect the contractor to have allowed for any changes to the contract sum for items such as inflation, etc., whereas fluctuating-price contracts do not. The relevant provisions are contained in the following clauses:

- Clause 37   Fluctuations
- Clause 38   Contributions, levy and tax fluctuations
- Clause 39   Labour and materials costs and tax fluctuations
- Clause 40   Use of price adjustment formulae

FLUCTUATIONS (CLAUSE 37)

Fluctuations can be dealt with in the conditions of contract in one of three ways:

- *Clause 38*: this allows only for the adjustments in price of contributions, levies or taxes; it is technically known as a 'firm price' contract.
- *Clause 39*: the assessment of labour and materials costs and tax fluctuations, using the traditional method of calculation.
- *Clause 40*: the calculation of price fluctuations by the formulae method.

Only one of these clauses may apply and this clause should be stated in the appendix. If this has been omitted then it is presumed that clause 38 will be used. The three clauses are contained in a separate booklet to JCT 98 detailing the various fluctuations provisions of each clause. In the approximate quantities version, clause 38 cannot apply. Note that a strictly firm price version is not now available, since each of the choices allows for fluctuations in cost, at least caused by changes in government legislation.

CONTRIBUTIONS, LEVY AND TAX FLUCTUATIONS (CLAUSE 38)

The contract sum is deemed to have been calculated as detailed by this clause. The adjustments that may therefore be necessary will occur where differences eventually arise because of the following provisions. The contract sum is based upon the types and rates of contribution, levy and tax payable by a person in the capacity of employer.

The prices in the contract bills are based upon those contributions that are current at the base date. Should these change or should the contractor's status be revised, then adjustment will become necessary. Although this clause is one of the trio of increased cost provisions, money may need to be repaid to the employer where the contributions decrease. Note, however, that levies payable under the Industrial Training Act 1964 are expressly excluded from any adjustment.

To enable the contractor to claim fluctuations in the costs of labour, the following provisions must apply. Each employee must have worked on the project for a minimum of two working days in any week for which the claim is applicable. The aggregation of days and parts of a day do not therefore apply. The highest properly fixed tradesperson's rate must be used, provided that such a tradesperson is employed by either the contractor or a domestic subcontractor. The clause is applied in the context of the Income Tax (Employment) Regulations 1973 (the PAYE regulations) under section 204 of the Income and Corporation Taxes Act 1970.

Where any of the tender types or tender rates change, then the net actual amount of an adjustment must be calculated. This sum is then paid to or allowed by the contractor. The change is measured as the difference between monies actually expended and those that were appropriate at the base date. It has already been noted that some changes are based upon a theoretical adjustment. For example, where an overseer works as a craftsperson any adjustments that may be due are based upon the craftsperson rate adjustment. In other circumstances where the workpeople are contracted out within the meaning of the Social Security Pensions Act 1975, the method of adjustment is on the basis that they are not contracted out employees.

The contributions, levies and taxes which are subject to adjustment are those which result because of an Act of Parliament or because of a change in an Act of Parliament. They are the amounts that a contractor must pay as a result of being an employer of labour, and therefore include sums such as the statutory insurances against personal injury or death. Originally the standard form of building contract (1963 edition) allowed only for fixed-price contracts and fluctuating-price contracts. Either the contractor was eligible to claim for increased costs, perhaps because of pay rises, or these were deemed to be included in the fixed-price tender. The mid 1960s saw the introduction of taxes such as selective employment tax, and strictly in the way the fixed-price contract was calculated this sum could not be recovered. Because of the expected further intervention by future governments, the clause was changed so that all contracts would not include the minimum provision of dealing with these items.

Materials – duties and taxes

The contract sum is deemed to have been calculated in the following manner in respect of materials, goods and fuels:

- The prices contained in the contract bills are based upon the types and rates of duty and tax applicable at the base date (excluding VAT). These include amounts payable on the import, purchase, sale appropriation, processing or use of the materials, goods, electricity and, where specifically mentioned in the bills, fuels.
- If any types or rates of duty, in the context of the above, are altered then the net difference between that actually paid by the contractor and what was presumed to have been allowed in the bills must be calculated. This sum will then be either paid to the contractor or refunded to the employer.

Landfill tax

The contract sum is calculated on the incidence and rate of landfill tax. The landfill site operator is accountable to HM Customs and Excise. Where waste materials arising from the works are deposited in such a landfill area and the landfill tax is changed, this will be adjusted within the terms of the contract. However, if the contractor could reasonably have foreseen such changes in such a tax then no reimbursement will be made to the contractor.

Fluctuations – work – sublet – domestic subcontractors

Where the main contractor has chosen to subcontract some of the works to domestic subcontractors then the following will apply. The subcontract must incorporate provisions to provide the same effect as clause 38 of the main contract, together with any percentage that may be appropriate. These adjustments will be paid to the contractor or refunded to the employer, whichever is appropriate.

Provisions relating to clause 38

The contractor must give a written notice to the architect of the occurrence of any of the events regarding the following provisions, for the purposes of this contract:

- Clause 38.1.2   Increases or decreases in rates of contributions, etc.
- Clause 38.1.6   Tender types and rates, and refunds
- Clause 38.2.2   Tender types and rates, and materials
- Clause 38.2.3   Landfill tax
- Clause 38.3.2   Domestic subcontractors' fluctuations

The written notice is a condition with which the contractor must comply in order to recoup any increased costs. The notice must be sent within a reasonable time, so that the architect is aware that further expenditure is likely. Presumably this also gives the architect the opportunity of rejecting a spurious claim that is outside the scope of these conditions.

The quantity surveyor and the contractor will calculate and agree the amount of fluctuation appropriate to each notified event. Any amounts that have been agreed are then added to or deducted from the contract sum. They will also be taken into account when calculating the contractor's determination payments under clause 28.

The contractor must provide the quantity surveyor with the necessary evidence, in order that the amounts of fluctuations may be calculated. This must be done within a reasonable time. The incentive necessary for the contractor to do this is that payment will be received within the interim certificates once they are agreed. When a fluctuation is claimed in respect of employees other than workpeople, the necessary evidence must include a certificate signed on behalf of the contractor each week, certifying the validity of the evidence. Fluctuations are always net under clause 38.4 and are thus exclusive of any profit to the contractor.

Where the contractor is in default over completion, the fluctuations will not be adjusted in line with actual expenditure. The amount of fluctuations will be technically frozen at the level applicable at completion date. The contractor will therefore still be eligible for increases, but not at the true level of expenditure. If the architect has already awarded an extension of time in accordance with clause 25, or subsequently does so, then the actual increases will be reimbursed to the contractor.

These fluctuation provisions are not applicable in respect of:

- Dayworks, because these will have been valued at current rates and therefore take into account the appropriate adjustments.
- Specialist works carried out by nominated suppliers or nominated subcontractors, as all fluctuations will be dealt with by their invoices.
- A contractor acting as a nominated subcontractor, where the above provision will apply.
- Changes in VAT, because these are exempt under this clause.

Definitions for use with clause 38

Clause 38.6 includes the following definitions that are appropriate both to this clause and to clause 39:

- *Base date*: in the first edition of JCT 80 this was described as the date of tender, which was defined as 10 days before the date fixed for the receipt of tenders. Representation to the tribunal suggested that this did not provide a firm enough date because employers infrequently revised the tender date, thereby necessitating recalculation and causing uncertainty. A firm date is now written into the appendix and called the base date, which brings the terminology in line with the formula rules for price adjustment. The necessity for the date is to avoid contractors having to make last-minute adjustments to their tenders.
- *Materials and goods*: this excludes consumable stores, plant and machinery, but can include timber used in formwork, electricity, and fuels if this has been specifically stated in the bills.
- *Workpeople*: persons whose rate of wages are governed by rules, decisions or agreements of the Construction Industry Joint Council or some other wage-fixing body. Overseers, for example, who work as tradespeople will be assumed to be tradespeople for the purpose and application of this clause. Note that workpeople

not only include those employed by the contractor on the project, but also those who may work off site in the contractor's workshops, e.g. joinery shop.
- *Wage-fixing body*: a body which lays down recognized terms and conditions of workers under the Employment Protection Act 1975.

Percentage addition to fluctuation payments or allowances

The percentage stated in the appendix must be added to fluctuations paid or allowed by the contractor. The percentage stated can only be recovered on certain defined fluctuations covered by clauses 38.1.2, 38.1.3, 38.1.6 and 38.2.2. The tendering documents will stipulate the percentage to be added. Although JCT discussed a figure of 20 per cent, the typical amount inserted is nil.

LABOUR AND MATERIALS COSTS AND TAX FLUCTUATIONS (CLAUSE 39)

This is the traditional full fluctuation clause, and therefore the provisions of clause 38 are included. The contractor is thus able to claim for increases in the costs of employing labour and the costs of purchasing materials in addition to the other statutory increases. The definitions of clause 39.7 are identical to those described in clause 38.6.

Labour

The contract sum is deemed to have been calculated as follows. The prices in the bills are based upon the rates of wages, and any changes will result in an adjustment to the contract sum.

- The adjustment is in respect of workpeople on site and of those employed directly upon the works.
- It also includes workpeople directly employed by the contractor who are engaged on the production of materials or goods in the contractor's workshops.
- The rules or decisions of the Construction Industry Joint Council or other appropriate wage-fixing body as applicable to the works will be used.
- Any incentive scheme or productivity agreement under the provisions of rule 1.16 of the Construction Industry Joint Council.
- The rates of wages also include other emoluments and expenses, such as holiday credits and insurances.
- Only wage changes unknown at the base date are eligible for adjustment. Promulgated changes are those that have already been agreed, and must therefore have been allowed for in the tender, even though they may not come into operation until some time after the contract has started. The actual amount of the changes must, however, be known. Proposed unagreed rates are therefore not promulgated within this definition.

- If the terms of the public holiday agreements change during the duration of the project, the appropriate costs of these changes can also be included under the heading of fluctuations. If such changes were promulgated at the time of tender then these are assumed to have been covered by the contractor.

- The contributions and levies of the CITB, for example, are excluded from adjustment and the contractor must therefore bear the costs of all increases. In practice considerable time is generally notified in respect of the future increases in these payments.

- Promulgated wage rate increases known before the base date would also mean that the contractor should have allowed for increases in the respective liability insurance. Only increases that are unknown at the base date can be claimed. Because the employer's liability insurance is often based upon the wages bill of the firm, this would be an accepted adjustable item in other circumstances. However, should the employer choose to provide a more expensive insurance then this claim would not be valid. Increases because of changes in the wages bill or because of inflation are therefore allowable where they were unknown at the time of tender.

- Increases in the cost of productivity bonuses resulting from increases in standard rates of wages are recoverable. Thus, where the appropriate wage-fixing body agrees upon an increase in the basic rate, the contractor will be able to reclaim this similar proportion paid by way of their productivity agreement.

- The contractor's employees, who are outside the scope of workpeople but who are engaged on the works, are subject to fluctuations as if they were craft operatives. An overseer working as a bricklayer would be reimbursed accordingly.

- The prices in the bills are based upon basic transport charges as submitted by the contractor and attached to the contract bills. Provision is therefore made for increases in travel allowances payable to workpeople under a joint agreement, including increased costs of employer's transport where used, or public transport fares.

- Clause 39.2 is largely a reproduction of the appropriate parts of clause 38, applicable to contributions, levies and taxes.

Materials

The contract sum is deemed to have been calculated in the following manner:

- The prices contained in the contract bills are based upon the market prices of materials, goods, electricity and, where specifically stated in the bills, fuels which were current at the base date.

- The above market prices are referred to as basic prices, and are included on a basic price list for these items. In practice only the major cost items are listed. This may be done by the quantity surveyor, in which case the list is determined at the time of tender, or alternatively the contractor is asked to prepare a list together with prices.

- Any changes in the market prices of these items are then subject to fluctuations and adjustments of the net changes. Only the items included on the basic price are subject to adjustment.
- Market price changes include any changes in duty or tax (other than VAT) on the import, purchase, sale appropriation, processing or use of goods, materials, electricity or fuels as specified.
- In order to avoid contention later, the quantity surveyor must approve the basic prices shown on the list. These are usually supported by invoices from a single supplier. It is usual then for the contractor to obtain the actual materials from this supplier. Any changes between quotation and invoice can therefore be calculated easily. Where it is necessary for the contractor to use an alternative supplier, the change in price may be calculated between the invoice and the price on the basic price list and adjusted for any changes due to the use of a different supplier.

Domestic subcontractors

The provisions relating to domestic subcontractors correspond generally to the provisions of clause 38.3, but of increased scope to cover wage fluctuations.

Clauses 39.5 to 39.8 generally also correspond to the similar provisions included under clause 38. Here they are set out briefly:

- Clause 38.5.1   Written notice of fluctuations required from the contractor
- Clause 39.5.2   Notice must be provided within a reasonable time after the occurrence of any price change
- Clause 39.5.3   Agreement of amount payable to be determined between the contractor and the quantity surveyor
- Clause 39.5.4   Amount of fluctuations to be added to the contract sum
- Clause 39.5.5   The contractor must provide the evidence for the computations
- Clause 39.5.6   Fluctuations must not result in any alteration to the contractor's profit
- Clause 39.5.7   Increased costs occurring after completion will only be paid for on the basis of those prevailing at the completion date; this may result in circumstances where the contractor has defaulted over completion
- Clause 39.5.8   Percentage addition to be included in the appendix where appropriate

Excluded work

- Work for which the contractor is paid daywork rates, since these will be current rates
- Work executed by nominated subcontractors or nominated suppliers
- Work executed by the contractor under clause 35.2
- Changes in the rate of VAT, since this is not dealt with under the contract

USE OF PRICE ADJUSTMENT FORMULAE (CLAUSE 40)

The traditional method of calculating fluctuations has many drawbacks:

- Disagreements over what is allowable
- Shortfall in recovery
- Delays in payment
- Extensive work in documenting and checking claims

An alternative procedure was therefore developed in the 1970s, based upon the idea of using index numbers. This method is known as the formula method of price adjustment and is based upon rules published by the Joint Contracts Tribunal. The indices which are used are those provided by the National Economic Development Office (NEDO) and the method has thus become known as the NEDO formulae. The rules are given contractual effect by incorporation in the main contract or subcontract. Two series have been devised:

- *Series 1*: 34 work category indices (March 1975)
- *Series 2*: 48 work category indices (April 1977, amended 1980)

The series 2 rates are in three sections:

- Definitions, exclusions, correction of errors
- Operation of work category and workgroup methods
- Application of formulae to the main contractor's specialist work

These have now been consolidated in the 'Formula Rules 1987'. In addition there are formula rules for subcontracts.

The use of clause 40 prescribes that the contract sum will be adjusted by use of the formula rules. The date of tender is chosen as the base date to which the indices will apply. Since the contract sum is exclusive of VAT, the operation of this clause will in no way affect the VAT agreement and clause 15.

The following definitions from rule 3 of the formula rules are to apply to clause 40:

- *Base date*: this is defined in the same way as clause 38.
- *Index numbers*: obtained from the monthly *Bulletin of Construction Indices* published by the Stationery Office.
- *Base month*: normally the calendar month prior to the date when the tenders are returned.
- *Valuation period*: the date of the valuation and the midpoint of the period when the fluctuations are calculated.
- *Work categories*: the 34 (series 1) or 48 (series 2) classifications of the contract work.
- *Workgroups*: the less detailed classification which may be used in preference to work categories. It operates by aggregating work categories into larger units. It reduces the number of calculations but may make the results untypical of the work being measured.

- *Balance of adjustable work*: some sections of the bills, e.g. preliminaries, are excluded from the work categories and are valued on the basis of averaging the index numbers which are used.
- *Non-adjustable element*: a proportion of the value of work may be excluded from the operation of the formulae under the local authorities edition of the JCT form.

The formula calculation adjustment is added to each valuation on the basis of the work carried out that month. Unlike the calculation of increased costs in clause 39, the amount determined under this clause is subject to retention. Initially the amount will be calculated using provisional indices, and three to four months later these will be firmed up, when the final indices can be determined.

Some articles may be manufactured outside the UK, and it would therefore not be realistic to apply the formula rules in such cases. If any variation in the market price occurs, they will be subject to fluctuation adjustment using similar provisions to those of clause 39.

The method of adjustment to nominated subcontracts depends upon the instructions, tender and agreement reached between the subcontractor and the architect. The available methods using a formula calculation are as follows:

- Electrical, heating and ventilating, air-conditioning, lifts, structural steelwork and catering; the relevant specialist rules are 50, 54, 58, 63 and 69.
- Where no specialist formula applies, use the formulae in part 1 of section 2 and one or more of the work categories set out in Appendix A to the formula rules.
- Some other appropriate method.

If the contractor decides to sublet any portion of the works, these provisions must be incorporated into the subcontract as appropriate.

The quantity surveyor and the contractor have the power to agree any alteration to the methods and procedures for the formula fluctuations recovery. This is presumably to help reduce the number of tedious calculations where an agreeable amount can be ascertained. The amounts calculated are then deemed to be the formula adjustment amounts. This is provided that such sums are a reasonable approximation to those that would have been calculated by the operation of the formula rules.

Where the contractor does not complete the works by the completion date, the following procedure is to be adopted. The value of work completed after that date is subject to formula adjustment on the basis of the indices applicable at the relevant completion date. Where a contract is therefore running behind schedule, the employer will consistently be paying higher amounts month by month than would otherwise be the case. But if the architect has granted an extension of time and fixed a later completion date, then formula adjustment will continue to run at the current dates.

CHAPTER 24

FINANCIAL MATTERS

Two clauses have been included under this heading dealing with matters relating to financial legislation:

- Clause 15   Valued added tax – supplemental provisions
- Clause 31   Construction industry scheme

VALUE ADDED TAX – SUPPLEMENTAL PROVISIONS (CLAUSE 15)

Definitions – VAT agreement

Value added tax (VAT) was introduced to the construction industry through the Finance Act 1972. It is administered through HM Customs and Excise. During the Chancellor's annual Budget statement there is the opportunity to amend both the extent and rate of this tax. This has been done many times since 1972.

Building work is either standard-rated (currently 17.5 per cent) or zero-rated. Examples of zero-rated works are residential buildings, which include children's homes, old people's homes, homes for rehabilitation purposes, hospices, student living accommodation, armed forces living accommodation, religious community dwellings and other buildings that are used for residential purposes. Certain buildings intended for use by registered charities may also be zero-rated. Buildings which are specifically excluded from zero-rating include hospitals, hotels, inns and similar establishments. The conversion, reconstruction, alteration or enlargement of any existing building is now always standard-rated, although this was not always the case when the tax was first introduced. All services which are merely incidental to the construction of a qualifying building are standard-rated, such as architects', surveyors' and other consultants' fees and much of the temporary work associated with a project. Items typically described as 'furnishings and fittings' (fitted furniture, domestic appliances, carpets, free-standing equipment, etc.) are always standard-rated irrespective of whether the project may be classified as zero-rated. The VAT guides provide further information.

Contract sum – exclusive of VAT

The contract sum referred to in article 2 and clause 14 is exclusive of VAT. Adjustments to the contract sum will also be exclusive of the tax, since the conditions specifically state that VAT will not be dealt with under the terms of the contract.

Possible exemption from VAT

If after the date of tender the goods and services become exempt from VAT, then the employer must pay the contractor an equal amount to the loss of the contractor's input tax. This will then equate with the amount that the contractor would otherwise have recovered.

The supplemental provisions are included into the contract by clause 15.1. The employer agrees to pay the appropriate tax to the contractor that is chargeable by the Customs and Excise office. The employer must therefore pay to the contractor the appropriate amount in respect of any positively rated items in the contract. The procedure for this payment is as follows:

- The contractor gives to the employer a written provisional assessment of the respective values of any goods and materials which have been included in a payment, and are subject to VAT.
- This should be done not later than the date for the issue of each interim certificate.
- The contractor must specify the rate of tax chargeable on these items. The rate is fixed by the Customs and Excise but it could change, and several different rates could be introduced.
- The employer, on receipt of the provisional assessment from the contractor, must calculate the amount of tax due.
- This amount is then included with the amount of the interim certificate and paid to the contractor.
- If the employer has reasonable grounds for objection to the provisional assessment, the contractor must be informed in writing within three working days of receipt.
- The contractor must then either withdraw the assessment, and thereby release the employer from the obligations, or confirm the assessment.
- When the certificate of making good defects has been issued, the contractor must then prepare a written final statement of the respective values of all the supplies of goods and materials. The contractor must specify the respective rates of tax on these items.
- This final statement can be issued either prior to or after the issue of the final certificate. For practical purposes it is advantageous to issue the statement after the final certificate has been accepted.
- The employer, on receipt of the final statement, must calculate the tax that is due. The balance to the contractor is paid within 28 days of receipt of the statement. In those circumstances where the employer may have overpaid the

contractor in respect of VAT, a refund of the appropriate amount will be due to the employer.

- The contractor must issue to the employer receipts under the certificates for the appropriate amount of tax that has been paid. These receipts must comply with regulations 2(2) of the VAT Regulations 1972, including the particulars as required by regulation 9(1), taking into account any of the amendments or re-enactments.
- The employer must disregard any set-off in respect of liquidated damages when calculating and paying the amounts of VAT due to the contractor. The contractor in a similar manner must ignore contra-charges of liquidated damages by the employer.
- If the employer and contractor disagree, it may require the decision to be made by the VAT commissioners. This should be done before the payment becomes due. The employer can also request the contractor to appeal further to the commissioners, should there still be disagreement with their findings. In these circumstances the contractor would be able to claim any costs or expenses involved from the employer. Before the appeal can proceed, the employer must pay to the contractor the full amount of the tax that has been charged.
- The commissioners' appeal decision is final and binding, unless they subsequently introduce a correction to the tax that has been charged.
- Arbitration is not applicable to VAT assessments by the commissioners.
- If the contractor fails to provide a receipt for the tax paid by the employer, then the employer is not obliged to make any further payments. This applies only if the employer requires a validated receipt for tax purposes, or where the employer has paid tax in accordance with the provisional assessments.
- If the employer has determined the contractor's employment, any additional tax which the employer may have to pay as a result of the determination may be set off against any payments to be made to the contractor.

CONSTRUCTION INDUSTRY SCHEME (CLAUSE 31)

The purpose of the original legislation was to deal with the problems of tax evasion by subcontractors, particularly the labour-only 'lump' subcontractors. Procedures were established whereby the main contractor collected tax on behalf of the Inland Revenue from those subcontractors who did not hold a tax certificate. Those subcontractors who held the appropriate 714 certificate were eligible for full payment on the basis that they dealt with the Inland Revenue direct. The existing statutory tax deduction scheme clause was replaced with the construction industry scheme (CIS) in June 1999.

Definitions included within clause 31

Act	Income and Corporation Taxes Act 1988, together with any statutory amendment or modification.
Authorization	Form CIS 4, CIS 5 or CIS 6, or a certifying document created on the contractor's headed stationery.

Construction operations	Those operations that are defined in section 567 of the Act.
Contractor	A person who is a contractor for the purpose of the Act.
Direct cost of materials	The direct cost of materials to the contractor.
Regulations	Income Tax (Subcontractors in the Construction Industry) Regulations 1993 SI 743, as amended by the Income Tax (Subcontractors in the Construction Industry)(Amendment) Regulations 1998, SI 2622.
Statutory deduction	Referred to in section 559 of the Act.
Subcontractor	Any person who is a subcontractor for the purposes of the Act and regulations.
Voucher	A tax payment voucher in the form CIS 25 or a gross payment voucher CIS 24.

Whether employer is a contractor

In the appendix to the form of contract the employer needs to be stated as either *a contractor* or *not a contractor*. Where the former applies, the employer is responsible for making the statutory deduction. Where the employer is *not a contractor*, the appropriate deductions must be made by the main contractor. If the responsible party fails to make the necessary deduction, they may find themselves liable to the authorities for the tax. The status of the employer for tax deduction purposes is also referred to in the fourth recital. An employer may automatically be a *contractor* to cover a wider statutory position, for example, in the case of local authorities. In the case of private employers it will largely depend upon whether the employer is engaged in *construction operations*. These are defined very broadly in the legislation.

Payment by employer – valid authorization essential

The employer must not make any payments under the contract until the valid authorization has been provided.

Validity of authorization – employer's query

Where an employer is not satisfied with the validity of the authorization, the contractor must be informed in writing. This will give the reasons why the validity is being questioned. This is to avoid any possibility of fraud taking place. No payments will be made until the authorization is considered to be valid. The contractor may be able to resubmit the original authorization if a letter from the contractor's tax office is able to confirm its validity.

Authorization

Where the authorization is a CIS 4 registration card, then 7 days prior to final payment, the contractor must give a statement to the employer showing the direct cost of materials. This is to enable the employer to make only a statutory tax deduction in respect of other items that are not defined as materials. Where the employer complies with this, then the contractor must indemnify the employer accordingly. Where the employer feels that the statement provided by the contractor is false, then the employer must make a fair estimate of the direct costs of materials. Where the authorization ia a valid CIS 5 or CIS 6 or an appropriate certifying document, then the employer must pay the amount due without making a statutory deduction.

Change of authorization

Where the contractor is subsequently issued with a change in CIS certificates then the employer should be informed immediately. Where the employer considers this to be valid, the appropriate action as described above will then be applied. Where the relevant authorization is withdrawn by the Inland Revenue, for any reason, the contractor must notify the employer, so that appropriate future action can then be taken. Following expiry of a CIS 5 or CIS 6 the employer must make no future payments to the contractor until the contractor provides the appropriate authorization.

Vouchers

Where authorization CIS 4 applies and the employer has already made payments to the contractor, the employer on the nineteenth day of the month provides the contractor with a copy of the CIS 25 voucher. This will have been sent to the Inland Revenue and will indicate the payments made and the tax deducted. Where authorization CIS 6 applies and the employer has made payments to the contractor, then the employer will add their tax reference and send the voucher to the Inland Revenue.

Correction of errors in making the statutory deduction

Where an employer makes an error or omission in the calculation of the statutory deduction, the error is usually corrected by repayment or by further deduction from payments due to the contractor.

Relation to other clauses

The conditions of this clause will prevail over other clauses in the contract should there be any differences in interpretation.

Disputes or differences

The relevant procedures that are outlined in the contract are to be used if there are any disputes or differences in interpretation. This is unless an Act, regulation or statutory instrument overrules such an interpretation.

CHAPTER 25

CLAUSES OF A GENERAL NATURE

The following clauses from JCT 98 have been grouped together for convenience but they apply to matters of a more general nature:

- Clause 1   Interpretations, definitions, etc.
- Clause 2   Contractor's obligations
- Clause 4   Architect's instructions
- Clause 5   Contract documents
- Clause 34   Antiquities
- Clause 41   Settlement of disputes – adjudication – arbitration – legal proceedings

Clauses 1, 2, 4 and 5 deal with matters of a broad nature that set the scene for the contract in general. Clause 1, for example, provides a considerable list of definitions which are then used throughout the contract. The inclusion of these avoids undue confusion both in the interpretation of the clauses and the use of the contract in practice. Clause 2 defines very briefly the contractor's obligation in connection with the works, and clauses 4 and 5 together cover those matters established as contractual and the procedures and provisions for instructions from the architect.

Clause 19A, dealing with fair wages payments to employees, has since July 1988 been deleted from the local authorities edition. Also clauses 32 and 33 are not used in JCT 98. These referred to outbreak of hostilities and war damage, respectively. In reading through contract documents some forty years ago, these seemed an unlikely occurrence in comparison to today's society. The clauses detailed the procedures to be followed in the event of such occurrences. These clauses were deleted in July 1992, with the provisions and conditions now included within the relevant clauses dealing with insurance provision.

The discovery of antiquities (clause 34) is not a remote possibility, since a number of archaeological finds have been made in recent years. It may be every builder's dream to unearth a unique Roman vase in one piece, but its value stays strictly with the landowner. Areas of known previous settlements may impose this condition more frequently upon developers and builders. Clause 41 is a clause which embraces arbitration good practice.

INTERPRETATIONS, DEFINITIONS, ETC. (CLAUSE 1)

The articles of agreement, the conditions and the appendix are to be read as a whole, and together they form the standard form of building contract. They cannot therefore be used independently, and in many instances they have an important bearing upon each other. For example, the details completed in the appendix will generally override the specific requirements indicated in each clause. The appendix will make recommendations for the contract, but the parties have some liberty on the amounts or times that they wish to insert against the appropriate clause (Chapter 14).

Definitions included in clause 1.3

A majority of the definitions listed also refer to the appropriate clause references, and these are discussed in detail where necessary.

Adjudication agreement	Reference to clause 41a for adjudication, clause 41b for arbitration and clause 41c for legal proceedings.
All risks insurance	Reference to clause 22.2 for the items to be covered.
Appendix	That part of conditions that are completed by the parties to the contract.
Base date	Referred to in clauses 38, 39 and 40 for example.
CDM Regulations	The Construction (Design and Management) Regulations 1994, including any amendments at the signing of the contract. The contractor will be required to comply with revisions that occur during construction but will be reimbursed accordingly.
Completion date	The date for completion as fixed and stated in the appendix or any date fixed either under clause 25 or in a confirmed acceptance of a 13A quotation.
Conditions	Clauses 1 to 37, either clause 38, 39 or 40, clauses 41 and 42 and the supplemental provisions (the VAT agreement) annexed to the articles of agreement.
Contract bills	The bills of quantities referred to in the first recital which have been priced by the contractor and signed by or on behalf of the parties to the contract.
Contract documents	The contract drawings, contract bills, the articles of agreement, the conditions and the appendix.
Contract drawings	The drawings referred to in the first recital which have been signed by or on behalf of the parties to the contract.
Contractor	The person (or firm) named as the contractor in the articles of agreement.
Date for completion	The date fixed and stated in the appendix.
Date for possession	The date stated in the appendix under reference to clause 23.1.

Defects liability period	The period named in the appendix under the reference to clause 17.2.
Employer	The person named as the employer in the articles of agreement.
Excepted risks	Ionizing radiations or contamination by radioactivity from any nuclear fuel or waste.
Final certificate	The certificate to which clause 30.8 refers, i.e. the last certificate to be issued under the contract.
Health and Safety plan	The plan referred to in the appendix to be provided by the principal contractor.
Numbered documents	These refer to the agreement in the articles of nominated subcontract agreement.
Planning supervisor	The architect or other person named in article 6.1 to perform this function.
Practical completion	When the works are complete as described in clause 17.1.
Specified perils	Fire, lightning, explosion, storm, tempest, flood, bursting or overflowing of water tanks, apparatus or pipes, earthquake, aircraft and other aerial devices or articles dropped therefrom, riot and civil commotion, but excluding excepted risks.
Works	The project to be constructed as described in the first recital and described in the contract documents.

Contractor's responsibility

This clause reinforces the responsibilities and obligations of the contractor under the terms of the contract. It essentially states that, regardless of any obligations the architect may have towards the employer or whether or not the employer appoints a clerk of works, the contractor will nevertheless be responsible for carrying out and completing the works in accordance with clause 2. This is so, irrespective of whether the architect or clerk of works inspects the works, workshops or any other place where work is being prepared. The architect is not made responsible under JCT 98 for the supervision of the works which the contractor is to carry out and complete. In essence the contractor's responsibility for work is in no way reduced by the conditions which allow for an architect or clerk of works' inspection.

Reappointment of planning supervisor or principal contractor – notification to contractor

If the employer replaces the planning supervisor or any contractor appointed as the principal contractor, the contractor must notified in writing.

Giving or service of notices or other documents

Notices are usually posted to the principal office of a firm unless this is otherwise stated in the contract documents.

Reckoning periods of days

Where action is required by a party under the terms of the contract, the period of time is calculated immediately after the date when the specified period begins. If a public holiday occurs within this period then it is excluded from the period stated.

Employer's representative

The employer may choose to appoint some other person to fulfil this function under the terms of the contract. The contractor must be appropriately informed. In order to avoid possible confusion over conflicting roles, neither the architect nor the quantity surveyor should be appointed to this position.

Applicable law

The laws of England are applicable to the contract unless otherwise stated. If the two parties agree to carry out the works under a different legal system then amendments to this clause (clause 1.10) should be stated.

Electronic data interchange

Where the appendix so states, the 'supplemental provisions for EDI' annexed to the contract shall apply.

CONTRACTOR'S OBLIGATIONS (CLAUSE 2)

Contract documents

The contractor will carry out and complete the works in accordance with the:

- Contract drawings
- Contract bills
- Articles of agreement
- Conditions of contract
- Appendix to the conditions of contract
- Numbered documents (subcontract agreements)

The above are collectively known as the contract documents. The contractor must use the materials and work standards that have been specified. The approval of the quality of materials or standards of work will be decided by the architect. These are to be to a reasonable satisfaction. The architect can, however, only require that standard described in the contract documents. In the selection of materials and the assessment of work, the architect and the documents must be precise. Although objectivity may achieve the appropriate and desired standards, some subjective analysis and interpretation by the architect will also be necessary.

Contract bills

The contract bills cannot override or modify either the application or the interpretation of these conditions of contract, with the exception of the rules for measurement. Unless it has been previously stated, JCT 98 assumes that SMM7 will have been used for the preparation of the bills. If other methods of measurement are preferred and have been used, this must be clearly stated in the contract bills. If a quantity surveyor chooses to measure items not in accordance with SMM7, this too must be stated. Any errors subsequently found in the contract bills resulting from their method of preparation must be corrected. The correction of these errors is automatic and does not require a variation instruction to be issued by the architect. The errors referred to cover descriptions, quantities or omission of items. They are errors made by the quantity surveyor during the preparation of the contract bills. Errors in pricing resulting from mistakes made by the contractor do not come within this definition, and will therefore not be corrected once the contract has been signed. A procedure for their correction prior to the contractor's tender being accepted is described in more detail in Chapter 8.

Discrepancies or divergences between documents

If the contractor finds any discrepancies or divergences between or within the following documents, the architect should be informed of such differences:

- Contract drawings
- Contract bills
- Architect's instructions
- Additional drawings or documents
- Numbered documents (nominated subcontracts)

The architect must then give instructions to the contractor in order to clarify the discrepancy or divergence, so that the works can proceed. The contractor cannot assume, for example, that the drawing will automatically take preference over the bills. However, it can be reasonably assumed that the conditions of contract will always take preference unless there is some clause to the contrary (clause 2.2.1). Note that the conditions of contract are not capable of discrepancy.

Discrepancies or divergences between the statement in respect of performance-specified work and an architect's instruction may also occur. If this occurs after the issue of the contractor's statement, the architect must be informed in writing in order that appropriate instructions can be given. If the contractor or architect find any discrepancy in the contractor's statement, the contractor must correct the statement to remove any discrepancy and inform the architect in writing of the correction that has been made. In this situation there will be no cost to the employer.

The conditions do not imply that the contractor must go looking for differences in the documents. If the compliance with the drawings later revealed that the bills were different but correct, the architect would need to issue further instructions for the correction of this work, if that was so desired. The contractor should, however, take all reasonable steps to avoid the occurrence of this situation.

ARCHITECT'S INSTRUCTIONS (CLAUSE 4)

Compliance with architect's instructions

The contractor must comply with all instructions issued by the architect, unless a reasonable objection in writing is made regarding the non-compliance. A reasonable objection may include the refusal to accept the nomination of a subcontractor, where an unsatisfactory relationship has existed on a previous contract.

Provisions empowering instructions

The contractor may also challenge the architect's authority to issue certain instructions. In these circumstances the architect may be requested to specify in writing the provision in the contract which empowers the issue of the particular instruction. The architect must comply with this request. If the two parties cannot agree upon this point, the matter may be referred to arbitration for a decision. If the contractor complies with the instruction on the basis of the architect's reply, then this shall be assumed to be an instruction under the terms of the contract.

The architect may write to the contractor requesting the immediate compliance with an instruction. If after 7 days from receipt of this notice the contractor does not comply, the employer may employ other firms to execute the work. The costs involved with this will be deducted from monies due to the contractor, or they may be recoverable as a debt by the employer. In more severe cases, particularly where the contractor persistently fails to comply with a written instruction, the employer may terminate the contractor's employment under clause 27.

Instructions in writing

All architect's instructions must be in writing to have any contractual effect.

Procedure if instructions given otherwise than in writing

If, however, oral instructions are given by the architect, the following procedure should be adopted:

- Confirmation should be given in writing within 7 days by the contractor to the architect.
- Unless the architect dissents in writing within a further 7 days, the instruction shall be accepted as an architect's instruction within the terms of the contract.
- Alternatively, if the architect confirms in writing the oral instruction within 7 days, then the contractor needs only to accept this as an architect's instruction.

■ If neither the architect nor the contractor confirms the oral instruction, but the contractor executes the work accordingly, then it may be confirmed by the architect at any time prior to the issue of the final certificate.

CONTRACT DOCUMENTS (CLAUSE 5)

The contract documents using the JCT 98 form of contract with quantities are as follows:

■ *Contract drawings* are the drawings which have been signed by or on behalf of the parties.
■ *Contract bills* are the bills of quantities which have been priced by the contractor and signed by or on behalf of the parties.
■ *Form of contract* is JCT 98 duly completed where required.

Where bills of quantities have not been prepared, either a specification or a schedule of rates may become a contract document. In these circumstances use the without quantities form of contract.

Custody of contract bills and contract drawings

The contract drawings and contract bills are to remain in the custody of the employer but must be available at all reasonable times for the inspection of the contractor.

Copies of documents

When the contract has been signed the architect must provide the contractor, free of charge, with:

■ One copy of the contract documents certified by the employer
■ Two further copies of the contract drawings
■ Two copies of the unpriced bills of quantities

Descriptive schedules, etc. – master programme of contractor

Within a reasonable period of time the architect should then supply the contractor with further information, which might include descriptive schedules and further drawings to amplify the contract drawings. These additional documents are not allowed to impose further obligations beyond those described in the contract documents.

It is the contractor's responsibility to provide the architect with two copies of the master programme. If this requires updating because of an extension of time then the contractor must supply a further copy of the revised master programme. A copy of the master programme should be retained on site for reference purposes. There is no requirement under the clause for the contractor to indicate the progress of the works.

Information release schedule

The introduction of an information release schedule is a new feature in JCT 98. This schedule informs the contractor when information will be made available by the architect. The schedule is not annexed to the contract. However, the architect must ensure that the information is released to the contractor in accordance with that agreed in the information release schedule. In practice the architect would need to coordinate this with the contractor's master programme for the works.

Provision of further drawings or details

The architect must supply to the contractor two copies of any additional drawings or details that may be necessary to explain the contract drawings. These may be necessary in order that the contractor can properly carry out the works to the appropriate and designated satisfaction of the architect. The contractor may in some circumstances have to request such information from the architect, giving reasonable notice of this intent.

Availability of certain documents

The contractor must also keep one copy of the contract drawings, one copy of the unpriced bills and other descriptive schedules, documents and drawings on the site for reference by the architect or the architect's representative at all reasonable times. Copies of all the information provided by the architect for the construction of works should be retained on site by the contractor for use by the contractor and the architect. A copy of the master programme must also be available together with a copy of any further drawings and details.

Return of drawings, etc.

When the contractor has received the final payment under the terms of the contract, request may be made to return all drawings, details schedules and other documents which bear the architect's name to the architect. The copyright of the design is vested in the architect. If the building owner or the contractor wishes to repeat the design, then the architect can request a further fee.

Limits to use of documents

None of the documents mentioned must be used for any purpose other than this contract. The contractor's rates must not be divulged to others, or be used for any purpose other than this contract. Where permission is sought from the architect or contractor, as appropriate, the contractual information may be reused for other purposes.

Issue of architect's certificate

Any certificates that the architect issues shall be sent to both the contractor and the employer concurrently.

Supply of as-built drawings, etc. – performance-specified work

Before the date of practical completion, the contractor, without charge, should supply the employer with drawings and other information that relate to the performance-specified work. This will show the work as built and also include maintenance and operating schedules, where appropriate.

ANTIQUITIES (CLAUSE 34)

Effects of find of antiquities

Antiquities include fossils and other items of interest or value which may be found on the site or during the excavation part of the works. These items become the property of the employer, and if such finds occur, the contractor should:

- Use the best endeavours not to disturb the object and, if necessary, cease construction operations within the vicinity until instructions are received from the architect. Where there is any reasonable doubt about a discovered object, this procedure should be followed as a precaution.
- Take all necessary precautions in order to preserve the condition of the objects from possible damage until they can be dealt with by experts if necessary.
- Inform either the architect or clerk of works of the discovery and its precise location on site.

Architect's instructions on antiquities found

The architect must issue instructions regarding the removal of antiquities. This may involve a third party, such as an archaeological society, examining, excavating or removing the object. Such a party will not be described as a subcontractor but as persons directly employed by the employer.

Direct loss and/or expense

The architect may consider that the above instructions involve the contractor in direct loss or expense, for which reimbursement is not provided elsewhere under the contract. In these circumstances the architect or usually the quantity surveyor shall ascertain the amount of such loss or expense. This amount will then be added to the contract sum. If the architect considers that an extension of the contract is appropriate then the removal of the antiquity becomes a relevant event under clause 25.

SETTLEMENT OF DISPUTES – ADJUDICATION – ARBITRATION – LEGAL PROCEEDINGS (CLAUSE 41)

Clause 41A: adjudication

This clause applies where a party refers any dispute or difference to adjudication. The adjudicator is a person to whom both parties agree. An appointment should normally be made within 7 days. Where the adjudicator becomes ill or unable to carry out the duties assigned then the parties may decide to appoint a new adjudicator.

The party making the referral should send to the adjudicator, within 7 days, a written statement of the contention or disagreement. A copy of this should also be sent to the other party. The adjudicator, acting under the Housing Grants, Construction and Regeneration Act 1996, i.e. not as an expert or arbitrator, should send the decision to the parties within 28 days. The adjudicator is not obliged to give reasons for the decision. In reaching the decision, the adjudicator should set out the procedure in ascertaining the facts and the law. The referral may include the following items:

- Using one's own knowledge and experience
- Within the terms of the contract, opening up, reviewing or revising any certificate, opinion, decision, requirement or notice
- Requesting further information from the parties
- Requiring the parties to carry out tests, additional tests or opening up the works
- Visiting the site and workshops
- Obtaining information and advice from employees or representatives
- Obtaining information and advice from others on technical or legal matters relating to the payment of interest

If a party chooses not to cooperate with the adjudicator, this will not invalidate the decision of the adjudicator. The parties may agree to meet their own costs or to allow the adjudicator to direct who should pay the costs involved in the whole proceedings. The decision of the adjudicator remains valid until the difference is determined by arbitration or legal proceedings.

Clause 41B: arbitration

The second alternative after adjudication is to refer a dispute to arbitration. This process is frequently longer and more costly than adjudication. The provisions of the Arbitration Act 1996, or any amendment, apply to such cases. The arbitration is to be conducted in accordance with the JCT 98 edition of the Construction Industry Model Arbitration Rules (CIMAR) that are current at the base date of the contract.

Clause 41C: legal proceedings

Where article 7B applies to any dispute or difference, these shall be determined by legal proceedings. Legal proceedings are really a last resort. They tend to be lengthy and expensive, and decisions usually take some time to be achieved.

PART 5

SUBCONTRACT CONDITIONS

CHAPTER 26

JCT NOMINATED SUBCONTRACT DOCUMENTATION

At the request of a number of its constituent bodies, the Joint Contracts Tribunal has reviewed the procedure for the nomination of subcontractors that was set out in the original version of JCT 80. It has done this in order to simplify the process that is involved. A new set of documentation was issued for this purpose in 1991.

The introduction of JCT 80 originally included two procedures, known as the basic method and the alternative method. The revised procedures have replaced these with a single method. The review has changed only the procedures involved with nomination. It has not changed in any way the liabilities of the various parties arising out of such a nomination. The new procedures are outlined in *Revised procedure for nomination of a subcontractor* (Joint Contracts Tribunal, 1991). This documentation provides a comprehensive set of information together with guidance notes.

DOCUMENTS

The following documents relate to revised procedures for the nomination of subcontractors:

- *NSC/T*: the standard form of nominated subcontract tender:
  - Part 1: The invitation to tender which is issued by the architect to prospective subcontractors
  - Part 2: the form of tender to be submitted by each subcontractor invited to tender
  - Part 3: the particular conditions to be agreed by the contractor and each subcontractor prior to entering into a nominated subcontract
- *Agreement NSC/A*: the nominated subcontract agreement to be entered into by the main contractor and each nominated subcontractor, subsequent to the latter's nomination by the architect.
- *Conditions NSC/C*: the conditions of subcontract. This was derived from nominated subcontract NSC/4. It is applicable to a subcontract entered into on NSC/A, which conditions are incorporated by reference in such agreement.

- *Agreement NSC/W*: the employer and nominated subcontractor agreement entered into between the employer and a nominated subcontractor. This was derived from NSC/2.
- *Nomination NSC/N*: it is essential that the architect uses this form when nominating a subcontractor in accordance with the main contract conditions.

ACTION UNDER THE 1991 PROCEDURE

The architect

- Completes the invitation to tender (NSC/T Part 1) and sends a copy with NSC/W, with page 1 completed, to each of the subcontractors from whom tenders are requested. These are submitted on NSC/T Part 2 together with drawings, specifications, bills of quantities, etc. (the numbered tender documents), as appropriate, that describe the work for which tenders are being sought, and the appendix to the main contract as it is anticipated it will be completed.
- When it is intended to nominate a subcontractor the architect:
  - Arranges for the employer to sign the applicable subcontractor's tender as approved and enters into the employer/subcontractor agreement.
  - Issues the nomination instruction on form NSC/N and sends this to the contractor together with a copy of:
    - invitation to tender and the nominated subcontractor's tender, i.e. the completed parts NSC/T Parts 1 and 2
    - the numbered tender documents including any amendments
    - the employer/subcontractor agreement as entered into
  - Sends a copy of the nomination instruction to the subcontractor together with a copy of:
    - the employer/subcontractor agreement as entered into
    - the completed appendix to the main contract
- When the subcontract has been entered into, the contractor will submit a copy of the applicable documents for record purposes.
- If the contractor and subcontractor are unable to enter into a subcontract within 10 working days of receipt of the nomination instruction, then the contractor is to advise the architect of the reasons involved:
  - If there are no fundamental reasons then the architect may extend the time for completing the formalities of entering into the subcontract.
  - If there are fundamental reasons then the architect has the options outlined in clause 35.9.2 of the conditions of contract.
  - If the reasons are not valid then the architect can insist upon a subcontract being formed.
- If before issuing the nomination instruction the architect wishes a proposed subcontractor to undertake design, ordering of materials or fabrication of components then this can be arranged as long as the employer/subcontractor agreement has been signed.

The subcontractor

- On receipt of the invitation to tender (NSC/T Part 1), the agreement (NSC/W) and the numbered documents (drawings, specifications, bills of quantities), the subcontractor:
  - Completes the tender on NSC/T Part 2.
  - Enters into the agreement (by signing or as a deed) and returns it to the architect.
- If the subcontractor is unaware of the main contractor when submitting a tender then such a tender may be withdrawn within 7 days after such notification, without any reasons being given.
- Upon receipt of a copy of the nomination (NSC/N) and a copy of the appendix from the main contract, a subcontractor:
  - Agrees with the contractor on the matters set out in NSC/T Part 3 and signs to signify such an agreement.
  - Checks that any differences between the appendix to the main contract and the invitation to tender are acceptable.
  - Enters into a subcontract with the contractor on NSC/A.
- Where the subcontractor cannot agree with the matters set out in NSC/T Part 3 then the main contractor should be informed in writing of the reasons involved.
- Upon receipt of the employer/subcontractor agreement, the subcontractor may be instructed to proceed with design work, ordering of materials or fabrication of components prior to being nominated. Arrangements are available for payment for this work prior to the issue of the nomination instruction, where this is desirable.

The contractor

- Upon receipt of the nomination instruction (NSC/N) and the accompanying documents from the architect, the contractor:
  - Agrees with the subcontractor the matters set out in NSC/T Part 3 and signs in agreement.
  - Enters into a subcontract on the NSC/A.
- If the contractor wishes to make a reasonable objection to the appointment of a subcontractor then this must be done within 7 working days from receipt of the nomination instruction.
- If the contractor is unable to enter into a subcontract within 10 working days from the receipt of the nomination then the architect must be advised of the reasons why this has not been done. The contractor must then subsequently follow any instructions that are issued by the architect.

TENDER NSC/T: PART 1

This is the architect's or contract administrator's invitation to tender that is sent to a proposed subcontractor. The three schedules provide the full details of the subcontractor's offer. These agreed terms, together with the other subcontract documents (drawings and bills of quantities or specification) will constitute the subcontract. The tender is the formal response to an enquiry by the architect. The subcontractor agrees to carry out the works in accordance with these documents. There are two possibilities:

- A VAT exclusive subcontract sum (clause 15.1)
- A VAT exclusive tender sum (clause 15.2)

A VAT exclusive tender sum is chosen when the work will be completely remeasured on completion. In both cases a 2.5 per cent cash discount is allowed to the main contractor. The offer is subject to the nomination procedures being completed. If either the main contractor or a subcontractor fails to agree upon the terms of the agreement, with good reasons, then the contract will not be formed. Where this situation arises then the employer may be faced with paying a subcontractor for design work, preordering of materials or the fabrication of some components, as instructed and agreed by the architect.

Part 1 provides the essential and general information relating to both the main contract and the subcontract conditions. The details of the main contract appendix are also provided, appropriately completed. Part 1 gives the following information:

- Numbered tender documents:
  - Drawings
  - Specifications
  - Bills of quantities
  - Schedule of rates
- Main contract works and location
- Names and addresses of employer, consultants and the main contractor
- Main contract information:
  - Form of main contract
  - Main contract alternative provisions
  - Any changes made to the standard form
  - Whether the main contract is executed under hand or as a deed
  - Facility for inspection of the main contract documents
  - Main contract appendix, to be attached
  - Any obligations or restrictions imposed by the employer that are not covered by the main contract conditions
  - Order of works; employer's requirements
  - Type and location of access
  - New completion date where this has been altered from the original date for completion stated in the main contract conditions
  - Other relevant information relating to the main contract
  - Subcontract; reference to clause 35.7 of the main contract; issues of retention and execution under hand or deed
  - Main contract appendix and entries therein
  - Earliest and latest starting dates

- Income and Corporation Taxes Act 1988 (Construction Industry Scheme, CIS)
- Attendance items: provided by the main contractor free of charge to a subcontractor:
  - *General attendance*: use of contractor's temporary roads, pavings and paths, standing scaffolding, standing power-operated hoisting plant, provision of temporary lighting and water supplies, clearing away rubbish, provision of space for the subcontractor's own offices and for the storage of their plant and materials and the use of messrooms, sanitary accommodation and welfare facilities.
  - *Other attendance*: this is in addition to the requirements for general attendance by the main contractor. The other attendance is specified separately for each individual subcontractor and may include:
    - Special scaffolding or scaffolding additional to the contractor's standing scaffolding
    - The provision of temporary access roads and hardstandings in connection with structural steelwork, precast concrete components, piling, heavy items of plant, etc.
    - Unloading, distributing, hoisting and placing in position, giving in the case of significant items the weight and size
    - Provision of covered storage accommodation including lighting and power
    - Power supplies, giving the maximum loads
    - Maintenance of specific temperature or humidity levels
    - Any other attendance
- Fluctuations
- Contributions, levy and tax fluctuations
- Labour and materials cost and tax fluctuations
- Formula adjustment

TENDER NSC/T: PART 2

This represents the tender, or offer, by a subcontractor. It states that the subcontractor will carry out and complete, as a nominated subcontractor, the subcontract works that form a part of the main contract works referred to in NSC/T Part 1. These will have been identified in numbered tender documents listed in NSC/T Part 1 and may also include entries that the subcontractor may have made in the tender. The amount stated by the subcontractor is exclusive of VAT, either as a contract sum or tender sum (see above). The tender should include a schedule of rates or prices for the measured work and daywork percentages for labour, materials and plant relevant to a particular daywork definition, e.g. RICS/CEC (Chapter 6). Reference is also made to the fluctuation provisions and the attendance to be provided by the main contractor.

Other matters to be included with the tender include the earliest and latest starting dates, the periods required for the submission and approval of all necessary subcontractor's drawings, etc., the execution of any work off site prior to commencement of the work on site and the period required for notice to commence work on site. Also covered are matters relating to insurances, VAT, special conditions or agreements on the employment of labour, limitation of working hours, etc., and the statutory tax deduction scheme.

PARTICULAR CONDITIONS NSC/T: PART 3

This covers any particular conditions agreed between the contractor and subcontractor. If, for example, any changes or additions to these items in NSC/T Part 1 may cause a subcontractor to reconsider a tender that has been submitted:

- Item 7   Obligations or restrictions imposed by the employer
- Item 8   Order of works: employer's requirements
- Item 9   Type and location of access

AGREEMENT NSC/A

This is the standard form of articles of nominated subcontract agreement between a contractor and a nominated subcontractor. It includes seven recitals and four articles. These are as follows.

Recitals

- *Recital 1*: the subcontractor has been invited to submit a tender on NSC/T Part 1 by the employer and architect (or contract administrator) and has submitted an offer on NSC/T Part 2 to supply and fix materials or goods or to execute work in accordance with the particulars that are set out in the numbered tender documents.
- *Recital 2*: the subcontract works are to be executed as part of the main contract works.
- *Recital 3*: the architect has selected the subcontractor's tender for carrying out the works in pursuant of clause 35.1 of the main contract conditions.
- *Recital 4*: the employer and subcontractor have entered into agreement NSC/W.
- *Recital 5*: under clause 35.6 of the main contract conditions the architect has issued an instruction on nomination NSC/N with a copy to the subcontractor.
- *Recital 6*: the contractor and subcontractor have agreed NSC/T Part 3 which has been signed by each of them.
- *Recital 7*: the subcontractor's offer is on the basis of a subcontract on the standard form of articles of nominated subcontract agreement NSC/A.

Articles

- *Article 1*: the subcontract consists of:
  - the articles of agreement
  - the completed NSC/T Part 1, Part 2 and Part 3
  - the numbered documents
  - the standard conditions of nominated subcontract (NSC/C)
- *Article 2*: the subcontractor's obligations, which are to carry out and complete the works.
- *Article 3*: payment, which shall be either a VAT exclusive contract sum or a VAT exclusive tender sum.
- *Article 4*: settlement of disputes; these shall be referred to adjudication, arbitration and then legal proceedings as necessary.

NOMINATION INSTRUCTION NSC/N

This is a form that is used for the final nomination of subcontractors, after all the preliminary procedures have been complied with. It is prepared under clause 35.6 of the main contract conditions.

AGREEMENT NSC/W

This is the agreement between a nominated subcontractor and an employer. In the JCT nominated subcontract form, the above agreement is a separate but integral part of the documentation. Whereas the bulk of the documentation is between the main contractor and the nominated subcontractor, NSC/W is an agreement between the employer and subcontractor. It is made in accordance with clauses 35.3 to 35.9 of the standard form of building contract (JCT 98).

When the architect has issued the preliminary notice of nomination and the conditions have then been signed by the respective parties, NSC/W is then issued. In NSC/W the nominated subcontractor warrants the employer that reasonable skill and care will be taken in:

- The design of the subcontract works, where the design has been undertaken by the subcontractor.
- The selection of the kinds of materials and goods for the subcontract works where these have been selected by the subcontractor.
- The satisfaction of any performance specification or requirement where appropriate.

The subcontractor may, prior to nomination, be requested by the architect to proceed within the agreement with any design work, the ordering of materials or the fabrication of components for the works. The employer may agree to pay for this work in advance of nomination, in which case the materials become the property of the employer.

The obligations of the employer to the subcontractor are largely financial:

- An agreement to notify the subcontractor of the amount included in interim certificates
- The possible early final payment under clause 35.17 of the main contract
- Payment to the subcontractor direct, should the main contractor fail to honour a certificate

The employer must be indemnified where payments are made direct to a subcontractor, after the insolvency of the main contractor.

NSC/W includes eleven clauses covering the following information:

- Completion of the subcontract and the subcontractor's obligations
- Design, selection of materials, satisfaction of performance specification
- Subcontractor's warranty and liabilities
- Architect's or contract administrator's instructions
- Interim and final payments
- Delay in the supply of information by a subcontractor
- Default in performance by the subcontractor
- Renomination
- Assignment
- Conflicts between the tender and agreement
- Arbitration.

SUBCONTRACT CONDITIONS NSC/C

These are described more fully in Chapter 27.

JCT NOMINATED SUBCONTRACT CONDITIONS (NSC/C)

This form of subcontract is intended to be used for nomination of subcontractors when JCT 98 is used as a main contract. It supersedes the existing standard form of subcontract (NSC/4) that was issued at the same time as the major revision to the main contract form, in 1980. The different subcontract forms (BEC, FASS and CASEC), the *green* form for nominated subcontracts and the *blue* form for domestic subcontracts have all now been replaced (Chapter 6).

A revision to the JCT form of subcontract was introduced in 1991, and became known as NSC/C. This has been subsequently revised to bring it into line with JCT 98. Whilst this covers the provisions and requirements set out in the main contract form, its format and numbering are considerably different. Main contractors must ensure that the obligations and responsibilities placed on them by the employer are appropriately and adequately transferred to the various nominated subcontracts. The form NSC/C is subdivided into sections, 1 to 9 but section 8 is not used. It is intended to be read and interpreted in conjunction with the other NSC documents that have been described in Chapter 26. In order for the subcontract to be binding on the parties it must be properly signed by completing the attestation, by the different parties concerned.

Reference should also be made to clause 35 of the main contract conditions, JCT 98, that provide for nominated subcontractors under the main contract.

SECTION 1: INTENTIONS OF THE PARTIES

Interpretations, definitions, etc.

This section outlines the different documents that are referred to in NSC/C clause 1.1. It also provides a series of definitions in clause 1.4, many of which are similar and refer to the definitions that are included in the main contract, JCT 98. It also includes a wide range of definitions, as one would expect, that relate to subcontracts and subcontractors. The various nominated subcontract documents are listed for reference purposes in Chapter 26. Clause 1.4 states how notices must be given and how to calculate the number of days in which action must be made.

It also makes reference to electronic data interchange (EDI). Clause 1.4D states that the subcontractor is not responsible to the contractor for works designed by the subcontractor.

The subcontract

Clauses 1.5 to 1.8 define the subcontract documents as the agreement (NSC/A), the (numbered) documents attached and the subcontract conditions. Clause 1.6 refers to the hierarchical nature of contract documents, making it clear that in the cases of differences, the terms of the main contract will overrule those of the subcontracts. Where any discrepancies occur, then a subcontractor should write to the contractor specifying the discrepancy or divergence. The contractor then informs the architect under clause 2.3 of the main contract conditions.

A subcontractor may be required to enter into a sub-subcontract for goods and services, with, for example, other suppliers. This is described as a sub-subcontract. If the sub-subcontract in any way restricts, limits or excludes the liability of such a supplier then the subcontractor must inform the contractor in writing of the limitation, exclusion or restriction. Under these circumstances the subcontractor's liability will not be expected to exceed the terms of the sub-subcontract.

Execution of the subcontract works

A subcontractor's obligations are to carry out and complete the subcontract works in compliance with the subcontract documents and in conformity with all reasonable directions and requirements of the main contractor (clause 1.9). The word 'reasonable' appears in all forms of contract and is generally understood. However, circumstances do occur where the meaning has to be interpreted by the courts. When agreed such interpretations can become case law. All materials and goods, so far procurable, should be:

- Of the kinds and standards described in the subcontract documents
- To the reasonable satisfaction of the architect

All work should be:

- Of the standards described in the subcontract documents
- Appropriate to the subcontract works
- To the reasonable satisfaction of the architect
- Carried out in a workmanlike manner
- In accordance with the health and safety plan

Subcontractor's liability under incorporated provisions of the main contract

A subcontractor must also observe, perform and comply with the provisions of the main contract, where these relate to the subcontract works. Subcontractors should also indemnify the contractor against and from any breach, non-observance, non-performance or any act or omission of their workers or agents. However, subcontractors are not liable for any defaults by the employer or contractor (clause 1.10).

Bills of quantities

If bills of quantities are provided as a subcontract document then these should have been prepared in accordance with SMM7 (clause 1.12).

Benefits under the main contract

The main contractor will request the subcontractor to obtain any rights or benefits of the main contract as they apply to the subcontract works (clause 1.13).

Strikes

The main contract or subcontract works may be affected by a local combination of workers, strike or lockout affecting those employed in any capacity with the project (clause 1.14). These include tradespeople on site and those engaged in the preparation, manufacture and transportation of goods and materials. Under the terms of the subcontract:

- Neither party will be able make claims on the other party.
- The contractor will try to ensure that the site is kept open for a subcontractor.
- The subcontractor will take all reasonable practical steps to continue with the subcontract works.

SECTION 2: COMMENCEMENT AND COMPLETION

Subcontract obligation

The subcontractor should carry out the works in accordance with the agreed programme details included in item 1 of NSC/T Part 3. Subcontractors should also carry out their parts of the works reasonably in accordance with the general progress of the main contract works. The subcontractor will start work on site in accordance with the notice detailed in item 1 of NSC/T Part 3. It is the main contractor's responsibility to give to subcontractors sufficient information on the general progress of the works, to enable them to fulfil their obligations under the terms of the subcontract (clause 2.1).

Delays and extension of time

If it becomes apparent that the regular progress of the subcontract works is likely to be delayed, the subcontractor should notify the contractor in writing of the circumstances causing the delay. The contractor then informs the architect. The subcontractor will always work through the main contractor, maintaining the correct levels of protocol. The subcontractor should also provide details on the expected effects of such a delay, estimate how the delay will affect the programme and generally keep the main contractor informed of any changes that might occur. Similar procedures, as outlined in the main contract, are then followed regarding whether the subcontractor should be granted an extension of the time. The various relevant events are listed in clause 2.6. These follow the same pattern and a similar content to those of the main contract conditions.

If a subcontractor feels aggrieved about not being granted an extension of time, the main contractor may be invited to join a subcontractor in arbitration proceedings (clause 2.7) to discuss the matter further. This may then lead towards litigation by allowing a subcontractor to use the contractor's name.

Where a subcontractor fails to complete the subcontract works on time then payment may become due to the main contractor from the subcontractor. This will be equivalent to any loss or damage suffered or incurred by the main contractor, that has been caused by the failure of a subcontractor in this respect.

Practical completion of the subcontract works

A subcontractor must give written notice to the main contractor when practical completion of the subcontract works has been reached. A copy of this notice is then passed to the architect, who decides on the date for the practical completion of the subcontract works. The architect then issues a certificate accordingly (clause 2.11). Clause 2.12 deals with defects that must be made good at the expense of the subcontractor, in accordance with the instructions of the architect, or on the directions of the main contractor. Clauses 17 and 18 of JCT 98 set out the liability of the main contractor in respect of defects. Subcontractors are similarly liable for defects in their work which occur before the end of the defects liability period before the expiry of the main contract. Such work includes:

- Defects
- Shrinkages
- Other faults
  - due to materials or work standards that were not in accordance with the subcontract
- Frost damage
  - occurring before the date of practical completion of the subcontract

Defects that arise through the design of the subcontract works are a separate matter between the employer and a nominated subcontractor (NSC/W).

Subcontractors upon practical completion of their subcontract works should clear the site of any property that belongs to them (clause 2.14).

Main contract conditions with sectional completion supplement

Where the main contract conditions provide for sectional completion of the works (Appendix: sectional completion supplement), this should be attached to item 6 of NSC/T Part 1 and enclosed with the copy of NSC/N.

SECTION 3: CONTROL OF THE WORKS

Instructions of the architect and directions from the contractor

Only the information contained in the subcontract documents imposes obligations on the subcontractor (clause 3.1). The subcontractor should keep on the site a competent person to whom the architect, through the contractor, can issue instructions (clause 3.2). For a typical nominated subcontractor this might mean the overseer of a gang of workpeople employed by the subcontractor.

The architect may issue further instructions to the subcontractor through the main contractor. The contractor must pass on such instructions efficiently in order to avoid the possibility of misunderstandings or delays. Subcontractors must then comply with such instructions, unless they make a reasonable written objection to their compliance. Reasonable objection may arise in connection with variations. For example, if a subcontractor was in the process of fabricating a component, a variation to change the size, shape or materials used may result in the nearly complete component becoming useless and worthless. Due perhaps to the subcontractor's other work in progress, time may not permit the manufacture of such a component caused by a variation.

Since instructions are issued through the contractor, they follow the principles laid down in JCT 98. To be effective, therefore, they must always be in writing or subsequently confirmed in writing by the architect, contractor or subcontractor.

Subcontractor's quotation

In the same way that the main contractor may be asked to provide a quotation against a variation instruction, so might a nominated subcontractor. It is referred to as a 3.3A quotation and includes the following items:

- The value of the adjustment to the contract sum
- Any adjustment to the time required for the completion of the subcontract works
- The amount to be paid in respect of any loss or expense
- A fair and reasonable amount in respect of the cost of preparing the 3.3A quotation

Work not in accordance with the subcontract

Where the subcontract works are not in accordance with the contract then a subcontractor can be given instructions to remove the work (clause 3.5). Alternatively the substandard work can be accepted and an appropriate deduction made from the contract sum (clause 3.6). Provision exists in the subcontract in the same way as the main contract for opening up the works for testing purposes (clause 3.7). The subcontractor is to indemnify the main contractor in respect of any liability and to reimburse the contractor for any loss as a result of direct compliance by the operation of these clauses.

Subcontractor's failure to comply with directions

If a subcontractor fails to comply with an architect's instruction within 7 days after its receipt, then the contractor may, with the architect's permission, employ others to comply with the direction (clause 3.10).

Architect's instructions – statement of authority

Subcontractors are able to object to architect's instructions where they consider that the architect is exceeding the powers granted under the terms of the contract (clause 3.11).

Right of access of contractor and architect

Both the architect and the contractor are to have reasonable access to the workshops of a subcontractor (clause 3.12).

Assignment and subletting

A subcontractor must not assign or sublet any part of the subcontract works without the architect's written consent (clauses 3.13 and 3.14).

General and other attendance

General attendance is provided to nominated subcontractors free of charge (clause 3.15). This also includes where the joint fire code applies. The main contractor may, however, have priced this item in the bills of quantities. Other attendance (clause 3.16) is provided individually to each different subcontractor depending upon their requirements and outlined in NSC/T. This will also be provided free to the subcontractor concerned. The costs of other attendance are normally recouped from the employer through pricing the items in the bills of quantities. Attendance items have been described in Chapter 23. Subcontractors should clear away their rubbish into places provided by the main contractor on site (clause 3.15). The eventual removal from site is provided by the contractor under general attendance. Subject to the above clauses, subcontractors are responsible for their own temporary workshops, sheds or other temporary buildings on site. The contractor must provide only for the space requirements on site for such buildings, as required by the general attendance requirements. Subcontractors are allowed the use of erected scaffolding. It is, however, a subcontractor's responsibility to ensure that as far as possible the scaffolding is fit for purpose and in a suitable condition.

Contractor and subcontractor not to make a wrongful use or to interfere with the property of the other

The main contractor and the subcontractors should not make wrongful use or interfere with the property of others working on the site (clause 3.19). Theft and the interference with a subcontractor's work can represent a major problem for the many different firms who may be working on a building site at the same time. One of the biggest problems is associated with the *borrowing* or theft of materials, plant and equipment by others.

SECTION 4: PAYMENT

Subcontract or tender sum

Either a subcontract sum (clause 4.2), based upon a lump sum, is agreed for work that has been clearly defined, or a subcontract tender sum (clause 4.3) that allows for remeasurement upon completion of the subcontract works. Where adjustments are made to the subcontract sum in preparation of the final account, then once these amounts have been agreed they can be included in the interim payments (clause 4.1).

Valuation of variations

The rules for the valuation of variations (clause 4.4) follow the same principles as the rules in JCT 98. These include variations resulting from architect's instructions, and work executed that was covered by an approximate quantity in the original bills. Work that is changed by the subcontractor can at a later date also become the subject of a variation if accepted by the architect. These rules for the valuation of variations are summarized as follows for work under a contract sum as NSC/A article 3.1:

- The rates and prices contained in the subcontract documents are used for work that is of a similar character and executed under similar conditions (clause 4.6.1.1).
- Pro rata rates to those contained in the subcontract documents are used where the work is of a similar character but is executed under different conditions (clause 4.6.1.2).
- Fair rates and prices to those contained in the subcontract documents are to be used for work that is not of a similar character (clause 4.6.1.3).
- The rates and prices contained in the subcontract documents are to be used for reasonably accurate forecasts of approximate quantities (clause 4.6.1.4).
- Pro rata rates to those contained in the subcontract documents are used where the approximate quantities are not a reasonably accurate forecast of the work (clause 4.6.1.5).
- Work that is omitted from the contract is valued at the rates and prices contained in the documents, provided the remaining work is not affected (clauses 4.6.2 and 4.6.5).
- Allowances can also be made to a reduction or addition in value to the preliminary items in the subcontract documents (clause 4.6.3).

- Work that cannot be properly measured or valued will use daywork rates (clause 4.6.4).
- If variations to the works substantially change the conditions under which any other part of the subcontract work is executed then this may necessitate revaluation under clause 4.6.

The subcontractor has the right to attend any remeasurement that may be necessary in connection with the valuation of variations (clause 4.8). Where the work is to be entirely remeasured under a tender sum as NSC/A article 3.2 then the above similar rules of valuation apply (clause 4.10 onwards). Alternatively a subcontractor may submit a price statement for carrying out the subcontract works.

Payment of subcontractor

Interim and final payments to nominated subcontractors are to be made in accordance with the provisions of clauses 4.14 to 4.25. These clauses incorporate the provisions of the main contract conditions of clause 35. Clause 4.15.1 makes it clear that the subcontractor must apply for payment through the main contractor. This must be done in writing, and whilst a letter will be accepted in practice, for accounting purposes an invoice is preferable. The invoice may include work completed, materials on site and materials off site where the provisions of clause 30.3 of JCT 98 (clause 4.15.3) have been observed. There are also provisions in JCT 98 (clause 35.13), repeated in NSC/W (clause 2.2.4), to allow an employer to pay a nominated subcontractor for design work, materials and goods prior to the instruction of actual nomination.

Notification by the main contractor to the subcontractor of the amount included in the certificate must be made within 17 days (clause 4.16.1) of the date of the interim certificate. This notification will also include for the appropriate amount of payment due, less the 2.5 per cent cash discount. The cash discount is allowable from all nominated subcontractors' invoices. Where this has not been allowed for at tender stage, it is usual then to add 1/39 to the total (clause 4.17.2.4). This envisages a cheque being posted to the subcontractor upon receipt of payment to the main contractor from the employer. In practice, where the employer defaults in payment, this will then cause a chain reaction of non-payment by the main contractor to the different nominated subcontractors. The subcontractor must issue a receipt to the contractor which can then be shown as reasonable proof of payment to the architect. Where the subcontractor is not paid the correct amount at the correct time then interest is to be added (clause 4.16.4). The rate of interest is 5 per cent above the Bank of England base rate.

Comparable provisions to those included in the main contract to allow an employer to pay a subcontractor direct, or for final payment to be made to a subcontractor, are also included.

Where the value of any of a subcontractor's goods or materials has been included in an interim certificate, and the employer has paid the main contractor for them, then such goods become the property of the employer (clause 4.15.2). The architect must inform the subcontractor of the amount of payment in this respect, in the usual way. This is an attempt to avoid the employer having to pay twice for these items, in the case of default by the main contractor. This issue has, in the past, caused disputes in the courts regarding the ownership or title to such goods.

The contract also provides for an early final payment to subcontractors, where subcontractors are prepared to indemnify the main contractor in respect of any future defects occurring in their work (clause 4.16.2).

The ascertainment of the amounts due under interim and final payments (clause 4.17) contain similar provisions to those of the main contract conditions (JCT 98 clause 30). Amounts in interim certificates are described as being subject to retention or amounts not subject to retention. The subcontractor should ensure that the main contractor has invoices for the work at least 7 days before the date of the interim certificate (clause 4.17). This is to allow time for inclusion in the quantity surveyor's valuation of the works. The gross valuation includes the following items:

- Value of work that has been properly executed
- Value of materials and goods delivered to the site
- Value of materials off site, where the architect has exercised discretion in this respect
- Amounts in pursuant of clause 4.1, as a result of payments or costs incurred by a subcontractor
- Amounts in respect of restoration, replacement or repair of loss or damage (clause 4.38) and the removal of debris which in clauses 6B.4 and 6C.4 are treated as if they were a variation
- Amounts in respect of fluctuations (clauses 4A or 4B)
- Amounts equal to 1/39 of the amounts in the last three items (the first three items will already include this cash discount amount)

Where a subcontractor is not satisfied with the amount of an interim payment, or the failure on the part of the architect to certify any amount, then the subcontractor together with the main contractor can instigate arbitration proceedings (clause 4.20). This may arise where:

- The main contractor has withheld payments under the set-off provision (clause 4.26 onwards), and these have been contested by a subcontractor.
- The employer has failed to operate the provisions for payments directly to a subcontractor, when the main contractor has defaulted in respect of interim payments.

A subcontractor can then choose to suspend the subcontract works until the matter is satisfactorily resolved. Notice must first be given to the employer and contractor of the subcontractor's intention (clause 4.21.1). This right must not be exercised unreasonably or vexatiously (clause 4.21.3).

The main contractor's interest in a subcontractor's retention money is fiduciary, as trustee for the subcontractor (clause 4.22). There is no obligation to invest the amount, i.e. where this is done any interest accruing will be for the benefit of the main contractor only. In some circumstances it may be necessary to provide a separate bank account to hold this money. This principle is the same as that used between the employer and the main contractor.

The adjustment of the subcontract sum or tender sum follows a procedure similar to the one for adjusting the main contract sum.

Contractor's right to set-off

The main contractor is entitled to deduct amounts from money that has been authorized for a nominated subcontractor (clause 4.26) provided that:

- The amount of set-off has been quantified in detail and with reasonable accuracy by the contractor.
- The subcontractor agrees that the amount is owed to the main contractor.
- The sum has been finally awarded in arbitration or litigation in favour of the main contractor.

The main contractor is also entitled to set off against any money due to the nominated subcontractor for loss and expense because of a failure on the part of a subcontractor to observe the provisions of the subcontract. The main contractor must be able to prove an actual loss or expense due to a subcontractor's breach of contract. Any amount ascertained in this respect can only be deducted from a subcontractor's outstanding payment on the judgment of an adjudicator, where the two parties cannot agree.

A subcontractor who disagrees with the amount specified in the contractor's notice of the intention to set off must within 14 days of the receipt of such notice send a written statement to the contractor (clause 4.30). This must be done by registered post or recorded delivery, setting out the reasons for such disagreement with the particulars of any counterclaim. At the same time the subcontractor should give notice of arbitration under section 9 of these conditions (clause 4.30.1). The subcontractor should also request the action by the adjudicator (clause 4.30.2).

Matters affecting regular progress

A subcontractor who feels that direct loss or expense will be incurred, for which adequate reimbursement will not be made, must inform the contractor accordingly. This must be done in writing, listing one or more of the reasons offered in clause 4.38.2. These reasons follow those outlined in clause 26.2 of the main contract conditions. A subcontractor's application under this clause should be made as soon as the subcontractor becomes aware that the regular progress of the works is likely to be affected. A subcontractor must then, as soon as possible, provide the main contractor with the details of the loss and expense incurred, in order that the architect and quantity surveyor can assess the likely amount of money that is involved. It is obviously important to the employer that any amounts to be added to the contract sum are known as soon as possible. This will enable the employer to be better aware of the overall financial commitment to the project.

The regular progress of the works may become materially affected because of an act, omission or default of the main contractor or any subcontractor. In these circumstances the subcontractor must give written notice to the main contractor of any possible claim that might be forthcoming. Any amount agreed therefore becomes a debt recoverable from the main contractor.

In clause 4.40 this represents the opposite position to the above. A nominated subcontractor is at fault and a claim is now due to the main contractor. During the preparation of this claim, the main contractor should also include in the calculations any liability towards other subcontractors that may be working on the site.

All of the provisions of clauses 4.38 to 4.40 are without prejudice to any other rights or remedies that the main contractor or nominated subcontractor may possess.

SECTION 4A/4B/4C: FLUCTUATIONS

This group of clauses follow the same pattern as the main contract provisions (clauses 37 to 40). Clause 4A (contribution, levy and tax fluctuations) will apply if no mention is made of either clause 4B (labour and materials cost and tax fluctuations) or clause 4C (formula adjustment). The use of clause 4C depends upon a suitable analysis of the original contract sum.

SECTION 5: STATUTORY OBLIGATIONS

Value added tax

Clauses 5A and 5B operate in a similar manner to clause 15 of the main contract conditions. They therefore provide for the subcontract sum to be exclusive of VAT. This is to be paid as an extra to the subcontract. Main contractors will in turn recover any tax that they have to pay as an input tax.

Income and Corporation Taxes Act 1988

Clause 5C refers to the Income and Corporation Taxes Act 1988 and the Construction Industry Scheme. These complement the provisions of clause 31 in the main contract conditions.

Construction (Design and Management) Regulations 1994 (5E)

The provisions of CDM, since they are in the main contract conditions, are replicated as far as subcontractors are concerned. These include reference to the planning supervisor, the development and compliance with the health and safety plan and the maintenance of a health and safety file.

SECTION 6: INJURY, DAMAGE AND INSURANCE

Injury to persons and property

A subcontractor is responsible for injury that may be caused to persons or property due to the negligent carrying out of the subcontract works. The subcontractor must therefore indemnify the contractor accordingly in connection with the possibility of this occurring. The indemnity excludes any act or neglect on the part of either the main contractor or another subcontractor working on the site.

The liability for loss or damage excludes injury or damage where clause 6C applies, where the employer is responsible. It is likely that an employer already has the existing premises insured. The liability also excludes the specified perils that have been defined in clause 1.3 of JCT 98.

Insurance against injury to persons and property

The subcontractor is responsible for maintaining such insurances that are necessary to cover liability in respect of personal injury, death and injury, or damage to property that might be caused whilst carrying out the subcontract works. The insurance in respect of personal injury or death should comply with the Employer's Liability (Compulsory Insurance) Act 1969 and any relevant statutory orders.

A subcontractor is not liable to indemnify the contractor or to provide insurance in respect of personal injury or death or damage to the works, materials or property by the effect of an excepted risk. These are defined in clause 1.3 of JCT 98.

Loss or damage to the works or subcontract works

Three options for insurance are offered. These follow the principles outlined in the main contract, JCT 98. The one retained in the subcontract will be the same as the one used on the main contract.

- Clause 6A   New buildings: contractor arranges joint names policy
- Clause 6B   New buildings: employer arranges joint names policy
- Clause 6C   Works in or extensions to existing structures: employer arranges joint names policy

Clause 6A: New buildings (note main contract conditions clause 22A)

The main contractor is responsible for insuring the works and materials against the specified risks (JCT 98 clause 1.3). This responsibility runs until the issue of the certificate of practical completion, and includes the subcontractor's work and materials. Any loss or damage occurring after this date to the subcontract works, unless it is due to negligence, breach of statutory duty, omission or default, is not the subcontractor's responsibility.

The subcontractor is also not responsible for any loss or damage that occurs due to the occurrence of one of the specified perils (JCT 98 clause 1.3). Neither is the subcontractor responsible for any loss or damage that might arise due to the fault of the main contractor or anyone for whom the main contractor is responsible.

Subcontractors are of course responsible for defects or damage caused by their own workers. They are also responsible for their own temporary buildings, plant, tools and equipment, and will need to insure them accordingly. Subcontractors should also consider whether any insurance should be provided, at their own expense, to cover any other possible risks that are not covered under clause 6A, such as impact, subsidence, theft, vandalism.

If any loss or damage occurs, either due to a specified peril or otherwise, then a subcontractor must inform the main contractor in writing of the extent, nature and location of the damage. A subcontractor, on the instructions of the architect or contractor, must restore any damaged subcontract works or materials and dispose of any debris that may have arisen. If the loss or damage is not the responsibility of the subcontractor, then it will be treated as a variation. The sum is then recovered from the contractor as a debt but it is not added to the contract sum. Main contractors must then claim the amount from their own insurers.

Clause 6B: New buildings (note main contract conditions clause 22B)

Under these conditions it is the contractor's responsibility to ensure that the employer insures the works in a joint policy that is referred to in clause 22B of the main contract (JCT 98). Subcontractors are either recognized as insured, under the joint names policy, or the insurers waive the rights of subrogation (i.e. substitution of one person for another, in regard to a legal claim) they may have against the subcontractor. The same procedures apply as under clause 6A, except the amount is treated as if it were a variation and then added to the contract sum. Employers would then seek to recover the amounts from their own insurers. The subcontractors will continue to insure their own personal effects, such as temporary buildings, plant and equipment and risks that are not otherwise covered by clause 6B, such as impact, subsidence, theft, vandalism.

Clause 6C: Works in or extensions to existing structures (note main contract conditions clause 22C)

These are similar provisions to clause 6B.

Policies of insurance, production, payment of premiums, default by contractor or subcontractor

The main contractor and the subcontractors, respectively, are all required to have the necessary insurance in compliance with the contract. The provision of such insurance by one party may affect another. There is therefore the provision to allow either party, at reasonable times, to inspect the policies or premium receipts of the other. Where the main contractor or a subcontractor fails to properly insure, the other party may do this on their behalf. The amounts of the premiums are then recovered from the party who is in default.

Subcontractor's plant, etc., responsibility of contractor

Any plant, tools or equipment brought on to the site by a nominated subcontractor are the sole risk and responsibility of the subcontractor. If any negligence, breach of statutory duty, omission or default of the contractor, or others for whom the main contractor has a responsibility, results in damage or loss of these items, the subcontractor can claim from the parties responsible. Subcontractors are, however, responsible for injury to persons and property resulting from the use of their plant or equipment. The main contractor will therefore require to be indemnified by a subcontractor against all possible claims and legal proceedings in this respect. Subcontractors are also responsible for any insurances that may be required for their own plant and equipment.

Joint fire code

All subcontractors must comply with the joint fire code and should ensure compliance by any person for whom a subcontractor is responsible under the terms of the contract.

SECTION 7: DETERMINATION

Determination of the employment of the subcontractor by the contractor

One of the remedies that the main contractor may request of the architect, if a nominated subcontractor defaults in one of the following ways, is the determination of the employment of the subcontractor (clause 7.1):

- Without reasonable cause the subcontractor suspends the works
- Without reasonable cause fails to proceed with the subcontract works in the manner provided in clause 2.1
- Refuses or neglects written notices from the contractor to remove defective work or improper materials or wrongly fails to rectify defects
- Fails to comply with the provisions of clause 3.13 or 3.14 (assignment and subletting)
- Fails to comply with the CDM Regulations

It is the contractor's responsibility to first inform the architect that one of the above has occurred. The contractor must also supply the appropriate details in respect of the incident. The architect may then instruct the main contractor to issue a notice of the impending determination of a subcontractor. The notice should be issued by registered post or recorded delivery. The subcontractor has 14 days in which to remedy the default, otherwise the employment is terminated.

In the event of a subcontractor becoming insolvent in accordance with the Companies Act 1989 or making arrangements to enter liquidation in accordance with the Insolvency Act 1986, the subcontract is terminated automatically.

If a subcontractor or any person employed, even without the subcontractor's knowledge, offers an inducement or reward in relation to the obtaining or execution of the subcontract, then the employer may determine the subcontract (clause 7.3). This clause may be extended to subcontracts on other projects being undertaken by the employer. When a subcontract has been terminated under this clause, the following items represent the respective rights and duties of the main contractor and subcontractor:

- The subcontractor's temporary buildings, plant and equipment may be used freely by the firm appointed to complete the subcontract works.
- The subcontractor must assign to the main contractor, without any payment, any benefit of any agreement for the supply of materials or goods or for the execution of any of the subcontract works.
- The contractor may be directed by the architect to pay for any goods or materials that have been delivered to site, but not paid for by the subcontractor.
- The subcontractor must remove personal effects such as temporary buildings, plant and equipment when requested to do so by the architect. If the subcontractor ignores this instruction, then the main contractor may, within a reasonable time, remove and sell them. The proceeds of such a sale are then held for the benefit of the subcontractor.
- The subcontractor must pay the main contractor for any direct loss or expense that has resulted from the determination.

Determination of employment under the subcontract by the subcontractor

If the main contractor makes a default for one of the following reasons then a subcontractor can determine their employment under the terms of the subcontract (clause 7.6):

- Without reasonable cause the contractor wholly suspends the works before completion.
- Without reasonable cause the contractor fails to proceed with the works so that the reasonable progress of the subcontract works is seriously affected.

A subcontractor may then take the following action (clause 7.7):

- Issue a written notice specifying the default, sent registered post or recorded delivery, with a copy to the architect.
- If 14 days after the notice of default has been given the contractor has not rectified the situation, then the subcontractor may terminate the employment under the subcontract.

The rights and remedies, without prejudice to other accrued rights, or to the subcontractor's liability to the contractor under clause 6 for injury to persons and property, are as follows (clause 7.7):

- The subcontractor removes personal items such as temporary buildings, plant, tools equipment, goods or materials. The subcontractor must do this with care to prevent injury, death or damage to those items for which the subcontractor is responsible under clause 6.
- The contractor must then pay the following items to the subcontractor:
  - The total value of work completed at the date of determination, ascertained under clause 4
  - The total value of any work in progress
  - Any sum ascertained in respect of direct loss and expense under clause 4.39
  - The costs of any materials or goods properly ordered for the subcontract works of which the contractor is bound to accept delivery
  - the reasonable cost of removal under clause 7.7.1
  - any direct loss or expense caused to the subcontractor because of the determination

Determination of the contractor's employment under the main contract

If the employment of the main contractor is determined under clauses 28 or 28A of the main contract then the employment of every nominated subcontractor is also automatically determined. The provisions of clause 7.7 will then apply to each subcontractor.

If the main contractor decides to terminate the contract because of one of the reasons listed in clause 28 of the main contract, the subcontracts are also automatically terminated. The main difference is that the subcontractors will receive a share of any payments that become due because of the termination of the contract. These amounts, in respect of subcontractors, are calculated under clause 28.2.2.6 of the main contract. The subcontractors will necessarily have suffered direct loss and expense, and they will need to substantiate their claims to the main contractor. The sums agreed and due to the subcontractors under interim certificates are not affected because of the main contractor's determination under clause 28.

SECTION 9: ADJUDICATION, ARBITRATION AND LEGAL PROCEEDINGS

Clause 9 incorporates the provisions of the main contract (JCT 98 clause 41) and gives the specific rules and powers to rectify a subcontract that has fallen into a dispute. The JCT arbitration rules (July 1988) are to apply.

ANNEXES

Annexes include the supplemental provisions for electronic data interchange which refers to clause 2.15 and the NSC/C sectional completion supplement.

CHAPTER 28

DOMESTIC SUBCONTRACT CONDITIONS

The construction industry changed rapidly during the latter part of the twentieth century in order to increase its efficiency, effectiveness and economic performance. Such changes continue to take place across a whole spectrum of activities. Many of these changes are not restricted to the UK practice alone but exist throughout the developed world. One of the important changes that occurred during this period was the widespread use of subcontracting firms in the construction industry, to undertake more work on behalf of the main contractor. Subcontracting has always existed, and it occurs in many different areas of life. However, the scope of change and growth amongst subcontracting firms has been considerable. There are now relatively few main contractors who carry out a comprehensive range of the different construction trades. The process of change is of course evolutionary in order to meet the demands placed upon the industry by employers and the need for contractors to manage their work in the best possible way. In a few years' time it is likely the position will have changed again in order to meet new methods of working and contracting generally. Some people believe that the wheel will again turn full circle. It may be that someone will *discover* the general contractor.

The main contract conditions do not generally recognize the main contractor's own subcontractors. These are often referred to as domestic subcontractors. As far as the contract is concerned, the work they carry out is deemed to be what is carried out by the main contractor. The architect, for example, will generally only deal with the main contractor's representative in the issuing of site instructions. It is the main contractor's responsibility to inform these subcontractors where these instructions relate to their work. For convenience and expediency the quantity surveyor may choose to remeasure or value the works of subcontractors with them and the main contractor in attendance.

Contractors at tender stage will decide on which work will be carried out by their own workers, if any. The remainder will then be subcontracted to specialist firms who may undertake just a single trade such as bricklaying or carpentry and joinery. In order to incorporate their prices within the main contract tender, they will in the first instance price the relevant extracts from the bills of quantities or work schedules. Several firms may be invited to submit prices to a contractor in this way. The main contractor will then use those prices that are the most advantageous, adding profit and overheads where relevant.

Under the JCT 98 form of contract the main contractor must first seek the approval of the architect before placing a contract with any of these firms (JCT 98 clause 19.2). Unless the architect has a reasonable objection, the consent will not normally be withheld. If the main contractor's tender is successful, a firm order will be placed with these firms. It is the main contractor's responsibility to ensure that such firms are able to carry out the work in accordance with the contract, to the appropriate standards and quality and in the time period shown on the contract programme. Wherever possible the main contractor should be satisfied with their performance on previous and similar projects, or through recommendation, and satisfied the price they have quoted is reasonable for the work undertaken. A very low quotation from a subcontractor, perhaps below cost, can create all sorts of problems during the construction of the project.

The terms of subletting will be agreed between the main contractor and the subcontractor. The main contractor may have preprinted conditions for this purpose. It is more likely that one of the standard subcontract forms will be used. The conditions outlined in these forms provide for fair and reasonable conditions of engagement to both parties, and their use is therefore to be encouraged. The use of such forms may also avoid possible friction arising between the main contractor and the subcontractor, where each may have made different assumptions and interpretations of non-standard conditions. Subcontracting can create many difficulties to each party, and ad hoc arrangements will not help either party to achieve satisfaction or to run the contract smoothly. The use of a standard non-nominated form is therefore recommended.

The following items refer to the subcontract conditions for use with the Domestic Subcontract DOM/1 articles of agreement. The form has largely replaced what was described as the *blue* form. It was called the blue form because it was printed on blue paper. The JCT 98 non-nominated form of subcontract comprises the following documents:

■ Articles of agreement for use with JCT 98 (DOM/1)
■ Subcontract conditions (DOM/1C)
■ Contra-charges and set-off under DOM/1

The following non-nominated forms are also available for use with other JCT forms:

■ Articles of agreement for use with Contractor's Design 1981 (DOM/2)
■ Articles of agreement for use in connection with IFC 84 (IN/SC)
■ Conditions for use with IN/SC

These include a range of amendments that are issued from time to time, often in conjunction with amendments with the main forms of contract. The different domestic subcontract conditions are also available for use with most of the major forms of contract (Chapter 6).

The following comments apply only to DOM/1 conditions for use with the standard form of contract. The articles of agreement follow the pattern of those in the main contract conditions. The agreement is in this case between the main contractor and a domestic subcontractor. A separate contract is formed between the main contractor and each of the domestic subcontractors. The name of the employer is provided for the subcontractor's information only. The employer is not a party to this contract, so they can neither sue nor be sued under the provisions of the contract. There are of course other remedies the employer or subcontractor can pursue in the case of a grievance that cannot be settled contractually.

- *Article 1* describes the subcontractor's obligation to become familiar with the provisions of the main contract and the overall aims and objectives of the employer. The subcontractor will, however, only be bound by those contract conditions that are written into the subcontract agreement. By becoming familiar with the main contract, the subcontractor will be able to gain a better understanding of any implications that may have an effect upon the tender price submitted and the proposed method for carrying out the works.
- *Article 2* states that the subcontract price is to be exclusive of VAT, in line with the main contract conditions. The agreed subcontract sum is entered here in the contract. This sum will include a 2.5 per cent cash discount for the main contractor, unless an alternative percentage has been inserted in Part 7 of these conditions. Article 2.1 refers to a subcontract sum where the subcontractor's tender sum is unlikely to be any different from the final account for the work. Where the work is to be remeasured, the subcontractor's tender sum will be inserted into the conditions. Only one of these alternatives is applicable, the other being deleted from the articles of agreement.
- *Article 3* refers to the settlement of disputes, and the requirement that all matters will be first referred to adjudication, then arbitration and finally to the courts if the dispute cannot be settled. The dispute will arise in the first instance because of a difference that may arise between the main contractor and a subcontractor. This may result in adjudication proceedings being instigated by the contractor against the employer. This will occur where the dispute is more of a matter to be resolved by the employer than a dispute directly with the contractor.

APPENDIX TO DOM/1

The appendix to DOM/1 provides information relevant to the subcontract conditions in fourteen parts.

Part 1

Part 1 contains information which is extracted from the main contract conditions, but which is relevant to the subcontract.

Section A

Section A includes the following information:

- Description of the works
- Form of main contract
- The main contract documents and where they can be inspected
- Whether the main contract has been executed under hand or under seal
- Which of the main contract conditions are to apply where an alternative is available, e.g.
  - architect or supervising officer
  - type of insurance for the works
- Any of the standard main contract conditions which have been amended or revised

Section B

Section B contains a copy of the completed appendix applicable to the main contract.

Section C

Section C is subdivided into three separate sections. The first section deals with any obligations or restrictions that have been imposed by the employers on the project. These will therefore have been covered in the preliminaries section of the bill of quantities. The first section may also include restrictions on working hours, the use of non-union labour, etc. Such matters will of course appertain to the contractor's domestic subcontractors. The second section outlines the employer's requirements affecting the order of the works where this is required. The third section covers a description of the location of the site and the mode of access.

Part 2

Part 2 describes briefly the particulars of the subcontract works that will be undertaken by this subcontractor. Any relevant documents – detailed drawings, specifications or bills of quantities, etc. – should be listed for identification and contractual purposes.

Part 3

Part 3 requires the subcontractor to insure for personal injury, damage to property and liability. The amount of insurance is to be specified in accordance with clause 7.2 of the subcontract conditions.

Part 4

Part 4 includes reference to clause 11.1 of the subcontract conditions regarding the time factor of the subcontract works, and provides for:

- The date for commencement of the subcontract works
  - this is designated between two calendar dates
- The subcontract period for this work
- The period required for notice to commence work on site
- The period for subcontract works off site and prior to commencement on site
- Any further details that may be appropriate to the time factor

Part 5

Clauses 16.2 and 17.2 refer to daywork undertaken by the subcontractor. The provisions of the main contract will apply in respect of the definition of prime cost of daywork regarding the analysis of the work. The appropriate RICS schedule will be used depending upon the nature of the subcontractor. Subcontractors have the option of including percentage additions for the daywork sections. However, the main contractor will only be eligible to reclaim from the employer, where this is appropriate, at the rates which the main contractor has inserted in the contract bills. The contractor will therefore endeavour to make sure they are no higher than those rates.

Part 6

Clauses 19A and 19B deal with the alternative VAT arrangements, one of which should be deleted.

Part 7

Part 7 covers the main contractor's cash discount on the subcontract invoices. The recommended discount is 2.5 per cent. Also listed is the retention percentage to be deducted from interim payments. This is normally in accordance with the provisions of the main contract, and 5 per cent is therefore considered to be the maximum.

Part 8

Clause 24 considers the contractor's claims against the subcontractor and the arrangement for set-off. This includes the same requirements as for nominated subcontractors. The name and address of the adjudicator and the trustee-stakeholder should be provided in accordance with this clause.

Part 9

Particulars of the subcontract's other attendance requirements should be described separately, initialed by the parties and attached to the appendix. The domestic subcontractor, in common with the nominated subcontractor, is entitled to the free provision of general attendance facilities. The main contractor may, however, make an allowance in their rates inserted in the bills for the appropriate costs associated with such items.

Parts 10 to 13

Parts 10 to 13 relate to the fluctuation provisions and follow a similar pattern to the main form. Clause 35 will apply if there is nothing to the contrary. Clause 34 expects, however, that one of the alternatives will be clearly identified. Where the formula rules are used, the base month appropriate to the subcontractor must be clearly indicated.

Part 14

If the main contractor desires to insist upon other stipulations in connection with a subcontractor, then this should be stated.

THE JCT DOMESTIC SUBCONTRACT CONDITIONS (DOM/1C)

These follow a similar pattern and content to the nominated subcontract conditions (NSC/C). However, the domestic subcontractor is not generally recognized in the same way by the employer as the nominated subcontractor. There are clear advantages in being appointed on the basis of the latter. For example, there are no direct payment provisions by the employer should the main contractor default in paying a domestic subcontractor. Neither is there any provision for an early release of retention. Where a nominated subcontractor feels aggrieved at receiving apparently unfair treatment from the main contractor, there is the possibility of an appeal to the architect to help settle the matter, since it was partially because of the architect that the subcontractor was nominated in the first place. The direct subcontractor does not easily have this redress, since the architect will only be responsible for approving this firm under clause 19 (assignment and subcontracts) of the main contract conditions. The architect, through the main contractor, will require that the work carried out by a domestic subcontractor meets the general requirements of the contract and is completed to the reasonable satisfaction of the architect. The domestic subcontractor must also comply with all instructions, make good any defective work and allow access for the architect to workshops if this is required. Such instructions will be issued through the main contractor.

PART 6

OTHER CONTRACT CONDITIONS

CHAPTER 29

AGREEMENT FOR MINOR BUILDING WORKS

The first edition of the JCT agreement for minor building works (now MW 98) was published in 1968, based upon the JCT 1963 standard form of building contract. This was subsequently extensively revised in January 1980, to bring it in line with JCT 80. It was revised again to bring it into line with the provisions of JCT 98. It is a simplified version of that form of contract and does not envisage the same contractual problems arising from the carrying out of construction works. The simplification inevitably, however, produces a less precise set of conditions. Since its publication in 1980 there have been a number of revisions and reprints of the form. The 1998 edition has been based on the printing of the 1980 edition incorporating amendments MW1 to MW8. These have also been extended to include MW9 to MW11 together with minor corrections.

MW98 uses the title 'contract administrator' in preference to 'supervising officer'. Architects have always been unhappy with the title 'supervisor', since they claim this is not a fair description of their activities; 'inspector of works' would be a preferred option and a better description. 'Contract administrator' is an improvement, since it represents more or less any profession's role within the scope of the contract. In fact, 'contract administrator' should replace other titles used in other forms of contract. Where the term 'architect' is used in this chapter it must also be interpreted to mean contract administrator.

The minor works form is not for use under the following circumstances:

- In Scotland
- Where bills of quantities have been provided
- Where the employer wishes to nominate subcontractors or suppliers
- On contracts of a duration that requires the full labour and materials fluctuations
- For works of a complex nature
- For works which involve complex services
- For works which require more than a short period of time for their execution

It is intended to be used only on minor building works, although no value has been stated. However, one supposes that in practice it is used on much larger projects. This is partially due to the reluctance of architects to get involved with the complexities of JCT 98. It has sold many thousands of copies, indicating its widespread acceptance and suitability for these sorts of works. Its use is restricted to where an employer appoints a professional consultant to advise and administer its terms. A professional consultant could include an architect or any other professional such as a surveyor or engineer, but not the contractor as in the case of design and build. The contract is carried out on an agreed lump sum on the basis of drawings and specifications. It can be an appropriate form for maintenance works where these can be readily defined.

The form of contract follows a similar pattern to JCT 98, but comprises a form of agreement and conditions of contract only. The agreement needs to be completed by the parties and it states:

- Works to be performed
- Name of the architect or contract administrator
- Drawings
- Contract sum
  - usually based on a specification, schedules or a schedule of rates
- Contract documents signed by the parties
- Quantity surveyor, if appointed

It also states that the Construction (Design and Management) Regulations 1994 apply.

A schedule of rates may be included should the project require remeasurement or the agreement of variations. A quantity surveyor may therefore still be used (possibly holding the title of contract administrator) to prepare the specification or schedule of rates and to agree the final account.

The articles of agreement include the following information:

- Article 1   The contractor's obligations to carry out the works in accordance with the contract
- Article 2   The amount of the contract sum
- Article 3   The name of the architect or contract administrator
- Article 4   The name of the planning supervisor, who is frequently the designer of the project
- Article 5   The principal contractor
- Article 6   Adjudication to be used in the case of disputes or differences
- Article 7A   Arbitration
- Article 7B   Legal proceedings

The conditions of contract include the clauses under eight headings. In addition there are supplemental conditions.

- Clause 1   Intention of the parties
- Clause 2   Commencement and completion
- Clause 3   Control of the works
- Clause 4   Payment
- Clause 5   Statutory obligations

- Clause 6   Injury, damage and insurance
- Clause 7   Determination
- Clause 8   Settlement of disputes

There are also five supplemental conditions. There is in addition a practice note for use with the agreement for minor building works.

The contents of these sections are now briefly described.

CLAUSE 1: INTENTION OF THE PARTIES

Clause 1.1 states that the contractor's obligations are to carry out the works:

- With due diligence and to complete the works
- In a good and workmanlike manner
- In accordance with the contract documents
- Using materials and work methods of the quality and standard specified
- To the reasonable satisfaction of the architect or contract administrator
- With due concern for matters of health and safety

Clause 1.2 states that the architect or contract administrator's duties include:

- The provision of the necessary information to the contractor to carry out the works
- Issuing certificates
- Confirming all instructions in writing

Clause 1 also includes reference to the planning supervisor and the CDM Regulations.

CLAUSE 2: COMMENCEMENT AND COMPLETION

These clauses state when the works will start on site and when they must be completed. The clauses also cover matters relating to the defects liability period, which is normally three months. In the event of the project not being completed on time, separate provisions for an extension of time, or damages for non-completion are included. The damages are assessed on a liquidated basis and the rate per week should therefore be inserted in the contract. The architect or contract administrator will certify the date when the works have reached practical completion.

CLAUSE 3: CONTROL OF THE WORKS

These matters deal with assignment and subcontracting, with the general provision that written permission must first be obtained from the architect or contract administrator. The contractor must at all reasonable times keep a competent person in charge on site to whom the architect or contract administrator can issue instructions. This is a less onerous position than JCT 98. Any person may be reasonably excluded from the site by the architect, for good reasons.

The architect's instructions to the contractor must be in writing, or confirmed in writing within 2 days. The provisions for covering non-compliance with an instruction are similar to JCT 98. The confirmation within 2 days is the period of time quoted for the confirmation of clerks of works' instructions under JCT 98. There is no provision for a clerk of works under this form of contract.

There are also provisions for variations to the contract. These are to be valued on a fair basis by the architect. Where a quantity surveyor is named in the agreement these duties will be carried out on behalf of the architect. Alternatively, a price can be agreed before the work is carried out, and this seems a sensible course of action to follow. Provisional sums are valued in the same way. Nomination is specifically excluded.

CLAUSE 4: PAYMENT

- Provision is made for the correction of any inconsistencies in the documents.
- Progress payments are normally to be made every four weeks and include the value of works properly executed and the value of materials and goods which have been reasonably and properly brought to the site. The form does not allow for the payment of materials off site. Retention is normally 5 per cent and payment should be made within 14 days to the contractor.
- If the employer fails to properly pay the agreed amount, then interest on outstanding payments will be added.
- This form of contract provides for a penultimate certificate. This is issued within 14 days after the date of practical completion of the works. Half the retention is released with this certificate.
- The contractor shall supply within three months from the date of practical completion, all documentation that is necessary to prepare the final account. The final certificate is then issued within 28 days of the receipt of such documentation, provided that all defects have been made good.
- Any changes in contributions, levy or tax charges can be recouped. However, these contracts are often of such a short duration that a footnote in the contract suggests this clause should be deleted from the contract.
- The contract is otherwise a fixed-price contract, with no allowances being made for changes to labour, materials, plant or other resource costs.
- Where the employer fails to pay the amount due, the contractor then has the right to suspend the works.

CLAUSE 5: STATUTORY OBLIGATIONS

Clause 5.1 requires the contractor to comply with all the statutory obligations, notices, fees or charges. Clauses 5.2 and 5.3 deal with the provisions of VAT and the statutory tax deduction scheme respectively; they are identical to the provisions of JCT 98. Clause 5.5 covers corruption, and although this may be incorporated into any contract it usually only applies where the employer is a local authority.

Where alternative A in the fifth recital is used then a number of clauses describe the issues that are associated with the CDM Regulations and the use of a health and safety file (Chapter 11).

CLAUSE 6: INJURY, DAMAGE AND INSURANCE

These clauses deal with matters of injury and damage and the requirements for appropriate insurance on the part of the contractor. The employer needs to be sure in the first place that if injury or damage occurs then the blame and responsibility fall on the contractor. This presumes that the injury or damage results because of the contractor's negligence, and not because of a fault in the design. Design faults may not necessarily provide the contractor with an adequate defence. The employer will therefore need to be satisfied that the contractor is properly insured (clause 6.4). As far as the works are concerned, alternative insurance provisions are available in the case of new works or existing structures. These are similar to clause 22 of JCT 80 and one of the alternatives therefore needs deleting. The contractor, and any subcontractors, will need to produce evidence of insurance if required.

CLAUSE 7: DETERMINATION

These clauses cover determination of the contract by either the employer or the contractor. The employer may determine if the contractor:

- Fails to proceed diligently with the works (clause 7.2.1)
- Suspends the works (clause 7.2.1)
- Becomes bankrupt (clause 7.2.2)

The contractor has the option of determining the contract where the employer defaults in respect of:

- Progress payments (clause 7.2.1)
- Interference or obstruction of works (clause 7.2.2)
- Failure to make the premises available (clause 7.2.2)

- Continuous suspension of the works for one month (clause 7.2.3)
- Bankruptcy or liquidation (clause 7.2.4)

The procedures for determination are described, and this act by either party should be considered as a last resort. This act often requires damages to be paid from one party to the other. Determination must be without prejudice to any other rights or remedies which the parties may possess.

CLAUSE 8: SETTLEMENT OF DISPUTES

The methods agreed for the settlement of disputes follow those of JCT 98.
They include adjudication, arbitration and legal proceedings.

SUPPLEMENTAL CONDITIONS

The supplementary memorandum referred to in this clause is in three parts.
The JCT's intention is that it should be used for reference and does not therefore need to be bound in with the executed agreement. The three parts are very similar to the five provisions of JCT 98:

- A   Contribution, levy and tax changes
- B   Value added tax
- C   Construction industry scheme (CIS)
- D   Adjudication
- E   Arbitration

PRACTICE NOTE M2

This states that the form is to be used where minor building works are to be carried out on a lump sum basis where the employer has retained a professional adviser. The definition of minor works is not given. This depends upon the size, nature, scope and complexity of the project. The form is recommended for use on projects of up to £70,000 at 1992 prices. In practice it is used for projects far in excess of this recommendation. A lump sum price will have been determined by the contractor on the basis of information supplied by the employer. The type of documentation supplied will vary depending upon different circumstances that are encountered. Where the work is complex a bill of quantities should be prepared. The contractor should be informed of the employer's status in respect of the construction industry scheme. Full labour and materials fluctuations provisions are not envisaged. There are no provisions for either nomination or the naming of subcontractors. Where this is desirable on the part of the employer, a separate contract with these firms is suggested.

Box 29.1 The minor works form compared with JCT 98

Minor works clause	Clause description	JCT 98 clause
1.1	Contractor's obligations	2, 8
1.2	Architect/contract administrator's duties	5
1.3	Reappointment of planning supervisor or principal contractor	6
2.1	Commencement and completion	17, 23
2.2	Extension of contract period	25
2.3	Damages for non–completion	24
2.4	Practical completion	23
2.5	Defects liability	17
3.1	Assignment	19
3.2	Subcontracting	19
3.3	Contractor's representative	10
3.4	Exclusion from the works	8
3.5	Architect/contract administrator's instructions	4
3.6	Variations	13
3.7	Provisional sums	13
4.1	Correction of inconsistencies	2
4.2	Progress payments and retention	16, 30
4.3	Penultimate certificate	–
4.4	Notices of amounts to be paid and deductions	30
4.5	Final certificate	30
4.6	Contribution, levy and tax charges	38
4.7	Fixed price	37
4.8	Right of suspension by contractor	28
5.1	Statutory obligations, notices, fees and charges	6
5.2	Value added tax	15
5.3	Construction industry scheme (CIS)	31
5.4	(Number not used)	
5.5	Prevention of corruption	27
5.6	Provisions for use where alternative A occurs in the fifth recital	–
6.1	Injury or death of persons	20, 21
6.2	Injury or damage to property	20, 21
6.3A	Insurance of the works – fire, etc. – new works	22
6.3B	Insurance of the works – fire, etc. – existing works	22
6.4	Evidence of insurance	22
7.1	Determination by employer	27, 28A
7.2	Determination by contractor	28, 28A
8.0	Settlement of disputes	41

Box 29.1   *(cont'd)*

Minor works clause	Clause description	JCT 98 clause
Supplemental conditions		
A	Contribution, levy and tax changes	38
B	Value added tax	15
C	Construction industry scheme (CIS)	31
D	Adjudication	41
E	Arbitration	41

Box 29.2   JCT 98 conditions specifically excluded from the minor works form

Clause description	JCT 98 clause
Levels and setting out of the works	7
Royalties and patent rights	9
Clerk of works	12
Materials off site	16
Partial possession	18
Loss and expense	26
Works by or on behalf of the employer	29
War damage	33
Antiquities	34
Nominated subcontractors	35
Nominated suppliers	36
Price adjustment formulae	40

CHAPTER 30

JCT INTERMEDIATE FORM OF BUILDING CONTRACT

Many of the professionals who are employed in the construction industry still regard the standard form of building contract (JCT 98) as too complex, at least for medium-sized projects. Others felt that the gap between JCT 98 and the agreement for minor building works was too great, and that some types of midway conditions were both desirable and essential for the efficient running of construction contracts. The *intermediate form* was thus conceived and its first edition was published in 1984. It soon became known as IFC 84. It was revised in 1998 (now referred to as IFC 98) to bring it into line with the conditions and the good practices developed in JCT 98.

It was originally envisaged that this form would be used on projects worth up to £250,000 at 1984 prices (now considerably more), but it has been used for projects of much larger values. One of the main differences between IFC 98 and JCT 98 is that there is no provision for nominated subcontractors, as is usually understood. There is provision, however, for the architect to name a subcontractor for whom the main contractor assumes a much greater responsibility. The main concept of naming is as follows. All of the contractors tendering are provided with detailed information regarding the named subcontractor's price and conditions, programme of work and the attendance requirements which the main contractor must provide. The contractor then assumes the same responsibility for the named subcontractor as with any normal domestic subcontractor. If, however, the named subcontractor defaults then there are provisions in the contract to safeguard the main contractor's interest.

IFC 98 generally adopts the layout of the minor works form, as the clauses are grouped under similar section headings. However, the clause content is more detailed and comprehensive, although much less so than JCT 98. Unlike JCT 98, IFC 98 is published in only one edition which allows for the differences envisaged in the contractual arrangements. In addition to the drawings, for example, either a bill of quantities, a specification or a schedule of works can be used. If the drawings are supported only by a specification then either a schedule of rates or a contract sum analysis must be provided in support of the tender.

THE AGREEMENT, ARTICLES AND RECITALS

The agreement is the page in the document which identifies the parties concerned with the contract. The recitals include:

- Brief description of the works to be constructed
- Documents which are to be included as contract documents
- The extent of application of the Construction (Design and Management) Regulations 1994
- Reference to the information release schedule
- Bonds.

There are now ten *articles* (originally there were five):

- Contractor's obligations
- Contract sum
- Architect or contract administrator
- Quantity surveyor
- Planning supervisor
- Principal contractor
- Completion of works by sections
- Dispute or difference: adjudication
- Dispute or difference: arbitration
- Dispute or difference: legal proceedings

INTENTION OF THE PARTIES

1.1   *Contractor's obligations.* To carry out and complete the works in accordance with the contract documents which have been identified in the recital above.

1.2   *Quality and quantity of work.* The quality is as specified in the contract documents, quantities depend upon type of contract documentation which has been prepared.

1.3   *Priority of contract documents.* A specification, schedule of works or bills of quantities cannot override the conditions of contract.

1.4   *Instructions as to inconsistencies, errors or omissions.* The architect must issue instructions regarding any inconsistencies which may arise in the documents. If these affect the contract sum then they are to be revalued under clause 3.7 (valuation of variations).

1.5   *Bills of quantities and SMM.* Contract bills are assumed to have been prepared under SMM7 unless otherwise stated. If and when the SMM is revised then this would be amended.

1.6   *Custody and copies of contract documents.* The contract documents are to remain in the custody of the employer, with the contractor being given one copy of the contract documents certified on behalf of the employer, and two further copies of drawings and other contract documentation, e.g. bills of quantities.

1.7 *Information release schedule.* The employer, through the architect or contract administrator, has provided the contractor with an information release schedule (fourth recital).

1.7 *Further drawings and details.* Other drawings and information are likely to be prepared during the course of construction and two copies of these are also to be made available to the contractor.

1.8 *Limits to use of documents.* The contractor must not use any of these documents for any other purpose than this contract.

1.9 *Issue of certificates by architect or contract administrator.* Certificates are normally issued to the employer with duplicate copies to the contractor.

1.10 *Unfixed materials or goods, passing of property, etc.* Goods and materials delivered to the site are not to be removed without the written permission of the architect or contract administrator. Where such items have been paid for by the employer they become the employer's property. The contractor, however, remains responsible for their protection from damage or theft.

1.11 *Off-site materials and goods, passing of property.* Materials or goods paid for by the employer which are stored off site become the property of the employer. Such items must not be used for purposes other than the works. The contractor is again responsible for their safe keeping.

1.12 *Reapppointment of planning supervisor or principal contractor – notification to contractor.* The contractor must be informed if the planning supervisor is changed.

1.13 *Giving or service of notices or other documents.* Unless otherwise specified, notices will be issued to the last known address.

1.14 *Reckoning periods of days.* Where an act is required to be done within a specified number of days, the period commences the following day and excludes public holidays.

1.15 *Applicable law.* Regardless of any issues associated with nationality, English law applies to this contract.

1.16 *Electronic data interchange.* Where the appendix so states, the supplemental provisions for EDI annexed to the conditions apply.

1.17 *Contracts (Rights of Third Parties) Act 1999 – contracting out.* The contract cannot be enforced by those who are not parties to it.

POSSESSION AND COMPLETION

2.1 *Possession and completion dates.* The date of possession is stated in the appendix and the contractor shall begin and regularly and diligently proceed with the works and complete them before the date stated in the appendix as the date for completion.

2.2 *Deferment of possession.* If this clause is to apply, i.e. it is stated in the appendix as such, then the employer may defer giving possession of the site to the contractor by the period stated. The period is not to exceed six weeks. The contractor is then allowed to claim for any direct loss or expense.

2.3 *Extensions of time.* These are clauses to be followed and interpreted in the event of the works being or likely to be delayed. The contractor must inform the architect or contract administrator that a delay is likely to occur and the cause of such. If the architect or contract administrator feels there are reasonable grounds for granting the contractor an extension of time then the contractor must estimate the amount of time that is appropriate and inform the contractor in writing of the decision.

2.4 *Events referred to in 2.3.* This includes the list of events which the contractor may refer to, to support a claim for an extension of time. The list includes similar items referred to in JCT 98.

2.5 *Further delay or extension of time.* This clause allows for further extensions of time as may be appropriate.

2.6 *Certificate of non-completion.* The architect or contract administrator must issue a certificate where the contractor has failed to complete the works, either by the date for completion or within any date extended under the terms of the contract.

2.7 *Liquidated damages for non-completion.* These are stated in the appendix as a rate of pounds per week or month to be paid by the contractor to the employer for delays which are attributed to the fault of the contractor.

2.8 *Repayment of liquidated damages.* If it is felt at some later date that the delays were not entirely the fault of the contractor then the damages, or a part of them, can be repaid to the contractor.

2.9 *Practical completion.* When the architect or contract adminstrator decides that the works are complete then a certificate is issued to that effect.

2.10 *Defects liability.* All defects, which are due to materials or work standards which are not in accordance with the contract, which are notified to the contractor within 14 days after the expiry of the defects liability periods are to be made good at the contractor's expense.

2.11 *Partial possession by the employer.* This is not a normal part of the contract. Practice note IN/1 allows for its incorporation if required.

CONTROL OF THE WORKS

3.1 *Assignment.* Neither the employer nor the contractor shall without the written consent of the other party assign the contract to someone else.

3.2 *Subcontracting.* The contractor cannot subcontract any part of the works without the written consent of the architect or contract administrator. The employment of a subcontractor is automatically terminated upon the determination of the main contract. A subcontractor is effectively treated in the same way as if they were the main contractor, so, for example, any materials or goods of a subcontractor which are brought to the site cannot be removed without written permission. If the value of a subcontractor's materials or goods is included in a certificate and this sum is paid by the employer to the main contractor then such items become the property of the employer.

3.3 *Named persons as subcontractors.* The contract documents may name a person or firm to undertake a section of the works as a subcontractor. Within 21 days of entering into the main contract, the contractor must communicate with the proposed subcontractor using the standard form of tender and agreement NAM/T. If the contractor is unable to enter into a subcontract in this way, the architect or contract administrator must be immediately informed of the reasons which prevent this. If the architect or contract administrator is reasonably satisfied that the particulars have prevented a subcontract being made then an instruction is given which may:

- Change the particulars so as to remove the impediment
- Omit the work
- Substitute the work with a provisional sum

In an instruction as to the expenditure of a provisional sum, the architect or contract administrator may require the work to be executed by a named subcontractor. This clause also deals with the situation where the employment of a named subcontractor is determined and the procedures that should then be followed.

IFC 98 is a work and materials contract. It is not a design contract, so the contractor is not responsible for any design work which may be required. If a named subcontractor has a design element then the main contractor is not responsible for any failures in the design but liability is limited only in respect of the goods, materials and work standards.

3.4 *Contractor's person in charge.* The contractor has to keep on the works a competent person in charge.

3.5 *Architect's or contract administrator's instructions.* This clause deals with the provision of instructions to the contractor, which must be in writing. There are no provisions for oral instructions. It covers matters of non–compliance and the action the contractor may wish to take in respect of challenging the validity of an instruction.

3.6 *Variations.* The architect or contract administrator may issue instructions requiring a variation and sanction in writing to those made by the contractor. The term *variation* is defined to include the alteration or modification of the design, quality or quantity of works as shown on the contract drawings. This may include the addition, omission or substitution of any work, the alteration of the kind or standard of materials or goods to be used, the removal from site of materials or work which are not in accordance with the contract and the addition, alteration or omission of any obligations or restrictions imposed by the employer.

3.7 *Valuation of variations and provisional sum work.* These broadly follow the provisions laid down in JCT 98 regarding similar work, using the priced schedules as a basis for valuation, fair valuations in the absence of any comparable items of work, dayworks and omitted work. The valuation where appropriate can include the adjustment to the values of preliminary items.

3.7.1   *Contractor's price statement.* The contractor must always comply with instructions that are legally required under the terms of the contract. Upon the receipt of an instruction, a contractor may prepare a price statement for the works, as described in JCT 98. The alternative approach is to value the work in accordance with the usual provisions for the valuation of variations (Chapter 17)

3.8   *Instructions to expend provisional sums.* The architect or contract administrator must issue instructions regarding the expenditure of provisional sums.

3.9   *Levels and setting out.* The architect provides the information required for setting out, the contractor is responsible for the actual setting out. Errors are amended in the same manner as JCT 98.

3.10   *Clerk of works.* The employer may choose to appoint a clerk of works whose sole responsibility is that of inspector.

3.11   *Work not forming part of the contract.* This allows the employer to employ persons or firms direct and concurrently with the main building work. The contractor must permit this work to be undertaken. Where the contract documents do not allow for this, the contractor must not unreasonably give the employer permission to arrange for the execution of such work.

3.12   *Instructions as to inspections and tests.* This allows the architect or contract administrator the authority to open up work for inspection. Where the work is in accordance with the contract then the costs of opening up are added to the contract sum. Where it is not then the costs together with the costs of rectification are borne entirely by the contractor. The substandard work may be allowed to remain, in which case an agreed reduction in cost is made to the contract.

3.13   *Instructions following failure of work, etc.* See also above.

3.14   *Instructions as to removal of work, etc.* The architect or contract administrator may issue instructions in respect of the removal from site of any goods or work which do not conform with the contract.

3.15   *Instructions as to postponement.* The architect or contract administrator may issue instructions in regard to the postponement of any work executed under the provisions of the contract.

PAYMENT

4.1   *Contract sum.* The contract sum is not to be altered in any way, other than within the terms of the contract. Any computation of errors is deemed to have been accepted by the parties.

4.2   *Interim payments.* Interim payments are to be made monthly unless the contract states otherwise. The employer must pay the contractor within 14 days of the date of the certificate. The majority of the work is subject to 5 per cent retention, i.e. work completed and materials and goods on or off site. Some items of work are not subject to any retention, e.g. tax, insurances.

4.3 *Interim payment on practical completion.* One-half of the retention fund is released at practical completion and the remainder with the issue of the final certificate.

4.4 *Interest in percentage withheld.* The employer acts as trustee of the retention fund but without any obligation to invest.

4.4A *Right of suspension by contractor.* Contractors have the right to suspend their obligations under the terms of the contract where the employer refuses to honour a certificate.

4.5 *Computation of adjusted contract sum.* Either before or within a reasonable time after practical completion the contractor should provide all the documents which are necessary to adjust the contract sum and to calculate the final account.

4.6 *Issue of final certificate.* When the contract sum has been agreed, the architect or contract administrator must issue the final certificate.

4.7 *Effect of final certificate.* Only the final certificate is conclusive except for matters which are the subject of proceedings, but note recent case law (see JCT 98).

4.8 *Effect of certificates other than final.* None of the interim certificates is conclusive evidence that any work, materials or goods included in them are in accordance with the contract.

4.9 *Fluctuations.* These allow for the possibility of fluctuations in cost on the main contract. The JCT fluctuation clauses and formula rules for use with JCT 98 and IFC 98 should be referred to for the specific details.

4.10 *Fluctuations, named persons.* This allows for fluctuations in cost in respect of any amounts for named subcontractors.

4.11 *Disturbance of regular progress.* This clause deals with any direct loss or expense which the contractor may have suffered and which will not be reimbursed by a payment under any other provision of the contract.

4.12 *Matters referred to in clause 4.11.* This clause identifies the matters which may give rise to a claim for loss and expense. These include delay in receipt of drawings, details, etc., opening up of work for inspection that is found to be in accordance with the contract, execution of work by the employer's own subcontractors, supply by the employer of materials or goods or a failure to supply postponement of the works, failure to provide access to the site, and certain architect's or contract administrator's instructions.

STATUTORY OBLIGATIONS, ETC.

5.1 *Statutory obligations, notices, fees and charges.* The contractor must comply with all notices which may be required by statute, statutory instrument, by-law, etc., and pay all the fees or charges which may be required. If such charges arise, for example, from a variation then they will be added to the contract sum.

5.2 *Notices of divergence from statutory requirements.* If the contractor finds any differences between the contract documents or that any of these documents are in conflict with any statutory requirement then a written notice of such divergence should be given to the architect. The contractor is not expected to search for discrepancies, but if they come to the contractor's notice then they should be notified to the architect.

5.3 *Extent of contractor's liability for non-compliance.* The contractor is not liable for work which does not comply with statutory requirements, if it has been carried out in accordance with the contract documents.

5.4 *Emergency compliance.* If the contractor is requested urgently to comply with a statutory obligation, before receiving appropriate instructions from the architect, then the minimum that is necessary should be carried out.
The architect must then be informed and the work treated as a variation.

5.5 *Value added tax, supplemental condition A.* The contract sum is exclusive of any value added tax and this is dealt with in the same way as in JCT 98.

5.6 *Construction industry scheme (CIS), supplemental condition B.* This clause is also comparable to the similar clause in JCT 98.

INJURY, DAMAGE AND INSURANCE

6.1 *Injury to persons and property and employer's indemnity.* The contractor is liable for and must indemnify the employer, against any action including costs and damages, in respect of injury or death of anyone. The contractor's only real defence is to show that such an occurrence was due to the neglect of the employer or of someone for whom the employer was responsible. The employer does have the option of accepting some of these risks and there is provision for such in clause 6.3.

6.2 *Insurance against injury to persons and property.* The provisions are broadly the same as JCT 98 in this respect, i.e. the contractor is responsible for the works and that of all subcontractors; the contractor should therefore seek to ensure that these subcontractors maintain relevant insurance, and if there is a default in this respect then the employer can insure on their behalf and deduct the premiums from the contract sum.

6.3 *Insurance of the works – alternative clauses.* The insurance will usually be taken out in the joint names of the employer and the contractor. There are three mutually exclusive clauses, 6.3A, 6.3B and 6.3C, and the appendix must indicate which is to apply. These provisions broadly follow the principles described in JCT 98.

6.3A *Erection of buildings – all-risks insurance of the works by the contractor or the employer.* These refer to the joint names policy for all risks insurance. They may be initiated either by the contractor or the employer, depending upon the type of project and what the contract requires. Clause 6.3.C relates to the insurance of existing structures. In addition there are clauses covering the insurance for the employer's loss in respect of liquidated damages and the compliance with the joint fire code.

DETERMINATION

7.1 *Notices under section 7.* These must be in writing and given by actual delivery, special delivery or recorded delivery.

7.2 *Determination by employer.* The employer may determine the contract if the contractor:

- Unreasonably suspends the works
- Fails to proceed regularly and diligently with the works
- Persistently refuses to comply with an architect's instruction to remove defective work, and this results in further damage to the works
- Fails to comply with the provisions of subcontracting and named persons

7.3 *Insolvency of contractor.* If the contractor becomes bankrupt, makes a composition or arrangement with creditors, or has a winding up order made against the company that is either voluntary or otherwise, then the contract is automatically determined. This may be reinstated and continued where the employer and liquidator or receiver agree.

7.4 *Corruption.* Where the employer is a local authority then determination of the contractor's employment can occur where a gift or inducement to seek favours has been made by the contractor. Certain of these would no doubt cause any employer to determine a contract.

7.5 *Insolvency of contractor – option to employer.* This clause follows the provisions outlined in JCT 98.

7.6 *Consequences of determination under clauses 7.2 to 7.4.* The consequences of determination can be summarized as follows:

- The contractor relinquishes the possession of the site.
- If the architect so instructs then the contractor must remove from the works goods, materials, plant, etc.
- The employer may engage others to complete the works and to use the goods, materials and plant. The benefits of these in practice may be limited, e.g. the right does not apply to hired plant, the goods and materials may be the subject of a retention of title clause, and materials or goods used will form a part of the final account.

7.7 *Employer decides not to complete the works.* This is an unusual occurrence. The costs outstanding to the contractor should be calculated within a period of six months.

7.8 *Other rights and remedies.* The provisions outlined are without prejudice to other rights or remedies.

7.9 *Determination by contractor.* The contractor under some circumstances is also able to determine the contract. These are as follows:

- The employer fails to honour a certificate.
- The employer interferes with or obstructs the issue of a certificate.
- If the works are suspended for a period of at least one month due to inconsistencies, variations, postponement, late instructions, delays or failure by the employer or those engaged directly by the employer or failure on the part of the employer in respect of site access.

7.10 *Insolvency of employer.* If the employer becomes bankrupt then the contractor can terminate the contract; this is similar to clause 7.2.

7.11 *Consequences of determination under clause 7.9 or 7.10.* These are as follows:
- The contractor removes temporary buildings, plant, tools, equipment, goods and materials and gives the subcontractors the same facilities.
- The contractor is paid in respect of the following:
  - value of work done but not yet paid
  - loss and expense ascertained
  - cost of goods and materials properly ordered and paid for
  - reasonable costs of removal under value of work done
  - direct loss and expense caused by the determination.

7.13 *Determination by employer or contractor.* This clause is now the same as in JCT 98 clause 28A.

7.14 *Consequences of determination under clause 7.13.* These are the same as where the contractor determines under clause 7.7.

INTERPRETATION OF THE CONTRACT

8.1 *References to clauses.* These are the clauses of IFC 98.

8.2 *Articles, etc. to be read as a whole.* This makes it clear that articles, clauses and items in the appendix are to be read as a whole.

8.3 *Definitions.* This clause contains a number of pertinent definitions, e.g. appendix, clause 6.3 perils, contract sum analysis.

8.4 *The architect or contract administrator.* If the person identified in article 3 is ineligible to be described as an architect under the Architect's (Registration) Acts 1931 to 1969 then the term 'supervising officer' is to be used.

8.5 *Priced specification or priced schedules of work.* This refers to the documentation in the second recital of the conditions of contract.

SETTLEMENT OF DISPUTES

9 These clauses refer to the adjudication, arbitration and legal proceedings outlined in articles 8 and 9.

ANNEXES

The annexes include the following items which need to be completed, by the respective parties to the contract, as necessary:

- Appendix
- Terms of bonds
- Schedule to advance payment bond

- Bond in respect of payment for off-site materials and goods
- Supplemental conditions (similar to JCT 98)
  - Value added tax
  - Construction industry scheme
  - Contributions, levy and tax fluctuations
  - Use of price adjustment formulae
- Supplemental provisions for EDI

CASES OF INTEREST

In some instances where the form of contract is thought to be either lacking or unclear, it is common practice to bring the matter before the courts for learned opinion. This may mean bringing the matter first to arbitration, then to the High Court, the Court of Appeal and finally the House of Lords, where leave to appeal along the way has been granted. The principles upon which these courts base their decisions are stated, and will then in future affect similar cases. This achieves a measure of consistency on matters on which the various parties to a contract may then rely. Where the higher courts reverse the decisions of lower courts, it is the higher opinions that will count in the future. The resulting body of opinion is then established as case law. It is published in the various law reports of the legal journals and, where relevant to the construction industry, in trade and professional journals. There are now also a number of construction law journals and other sources that record the considerable case law of this industry.

The following cases represent just a sample of some of the disagreements that have reached the courts over a number of years. The complete list would form several books alone. Some of the cases described have become household names for the student of building contracts.

Amalgamated Building Contractors v. Waltham Holy Cross UDC (1952)

A contractor had applied for an extension of time because of labour and material difficulties. The architect acknowledged the request but did not at this time grant an extension to the contract period. Some time after the completion of the works, the architect granted the contractor an extension in retrospect. It was held that this was valid on the grounds it was a continuing cause of delay. The architect was unable to determine the length of extension until after completion. The parties must therefore have envisaged the retrospective application of the extension clause. The Court of Appeal held that granting of an extension of time after completion would therefore seem to be an explicit possibility.

AMF International Ltd v. Magnet Bowling Ltd and Another (1968)

Bowling equipment had been installed for Magnet Bowling Ltd by AMF International Ltd. This equipment was damaged after a flood had been caused by exceptionally heavy rain. The bowling centre was only partially completed at this time. AMF therefore claimed £21,000 in damages against Magnet Bowling Ltd and the main contractor. The point at issue in this case was whether the indemnity clause included in the contract would protect the employer and enable him to claim successfully against the contractor. It was held that:

1. The employer was liable in tort due to their negligence in failing to check that the site had been made safe for AMF.
2. The contractor was liable under the Occupiers Liability Act for failing to take reasonable care.

The damages were apportioned on a 60/40 basis between the main contractor and the employer. The question of whether the employer could recover from the contractor then arose. It was held that:

1. The employer, because they had been found guilty of negligence, could not therefore claim under the indemnity clause. This is based upon a legal principle that the indemnity clause will not protect negligence on the part of the one who relies upon it.
2. The employer was able to recover on contractual grounds since the contract included a clause requesting the contractor to protect the works from damage by water.

Anns v. London Borough of Merton (1978)

It was alleged that in 1962 a builder had erected a block of maisonettes with defective foundations. It was further suggested that the defendant local authority had, through its inspector, either failed to inspect the foundations or had inspected them carelessly. The plaintiffs had taken long leases on some of the maisonettes. In February 1972 the plaintiffs issued writs against the builder and against the local authority. This action was brought to decide whether the proceedings against the local authority were barred under the Limitation Act of 1939, which was then in use. Obviously the plaintiffs' cause of action against the council accrued in 1962. It was therefore too late, under this Act, to start an action in 1972. In fact although the appeal was nominally on the limitation issue, the speeches in the House of Lords were mainly concerned with the more fundamental question of whether an action would lie against the local authority at all on such facts. In the House of Lords it was held that such an action would lie, and it would thus not be statute barred.

If the fact that the defects to the maisonettes first appeared in 1970, and the writs were issued in 1972, then the consequences must be that none of these actions were barred by this Act.

Architectural Installation Services Ltd v. James Gibbons Windows Ltd (1989)

The plaintiffs were a specialist labour–only subcontractor installing window units to the defendants as main contractors. The subcontract included a clause for termination by notice on specified grounds, including default of the subcontractor for wholly suspending the works or failing to proceed with them expeditiously and then remaining in default for 7 days after being given written notice of the default by the main contractor. The plaintiffs sued the defendants for monies alleged due and for wrongful determination of the contract. The defendants counterclaimed stating that they were entitled to determine the contract and to seek damages for defective work. It was ordered that two preliminary issues be tried, each of which assumed that the plaintiffs had wholly suspended or failed to proceed with the works in breach of clause 20 of the subcontract. It was held that:

1. On the assumptions set out in the preliminary issues it was agreed that it was not a rightful termination of the subcontract, because there was no sensible connection between the defendants' notice of default and the notice of determination.
2. However, the defendants' telex was a rightful termination of the subcontract at common law since clause 8 did not exclude common law rights expressly or by implication, but existed side by side with the common law right to terminate.

Bacal Construction v. Northampton Development Corporation (1975)

This was a contract for a housing development. The development corporation had provided borehole data on which the contractor had based their substructure design. During the course of the work, tufa was discovered in several areas and this required a redesign of the foundation. It was held that there should be an implied term that the ground conditions would accord with the hypotheses upon which the contractor had been instructed to design the foundation. Because the client had provided the borehole data that was now shown to be inaccurate, the differing costs were to be borne by the employer.

Percy Bilton v. Greater London Council (1982)

Bilton contracted with the Greater London Council (GLC) to erect a housing estate. The contract was dated 25 October 1976, and was substantially JCT 63. The original completion date was January 1979. The nominated subcontractor for the mechanical services went into liquidation on July 1978. By this time the contract was already running 40 weeks behind the programme. A new subcontractor was nominated, but subsequently withdrew before starting work. A new nominated subcontractor was not appointed until December 1978. The programme for this meant that the work would not now be completed until January 1980. Various extensions of time were granted and the extended completion date became February 1980. However, the contract was not completed by that date and thereafter the GLC deducted liquidated damages. The House of Lords held that the withdrawal of a subcontractor is not a fault or breach of contract on the part of the employer, nor is it covered by JCT 63 clause 23. The architect's clause 22 certificate was valid and the GLC was entitled to deduct liquidated damages.

Bottoms v. Lord Mayor of York (1812)

The contractor in carrying out work for a sewage works had intended to use poling boards to support the excavations. However, the nature of the soil turned out to be different from what was expected and necessitated extra works. The engineer refused to authorize this as a variation. The contractor then abandoned the works and sued the client for the value of the work completed. Neither party had sunk boreholes. Before the signing of the contract, the defendants had been advised that the contractor's price was such that the contractor was bound to lose money on the type of soil conditions that were anticipated. It was held that the plaintiff's claim must fail. The plaintiff was not entitled to abandon the works on discovering the nature of the soil or because the engineer declined to authorize extra payments.

Bower v. Chapel-en-le-Frith Rural District Council (1911)

The plaintiffs were the successful tenderers for the erection of a waterworks for the council. The contract was for a lump sum, based on specifications and bills of quantities. The plaintiffs were required by the council to purchase a windmill tower and pump from a named supplier and to fix them. The windmill did not work as expected. It was not argued that the default was in any way due to defective installation by the contractor. The council then requested the plaintiffs to replace it with one that would work efficiently. The plaintiffs argued that they were not responsible, as they had played no part in the choice of the windmill. It was held that the plaintiffs were not liable.

Brightside Kilpatrick Engineering Services v. Mitchell Construction (1973) Ltd (1975)

The plaintiffs were nominated as subcontractors under JCT 63. The subcontract was placed by the defendants which stipulated that the subcontract documents should consist of a 'standard form of tender, a specification, conditions of contract and a schedule of facilities'. The order continued, 'the form of contract with the employer is the RIBA 1963 edition', and concluded 'the conditions applicable to the subcontract shall be those embodied in the RIBA as the above agreement'. Printed references to the 'green form' of subcontract had been deleted. In the Court of Appeal it was held that the words should be read as referring to those clauses in the main contract which referred to matters concerning nominated subcontractors. Clause 27 of JCT 63 regulated the nominated subcontract relationship, and the only way to give a sensible meaning to the relationship was to read into it the terms of the 'green form'.

British Crane Hire Corporation Ltd v. Ipswich Plant Hire Ltd (1974)

Both of these parties were involved in the plant hire business. A drag line crane was hired over the telephone. The hire charges were agreed, but nothing was discussed about the terms and conditions of hire. After delivery the plaintiff owners sent to the defendants the then current General Conditions for Hiring Plant. The defendants failed to sign and return the acceptance note. In the Court of Appeal it was stated that, since both parties were in the plant hire business, and that both used the same standard conditions when hiring plant, it must be assumed that the contract was made on those terms.

Courtney v. Fairbairn Ltd v. Tolaini Brothers (Hotels) Ltd and Another (1975)

In 1969 Mr Tolaini, a hotel owner, decided to develop a site in Hertfordshire. He contacted Mr Courtney, a property developer and building contractor. Mr Courtney wrote to Mr Tolaini stating that if he (Mr Courtney) found suitable sponsors for the scheme, he would undertake the building work. Mr Courtney found sponsors who were able to reach a satisfactory financial arrangement with Mr Tolaini. Mr Tolaini then instructed his quantity surveyor to negotiate a building contract for the project based upon cost plus 5 per cent. The financial aspects were quickly concluded, but negotiations between the quantity surveyor and the building contractor broke down. Mr Tolaini therefore decided to let the contract to another firm, and was promptly sued for breach of contract by Mr Courtney.

It was held, on appeal, that the exchange of letters did not constitute a contract. Because price was of such fundamental importance to the contract, no contract could be formulated until this was agreed, or there was an agreed method of ascertaining it.

Crowshaw v. Pritchard and Renwick (1899)

The plaintiff wrote to the defendant enclosing a drawing and specification, and inviting a tender for some alteration work. The defendant replied, 'Our estimate to carry out the sundry alterations to the above premises, according to the drawings and specification, amounts to £1,230'. The plaintiff wrote next day accepting the defendant's 'offer to execute' the works. Later the defendant refused to go ahead. They contended that by using the word 'estimate', they did not intend it as an offer to do the work, and they contended there was a trade custom that a letter in this form was not to be treated as an offer. It was held that the estimate was an offer which had been accepted by the plaintiff and the defendant was liable in contract. The judge in the Queen's Bench Division stated, 'It has been suggested that there is some custom or well-known understanding that a letter in this form is not to be treated as an offer. There is no such custom, and, if there is, it is contrary to law'.

Davies & Co. Shopfitters Ltd v. William Old (1969)

A nominated subcontractor submitted a tender which was accepted by the architect. In placing the order for the work, the main contractor added new terms that had been absent from the earlier documentation. One of those terms stated that no payment would be made to the subcontractor until the main contractor had been paid. In spite of this the nominated subcontractor started work. It was held that the order from the main contractor constituted a counter-offer which had been accepted by the subcontractor prior to starting work.

Davies Contractors Ltd v. Fareham UDC (1956)

A contractor undertook to build 78 houses in 8 months for a fixed price. They attached a letter to their tender with the proviso that the price was on the basis of adequate supplies of labour being available. The unexpected shortages of labour on a national scale increased the contract period to 22 months. The letter, however, failed to be incorporated into the contract. The courts therefore held that the employer must suffer the delay, but that the contractor must bear the expense.

Dawber Williamson Roofing v. Humberside County Council (1979)

A main contractor had, with approval, sublet the roofing part of the project to a domestic subcontractor. The slates were delivered to the site and their value was included in an interim certificate as materials on site. The certificate was honoured and the sum for these materials was paid to the main contractor. Prior to paying this to the domestic subcontractor the main contractor went into liquidation. The subcontractor therefore sought permission from the council to remove the slates, but this was refused. The subcontractor sued for the value of the slates on the grounds that they remained the subcontractor's property until fixed. It was held that the title rested with the subcontractor. This resulted in the employer paying twice for these materials.

Dawneys v. F.G. Minter Ltd and Another (1971)

This case involved certified payments to a nominated subcontractor being withheld by the main contractor as set-off against damages. Lord Denning summed up this case as follows:

> When the main contractor has received sums due to the subcontractor – as certified or contained in the architect's certificate – the main contractor must pay those sums to the subcontractor. He cannot hold them up so as to satisfy his cross claims. Those must be dealt with separately in appropriate proceedings for the purpose. This is in accord with the needs of business. There must be a cash flow in the building trade. It is the very lifeblood of the enterprise. The subcontractor has to expend money on steelwork and labour. He is out of pocket. He probably has an overdraft at the bank. He cannot go on unless he is paid for what he does as he does it. The main contractor is in a like position. He has to pay his men and buy his materials. He has to pay the subcontractors. He has to have cash from the employers, otherwise he will not be able to carry on. So, once the architect gives his certificates, they must be honoured all down

the line. The employer must pay the main contractor; the main contractor must pay the subcontractor; and so forth. Cross claims must be settled later.

Dodd v. Churton (1897)

A contract provided for the whole of the works to be completed by June 1892. Liquidated damages were included in the contract. There was a provision in the contract that any authority given by the architect for additional work would not vitiate the contract. There were thus no provisions for an extension of time. However, additional works were ordered that delayed the works beyond the completion date. The employer decided on a reasonable amount of time for doing the additional works and then claimed the amount of liquidated damages mentioned in the contract. The Court of Appeal held that by giving the order for additional works, the employer had waived the stipulation for liquidated damages in respect of non-compliance of the works by the stipulated time. It was stated that where one party to a contract is prevented from performing it by the act of another, they are not liable in law for that default.

Dodd Properties v. Canterbury City Council (1980)

Canterbury City Council built a multi-storey car park next to a building owned by Dodd Properties. Piling operations caused damage to the building, and Dodd Properties claimed damages for both the cost of necessary repairs and for the loss of business during the carrying out of the repairs. When the case came to court, the work had not been done since Dodd Properties were awaiting the damages to pay for the repairs. They claimed current rates, whereas the council suggested the rates should be based upon what was relevant at the time the damage was done. It was held that there had been no failure on the part of Dodd Properties to investigate the loss and their claim was therefore upheld.

Dutton v. Bognor Regis UDC (1971)

The foundations of a house were built upon land that had been used as a refuse disposal tip. The foundations were passed and approved by the local authority building inspector. Some time later the plaintiff purchased the house from the original owner, and cracks that began to appear in the wall were discovered to be due to settlement. The builder settled the matter for an agreed sum. The local authority, however, was held to be liable for negligence, since it owed a duty of care and this had not been reasonably given. Lord Denning expressed the view that the cause of action arose when the defective foundations were laid.

East Ham Corporation v. Bernard Sunley and Sons Ltd (1965)

The corporation claimed damages for defective work in the construction of a school in East Ham. The building contract was in the RIBA form (revision of 1950). The contractor had completed the work and the architect had issued his final certificate three years after practical completion. This final certificate issued by the architect at the end of the job was conclusive evidence of the adequacy of the works. Two years later some cladding panels fell down due to faults in the fixings. The courts had to consider two questions:

1. On the true construction of the contract, was the final certificate issued by the architect conclusive evidence as to the sufficiency of the works subject to the exceptions mentioned in clause 24(f)?
2. Did the words 'reasonable examination' mean an examination carried out during progress or at the end of the defects liability period?

It was held by the courts that the final certificate was conclusive evidence that the works had been properly carried out, subject to the provisions of the final certificate clause.

The stone panels fell off as a result of the careless and incompetent fashion in which they had been attached. It was extremely fortunate that no one was killed or injured. Many contractors would have been not only too willing but anxious to remedy such defects. Not these contractors. They had relied on an escape clause in the contract and sought to throw the whole of their defective work on to the council. The Court of Appeal concluded that the contractors escaped liability under the contract provision. However, on leave to appeal to the House of Lords, the decision was reversed. The employer was therefore awarded damages based upon the costs at the time of performing it.

English Industrial Estates Corporation v. George Wimpey and Co. Ltd (1972)

This case was based upon the JCT conditions relevant at the time. In addition, a special condition allowed for the employer's tenant to install machinery and store materials during the construction of the works. The contractor was, however, to be responsible for insuring the works. During the construction period an extensive fire damaged the works with a loss of approximately £250,000. The issue then arose as to whether this loss should be borne by the contractor's insurers or by those of the employer. The contractor claimed that clause 12 of the conditions of contract excluded the possibility of special provisions, particularly where they were in conflict with standard conditions, in this instance clause 16 (clauses 18 and 2 of JCT 98, respectively).

The courts held in favour of the employer, because they were not satisfied that there had been a sufficient taking of possession for clause 16 to apply. It would appear that since, at the outset, both parties envisaged this early occupation by the employer, the insurance risk would remain with the employer.

A E Farr v. The Admiralty (1953)

The plaintiffs contracted to construct a jetty at a naval base. Clause 26 (2) of the contract stated, 'The works and all materials and things whatsoever including such as may have been provided by the Authority on the site in connection with and for the purpose of the contract shall stand at the risk and be the sole charge of the contractor, and the contractor shall be responsible for, and with all possible speed make good, any loss or damage thereto arising from any cause whatsoever other than the accepted risks'. After the plaintiffs had done a substantial part of the work, the jetty was damaged by a destroyer (owned by the employer) which collided with it. The plaintiffs claimed the costs of repairing the jetty. It was held because of the clause that the plaintiffs were liable for the repairs to the jetty.

A E Farr Ltd v. Ministry of Transport (1965)

This was an appeal by A E Farr Ltd to the House of Lords which reversed a decision of the Court of Appeal. The case related to the payment of working space on a civil engineering contract using the relevant documents. The method of measurement at the time allowed for separate bill items to be measured for working space which would include their consequent refilling. Although the ICE conditions of contract at the time included a clause, 'The contractor shall be deemed to have satisfied himself before tendering as to the correctness and sufficiency of tender for the works', this could not be extended to include errors made by the client's advisers. The court therefore accepted that separate items should have been measured in accordance with the appropriate method of measurement. All bills of quantities are prepared in accordance with a method of measurement and strict compliance with this should be made, or specific amendments stated where necessary.

Re Fox ex parte Oundle & Thrapston RDC v. Trustee (1948)

A building contractor became bankrupt and the question the courts had to decide was whether the materials not incorporated into the works were vested in the contractor's trustee. The basis for this was the reputed ownership clause contained in the Bankruptcy Act: 'All goods being at the commencement of the bankruptcy in the possession, order or disposition of the bankrupt in his trade or business by consent or permission of the true owner, he is the reputed owner thereof and his ownership passes to the trustee'. It was held that the trustee had no claim to materials on site which had been paid for in interim certificate. He did, however, have a good claim to materials paid for, but retained in the contractor's yard.

Gilbert Ash (Northern) Ltd v. Modern Engineering (Bristol) Ltd (1973)

This case involved a contractor who had deducted sums due to a nominated subcontractor for set-off, because of a breach of warranty by the subcontractor. The Gilbert Ash form of subcontract had been used, which included a set-off clause. A sentence in this clause allowed for set-off for any breach of the subcontract, however minor, and unrelated to the actual damages actually suffered. This was held to be quite outside any rights which the courts will enforce and declared to be a penalty. Another sentence did, however, allow set-off for *bonafide* claims of the contractor, and the court held that these sums were capable of being deducted.

M J Gleeson (Contractors) Ltd v. London Borough of Hillingdon (1970)

Gleeson's contract for the erection of 300 houses and associated buildings was based upon drawings and bills of quantities. The bills allowed for the completion of works in various stages, but the appendix to the conditions of contract provided for a single completion date only. It was held that liquidated damages could not be deducted until the date for completion shown in the appendix was exceeded.

The decision was based upon the wording of the clause: 'Nothing contained in the Contract Bills shall override, modify or affect in any way whatsoever the application or interpretation of that which is contained in these conditions'. The court expressed the view that it would have been an easy matter to add words to that clause, to the effect that it was subject to conditions included in the bills. This, however, is contrary to recommended practice and could lead to confusion and uncertainty.

M J Gleeson v. Taylor Woodrow Construction Co. Ltd (1989)

This case concerned a management contract for building works on the Imperial War Museum. Taylor Woodrow were to organize, manage and supervise the works and Gleeson were to carry out their portion of the subcontract works. The subcontract works were delayed. Gleeson were awarded 39 days against a claim for 73 days. Taylor Woodrow argued that because of this delay, they had been subjected to claims from other subcontractors and they therefore deducted 'set-off claims' from Gleeson and in addition liquidated damages. Gleeson accepted that the dispute concerning the extension of time and the liquidated damages aspect would be dealt with under the arbitration procedures. However, they considered that the set-off sums had been wrongfully deducted. Taylor Woodrow argued that these should also be dealt with under arbitration. The court held that a prerequisite of setting-off was to quantify it precisely. This had not been done. The court also believed that the set-off and the liquidated damages might involve some duplication, since both were concerned with the alleged delay by the subcontractor.

Accordingly, the court found that there was no defence to Gleeson's claim and no dispute that could be referred to arbitration as far as the set-off amount was concerned. Judgment was therefore given for the plaintiffs, with the remaining matters of liquidated damages and the extension of time being referred to arbitration.

Gloucester County Council v. Richardson (1968)

This contract was based upon the 1939 form of contract, which is considerably different from that in use. It illustrates the problems of warranty agreements which the new form seeks to overcome. The main contractor was instructed by the architect to accept a contract from a nominated supplier. This was for the supply of precast concrete columns, on terms and at a price agreed by the employer. After erection, cracks appeared in the units and the question of the contractor's liability arose. It was held that the contractor was not liable since they had been directed to enter into a contract which severely restricted their right of recourse against the supplier in the event of defects. Note that the 1939 form did not give the contractor the right to object to nomination of a supplier or to insist upon indemnity.

Gold v. Patman & Fotheringham Ltd (1958)

This contract contained a provision whereby the contractor sought to indemnify the employer against claims in respect of damage to property arising out of the works. The indemnity only occurred where it could be shown that the contractor had been negligent in carrying out the works. Furthermore, the contractor had only insured the works in their own name, and not that of the employer since this was not a requirement of the contract. (The current JCT form of contract requires a joint insurance in the names of the contractor and the employer.) During construction, piling operations damaged adjoining property, but no question of negligence arose since the contractor was carrying this out correctly within the terms of the contract. The owners of the adjoining property therefore sued the employer, and they also attempted to bring an action against the contractor for failure to safeguard the employer's interest. It was held that the contractor's obligation was only to insure themselves and not the employer as well.

Martin Grant & Co. Ltd v. Sir Lindsay Parkinson & Co. Ltd (1985)

Martin Grant & Co. were subcontractors for formwork on a large number of local authority housing projects. The subcontract was in non-standard form and contained no provisions for the risk of delay. Substantial delays arose in the performance of the main contracts. Martin Grant & Co. suffered considerable loss, since they had to carry out their work several years later than they anticipated. Martin Grant & Co. contended that the main contractor would make sufficient work available for them in order that they might maintain a reasonable progress of the works in an economic and efficient manner.

The trial judge decided that the express terms of the subcontract left no room for the kind of implied term for which Martin Grant & Co. contended. The case went to the Court of Appeal, where it was eventually dismissed. The express terms of the subcontract left no room for the term contended for to be implied. There is no general rule of law implying such a term in a building subcontract, either because of the relationship of the parties or otherwise.

Greaves & Co. (Contractors) Ltd v. Baynham Meikle and Partners (1976)

A firm of consulting engineers were responsible for the design of a warehouse. The two-storey building would be used for the storage of oil drums, and forklift trucks would be used for transporting them around the building. Within a few weeks, cracks occurred in the building. The engineers were sued for the costs of the remedial work that was necessary, on the basis that the design was unsuitable for the purpose intended. It was held by the Court of Appeal that because the engineers knew of the building's proposed use beforehand, they were liable for the costs of the remedial work.

Hadley v. Baxendale (1854)

The principles for assessing damages resulting from a breach of contract were stated in this case as that arising naturally from the breach, and that which may reasonably be supposed to have been in the minds of the parties at the time they entered into contract.

Hedley Byrne & Co. Ltd v. Heller and Partners Ltd (1964)

A firm of advertising agents gave credit to a client in reliance upon a banker's reference, and suffered loss when the client became insolvent. The reference had been given carelessly, but since the bank had expressly disclaimed liability when giving it, the action failed. Nevertheless, the House of Lords stated that, contrary to what had previously been believed, liability for negligence may extend to careless words as well as to careless deeds and that damages may be awarded for financial loss as well as for physical injury to persons and property.

W Higgins Ltd v. Northampton Corporation (1927)

Higgins contracted with the local authority to erect 58 houses. However, Higgins had completed their tender incorrectly. They thought they were tendering to erect the houses at £1,670 a pair, when in fact it was £1,613, because of the way in which Higgins had made up the bills. It was held that Higgins were bound by this mistake and could not claim to have the contract set aside or rectified.

Howard Marine v. Ogden and Sons (1978)

Ogdens, a firm of contractors, were one of the tenderers for the construction of a sewage works for the Northumbrian Water Authority. Surplus excavation material was to be loaded into barges, shipped downstream and dumped at sea. Howards were a firm who owned barges that were capable of doing this work. A quotation for the hire of the barges specified the volume of material that each barge could carry, and this was 850 cubic metres. On this basis the contractor could establish the number of trips necessary and their effect upon the programme. But one fact was overlooked. Each vessel travelling on water has a safe loading line which depends upon the weight of the material. The amount carried was therefore at the most 850 cubic metres, but because the spoil was clay this in practice turned out to be considerably less. Various telephone conversations took place between Howards and Ogdens, but misunderstandings occurred. To further complicate matters, the payload figure in *Lloyd's Register* was incorrect. This was stated as 1,600 tonnes but was in fact only 1,055 tonnes. Although Ogdens had prudently based their calculations on a lower figure than the agreed 850 cubic metres, the barges failed to carry even this volume. Ogdens then refused to pay for any more barges, and Howards withdrew their transport. Howards claimed for their outstanding hire, and Ogdens counterclaimed for misrepresentation and damages.

The moral of the story is that, when making enquiries, it is unsafe to rely upon telephone conversations alone. Such answers may be too casual and of little legal consequence. Either independent advice should be sought or, preferably, the salient facts should be in writing.

The principle of *Hedley Byrne* v. *Heller and Partners* (1964) came up for consideration by the Court of Appeal. This stated: 'When an enquirer consults a businessman in the course of his business and makes it clear to him that he is seeking considered advice and intends to act on it in a particular way, a duty to take care in giving such advice arises and the adviser may be liable in negligence if the advice he gives turns out to be bad'.

IBA v. EMI and BICC (1980)

EMI were the main contractors on a project for the Independent Broadcasting Authority (IBA). BICC were nominated subcontractors for an aerial mast. The mast collapsed because of BICC's failure to consider the effects of asymmetric ice loading on the struts. Although EMI took no part in the design process, they had accepted the contractual responsibility for the adequacy of the design. It was stated that one who contracts to design and supply an article for a known purpose, must ensure that it is fit for that purpose.

Owing to the complexity of modern buildings, it is often necessary for the architect to employ specialists. Some of these specialists may be independent consultants or subcontractors. One of the functions of the architect is to coordinate the design to ensure it is compatible with the overall scheme and is reasonably fit for the purpose intended (see *Moresk Cleaners Ltd* v. *Hicks* 1966).

Ibmac Ltd v. Marshall (Homes) Ltd (1968)

The defendants, who were building a council housing estate, employed the plaintiffs to build a roadway to serve it. The plaintiffs quoted a price for the work, supplying a bill of quantities with rates for the work. The plaintiffs found that the work was more difficult than they had envisaged, since the site lay at the bottom of a steep hill. There were also serious difficulties with water. The plaintiffs abandoned the works and claimed payment for the work that they had done. Their claim failed in the Court of Appeal.

Killby & Gayford Ltd v. Selincourt Ltd (1973)

In a letter the architects asked a firm of contractors to price alteration work. The letter concluded, 'Assuming that we can agree a satisfactory contract price between

us, the general conditions and terms will be subject to the normal standard form of RIBA contract'. The contractors submitted a written estimate. The architect replied that he was accepting the estimate on behalf of the client and he wanted the contractors to accept his letter as their formal instruction to proceed with the work. No JCT contract was ever signed, but the contractors proceeded with work. In the Court of Appeal it was held that the exchange of letters incorporated the current JCT form.

William Lacey (Hounslow) v. Davis (1957)

The plaintiff tendered for the reconstruction of war-damaged premises belonging to the defendant, who led the plaintiff to believe that they would receive the contract. At the defendant's request, the plaintiff calculated the timber and steel requirements for the building, and prepared various further schedules and estimates which the defendant made use of in negotiations with the War Damage Commission. Eventually the plaintiff was informed that the defendant intended to employ another builder to do the work. In fact, the defendant sold the premises. The plaintiff claimed damages for breach of contract, and alternatively, remuneration on a *quantum meruit* basis in respect of the work done by it in connection with the reconstruction scheme. It was held that although no binding contract had been concluded between the parties, a promise should be implied that the defendant would pay a reasonable sum to the plaintiff in respect of the services rendered. Judgment was entered for the plaintiff.

Lewis and Brass (1877)

An architect invited tenders for building works. The defendant's tender read, 'I hereby agree to execute complete, within the space of 26 weeks from the day receiving the instructions to commence, the whole of the work required to be done . . . for the sum of £4,193'. The architect replied that he had been instructed to accept the tender and a contract would be prepared, ready for signature within a few days. The defendant had made a mistake in their tender and tried to withdraw from the contract. In the Court of Appeal it was held that the tender and acceptance formed a contract. A tender and its acceptance may therefore amount to a contract, even though the acceptance refers to a formal contract to be drawn up afterwards.

Sir Lindsay Parkinson & Co. Ltd v. Commissioners for Work (1949)

A construction contract was agreed on the basis that the contract price would be the cost of the works, plus a net profit not exceeding £300,000. The project was a typical cost-plus contract carried out at this time. Additional works were ordered which extensively increased the size of the project. The court held that the contractor was entitled to a *quantum meruit* payment for this additional work, since this work could not have been envisaged at the start of the project.

London Borough of Newham v. Taylor Woodrow (Anglian) Ltd (1980)

This infamous case was for a contract to erect a 22-storey block of flats, known as Ronan Point. The building was of a prefabricated concrete construction. One morning, gas caused an explosion in a flat on the eighteenth floor. This resulted in the collapse of the south-east corner. The local authority claimed for the cost of repairing and strengthening a number of similar blocks. The court acquitted the contractor of negligence, largely on the basis that several other local authorities had approved and accepted the design. The court held, however, that the contractor was liable for a breach of contract, on the basis that the flats should have been designed and constructed so that they would be safe and fit for their purpose. In addition the building contract included a provision that the contractor would be responsible for any faults and repair work at their own expense. This case was again reopened in 1984 but the final decision on liability has still to be resolved.

London County Council v. Wilkins, Valuation Officer (1956)

This case established that the provision of temporary accommodation on site by a contractor was rateable if erected for a sufficient period of time.

London School Board v. Northcroft (1889)

The quantity surveyor was responsible for measuring buildings up to a value of £12,000. The employers brought an action for negligence resulting from two clerical errors in the quantity surveyor's calculations. This involved overpayments to the builder of £118 and £15, respectively. It was held that the quantity surveyor, who had employed a skilled clerk who had carried out hundreds of intricate calculations correctly, was not liable, since he had done what was reasonably necessary.

Marston v. Kigrass Ltd (1989)

The plaintiffs claimed for preparatory works over and above the costs and work of preparing their tender for a design and build contract to provide a factory to replace one that had burnt down. The contract was never placed since the funds from the insurance company were insufficient to cover the costs of rebuilding. The plaintiffs' tender was the best value for money but because of the tight timescale it needed to be supplemented with further details. The tender was followed by a vital meeting when the plaintiffs were the only tenderers invited to discuss their tender. The chairman of the company said that no contract could be entered into without the insurance money for rebuilding. Nothing was suggested that any preparatory work would be at the contractor's own risk, in the event that the insurance money did not materialize. The defendants were well aware that the plaintiffs would need to start the preparatory work before the contract was signed. It was also made clear that subject to this work being satisfactory the contract would be given to the plaintiffs.

It was held that there was an express request made by the defendants to the plaintiffs to carry out a small quantity of design work and there was an implied

request to carry out preparatory works in general. Both express and implied requests gave rise to a right to payment of a reasonable sum.

Minter v. Welsh Health Technical Services (1980)

This case was based upon a claim for loss and expense, the equivalent of clause 26, but on the 1963 form. The amounts paid were challenged on the basis that they had not been certified and paid until long after the contractor had incurred the loss. The courts held that direct loss and/or expense were to be read as conferring a right to recover sums on the same principle as common law damages. In essence, therefore, this might include compensation for the loss of the use of capital.

Molloy v. Liebe (1910)

A contract provided that no works beyond those that were included in the contract would be paid for by the employer, unless it was in writing from the employer and the architect. During the progress of the works, the employer required certain work to be carried out. The employer insisted that these were within the terms of the contract. The contractor insisted that they represented extra works and must be paid for separately. It was held that it was open to the arbitrator to decide. If the arbitrator concluded that the works were not part of the original contract, then the employer by implication agreed to pay for them.

Re Newman ex parte Capper (1876)

This contract included a provision of a large fixed amount of liquidated damages payable by the contractor in the event of any breach occurring. The court held that this was in effect a penalty since it could not relate with all the varying amounts of loss which might arise. It was not therefore enforceable. The courts did, however, fix an appropriate amount that was upheld as being damages sustained.

North West Metropolitan Regional Hospital Board v. T A Bickerton Ltd (1976)

This case placed the responsibility of renomination with the architect rather than the contractor, after a nominated subcontractor had defaulted. A nominated subcontractor went into liquidation before they had completed their work. The main contractor then requested the architect under the terms of the contract (1963 edition) to appoint a new firm as a successor to this nominated subcontractor. Because a new price was required, and that this was likely to be a higher price, the employer suggested that the contractor was responsible for completing this work in any way, but to the approval of the architect. This may result in the contractor doing the work themselves or subletting with approval, but that any extra cost must be borne by the main contractor.

The 1963 edition of the form of contract is unclear on this point, but the case was decided in the contractor's favour. It is the architect's responsibility to renominate in the event of default, and the employer is therefore obliged to pay the additional cost of the new firm employed. JCT 80 clarified this point along the lines of this decision.

Oram Builders Ltd v. M J Pemberton and C Pemberton (1985)

This project was on the minor works agreement (JCT 1980 edition). Article 4 is an arbitration agreement in general terms. The supervising officer issued instructions to the plaintiffs which constituted a substantial variation to the contract. The plaintiffs claimed to have done the extra work and so required a certificate for its payment. This was not forthcoming and the plaintiffs claimed extra costs for the work or alternatively damages. The defendants disputed the plaintiffs' claim and counterclaimed for alleged breaches of contract. The preliminary issue, with which this judgment is concerned, was whether the courts had powers to review the exercise of the supervising officer's discretionary powers under the contract and open up certificates. It was held that the courts had no jurisdiction to go behind a certificate of the architect or supervising officer. Where there is an arbitration clause in general terms referring any dispute or difference between the parties concerning the contract to the arbitrator, then even on a narrow interpretation of the reasoning of the Court of Appeal in *Northern Health Authority* v. *Derek Crouch Construction Co. Ltd* (1984), the High Court has no jurisdiction to go behind an architect's certificate.

Peak Construction Ltd v. McKinney Foundations Ltd (1970)

This case considered two separate matters concerning a project that overran its contract period. The first problem was concerned with an extension of time. McKinney Foundations Ltd were the nominated subcontractors for piling work. Defects were found upon completion of this work, but there was a prolonged delay on the part of the employer in both obtaining an engineer's advice and deciding upon the remedial work to be carried out. When the project overran, the employer levied liquidated damages on the main contractor. The main contractor then attempted to recover these damages from the subcontractor. Furthermore, the above was not a reasonable cause for granting an extension of time within the terms of the contract. It was held, however, that because the delay was due largely to the employer's default, liquidated damages could not be deducted. This was the case if the extension of time clause did not make provision for such a delay, or even if there was failure to extend the time.

The second matter was concerned with the price fluctuation clause. The court held that this clause continued to operate after the time for completion was past. This matter was clarified in JCT 80.

Porter v. Tottenham Urban District Council (1915)

The plaintiff contracted to build a school for the council upon land belonging to them. The contract provided that the plaintiff should be entitled to enter the site immediately, and that the works should be completed by a specified date. The only access to the site was from an adjoining road and to lay a temporary sleeper road and then subsequently a permanent pathway. The plaintiff began work but was forced to abandon it because of a threatened injunction from an adjoining owner, who claimed that the road was their property. The third party's claims were held to be unfounded. The plaintiff completed the works then claimed damages against the council in respect of the delay caused by the third party's action. In the Court of Appeal, the plaintiff's claim failed. It was stated that there was no implied warranty by the council against wrongful interference by third parties with free access to the site.

Roberts & Co. Ltd v. Leicestershire County Council (1961)

A contractor submitted a tender which specified a completion of 18 months. The county architect, unknown to the contractor, decided that 30 months was a more appropriate contract period. Prior to the signing of the contract, there were two meetings at which the contractor referred to their plans to complete the work according to a progress schedule in 18 months. It was held that the contractor was entitled to rectification on the grounds that the defendants were estopped by their conduct from saying that there was no mistake.

Scott v. Avery (1856)

This case concerned an insurance policy which indicated that only arbitration proceedings were permissible in the event of a dispute occurring. It therefore sought to prohibit direct legal actions in the courts. It was held that this provision was valid. The majority of the standard forms of building contract contain arbitration provisions. This course of action is agreed to between the parties, who may consider arbitration more appropriate. If one of the parties wishes to hold to arbitration, they may apply for a stay of proceedings on the basis that the contract follows the *Scott* v. *Avery* principle. In practice the courts in deciding whether to grant this are likely to take into consideration any hardship that may be caused.

Sutcliffe v. Thackrah and Others (1974)

The facts of this case were that Mr Sutcliffe employed Thackrah and Others, a firm of architects, to design and supervise the construction of a house. During construction two certificates (Nos 9 and 10) were issued and duly paid by Mr Sutcliffe. Shortly afterwards the contract with the builders ended when the builders became insolvent and another firm completed the work.

It was then discovered that the two certificates had included defective work. Because it could not be recovered from the builders, Mr Sutcliffe sued the architects for his loss because of their negligence in issuing the certificates. The negligence was not due to a failure to detect the defective work, since one of the architects was aware of this; it was due to a failure to pass the information to the quantity surveyor who had valued it, assuming that it was in accordance with the contract.

The official referee who tried the case in arbitration, held that the architect was negligent and awarded damages to Mr Sutcliffe. The Court of Appeal reversed this decision but this was finally overruled by the House of Lords. It was held that the architect was discharging a duty under the contract between the parties, but that there was no dispute over the subject matter and so it could not be contended that he was acting in arbitration.

In order for quantity surveyors to safeguard themselves from possible repercussions in the future, they now include on their valuations words to the following effect: The valuation assumes that the work is in accordance with the specification of materials and work standards. A quantity surveyor would clearly not include in a valuation work that was obviously defective, but the matter of quality control is more rightly the responsibility of the architect. The case of *Chambers* v. *Goldthorpe* was cited, but this was overruled by the House of Lords.

George E. Taylor & Co. Ltd v. G. Percy Trentham (1980)

Taylor were nominated subcontractors to Trentham. They also had a contract with the employer whereby they warranted due performance of the subcontract works so that the main contractors should not become entitled to an extension of time. The employer paid Trentham only £7,526 against an interim certificate of £22,101. The amount withheld was the balance payable to the subcontractors after deduction of the main contractor's claim against them for delay. It was held that the employer was not entitled to withhold the money, the contract between the employer and the subcontractor being *res inter alios acta*.

Trollope & Colls and Holland and Hannan and Cubitts Ltd v. Atomic Power Construction (1962)

The contractor tendered for work under a contract which included provisions for variations and fluctuations. After a delay of 4 months the contractor was requested to commence work upon the site. This request included a letter of intent to enter into a formal contract when the terms of the contract were settled. Substantial changes to the scheme had been notified to the contractor and this continued to occur until the terms of the contract were agreed 10 months later. The contractor therefore claimed that they should be paid on a *quantum meruit* basis for the work that had already been completed prior to the signing of the formal agreement. This they claimed was more equitable than on the basis of their original tender adjusted by variations, since at that time no contract was in existence.

It was held, however, that both parties had contemplated entering into a contract which would affect all the work associated with the contract. The contract was formulated in the light of the work going on at the time and it therefore had a retrospective effect.

Trollope & Colls Ltd v. North Western Metropolitan Regional Hospital Board (1973)

This was a very large building contract where the Court of Appeal reversed the decision of the trial judge, but the House of Lords confirmed the trial judge's opinion. The Hospital Board wanted the project completed in three phases, using three largely similar sets of conditions. Phase 1 was delayed by 59 weeks and the architect granted an extension of time by 47 weeks. Phase 3 was to start contractually 6 months after Phase 1 had received its certificate of practical completion. Phase 3 did, however, still retain its original completion date which then envisaged a 16 month contract period rather than the previously accepted 30 month contract period.

In these circumstances, could the time for completion of phase 3 be extended by the period of 47 weeks granted on phase 1? Unexpectedly it was the contractor who suggested that it could not, but for their own advantage. The contractor professed to being able to complete in time, and they therefore called upon the Hospital Board to nominate their appropriate subcontractors. This they were unable to do. In this situation the contractors would require new prices to be agreed relevant to the prevailing date. The contractors therefore sought a declaration in the High Court that the date for completion of phase 3 was unaffected by phase 1 being behind schedule.

The decision was made in favour of the contractor on the basis that the business efficacy of this large contract did not necessitate implying the kind of term the board required. The House of Lords was not satisfied that the parties had overlooked the effect of delays on earlier phases, in fixing the completion date for phase 3.

Tyrer v. District Auditor for Monmouthshire (1974)

A firm of contractors who went into liquidation had been employed upon several contracts by a local authority. The local authority found that this firm had been overpaid on interim certificates. The district auditor established that overpayment was due to the negligence of the quantity surveyor, in accepting excessively high rates for work carried out and failing to check simple arithmetic. The auditor then surcharged the quantity surveyor for the sums that he was unable to recover from the liquidator. The quantity surveyor appealed on the grounds that he had undertaken his duties in the capacity of an arbitrator. The court, however, rejected this appeal stating that 'there was nothing to show that the appellant was in quasi–judicial position when carrying out his duties here'.

Victoria Laundry Ltd v. Newman Ltd (1948)

This case provides an interesting contrast with the case of *Hadley* v. *Baxendale*. The plaintiffs were launderers and dyers and required a large boiler to extend their plant and to help them win some lucrative contracts. A firm of engineers contracted to sell them such a boiler, but certain faults arising meant its delivery was seriously delayed. The plaintiffs claimed damages: first, equivalent to the estimated loss of the increased profits the use of the boiler would have acquired for them; and second, the amount that they would have earned from dyeing contracts during the same period. The court held that the engineers were liable for the ordinary losses which they must have known from the particular circumstances. They were not liable in the second instance since this would have required special knowledge which they did not have.

University of Warwick v. Sir Robert McAlpine (1989)

The buildings at Warwick University had ceramic tile cladding on a concrete frame. The cladding which had been carried out by subcontractors began to fail. The architects blamed bad work and McAlpine, the main contractors, blamed the design. The university decided to remedy the defects by a resin injection process that was an innovation. The sole licensees of the process were a firm known as CCL. CCL were recommended by the architect and employed by McAlpine to carry out the remedial works. The main contract conditions were varied, and whilst CCL were technically not nominated subcontractors they were employed by McAlpine

McAlpine had considerable reservations about the use of this new process. The remedial works were not successful and the university alleged that McAlpine were in breach of an implied fitness for purpose. It was held that a term that resin be fit for its purpose could only be implied in the main contract if the university had relied on McAlpine. As they had not done so, no such term was to be implied.

City of Westminster v. J. Jarvis & Sons Ltd and Peter Lind (1970)

Jarvis contracted with Westminster for the erection of a multi-storey car park, with flats, offices, showrooms and ancillary works. The contract was in JCT 63 form. Lind were nominated subcontractors for the piling work. Lind carried out the work and purported to complete by the subcontract completion date. Lind left the site. Some weeks later it was discovered that many of the piles were defective, either as a result of bad work standards or poor materials. Lind carried out the remedial works. The main contract works were delayed for some 21.5 weeks. Jarvis, the main contractor, claimed an extension of time under clause 23(g). In the House of Lords it was held that Jarvis was not entitled to an extension of time since the works had apparently been completed by the due date. A 'delay' within the meaning of the clause occurs only if, by the subcontract completion date, the subcontractor has failed to achieve such completion of their work that they cannot hand it over to the main contractor.

Re Wilkinson ex parte Fowler (1905)

This contract allowed for the right to make direct payment to certain firms, where the main contractor defaulted or delayed in proper payment to them. The contractor became bankrupt and the employer decided to make payments out of the retentions owing to the contractor. It was held by the courts that such payments were valid against the trustee.

William Sindall v. North West Thames Regional Health Authority (1977)

The main contractor introduced a bonus incentive scheme on a site in accordance with the principles recommended by the National Joint Council for the Building Industry. The contract was on a fluctuations basis incorporating clause 31A. An increase in the basic rates of wages occurred and this, in a bonus scheme, was a voluntary decision and not as an unavoidable consequence of following the obligatory rules or decisions, and the cost of its operation was therefore outside the scope of clause 31A. This was in spite of the fact that once the contractor had introduced their bonus scheme on the basis of the NJCBI rules, they had also to pay the eventual increase.

Williams v. Firzmaurice (1859)

This involved the contractor in building a house in accordance with the drawings and specification supplied by the architect. The specification included a clause that the contractor undertook 'to provide the whole of the materials necessary for the completion of the works and to perform all the works of every kind mentioned'. Floorboarding, although shown on the drawings, was not included in the specification. The contractor therefore refused to carry out this work unless it was paid for as extra works. It was held that the boarding was necessary and was therefore included in the contract price, even though it had been omitted from the specification.

BIBLIOGRAPHY

Ashworth A 1996 *Precontract studies* Longman, Harlow

Ashworth A 1998 *Civil engineering contractual procedures* Longman, Harlow

Ashworth A 1999 *Cost studies of buildings* Longman, Harlow

Ashworth A and Hogg K I 2000 *Added value in design and construction* Pearson, Harlow

Banwell P 1967 *The placing and management of contracts for building and civil engineering* HMSO, London

Barlow J 1996 *A statement on the construction industry* Royal Academy of Engineering, London

Bennett J and Jayes S 1998 *The seven pillars of partnering* Reading Construction Forum and Thomas Telford, London

Centre for Strategic Studies in Construction 1996 *Design and building: a world-class industry* University of Reading, Reading

Chappell D 1991 *Which form of building contract?* Architectural and Design Technology Press, London

Construction Industry Board 1996 *Towards a 30 per cent productivity improvement in construction* Thomas Telford, London

Construction Sponsorship Directorate 1995 *Fair construction contracts* Department of the Environment, London

Cornes D L 1994 *Design liability in the construction industry* Blackwell Science, Oxford

Davies W H 1982 *Construction Site Studies* Butterworth, London

Davis Langdon and Everest 2000 *Contracts in use survey during 1998* Davis Langdon and Everest, London

Dearle and Henderson 1988 *Management contracting: a practice manual* E & F N Spon, London

Department of the Environment 1993 *Private finance and public money* HMSO, London

Egan J 1998 *Rethinking construction* Department of the Environment, Transport and the Regions, London

Flanagan R 1999 *Linking construction reseaerch and innovation to research innovation in other sectors* CRISP, London

Graves A 1998 *The Government client improvement study* Agile Construction Initiative, Bath University

Gray C 1996 *Value for money* Reading Construction Forum, London

Harvey R C and Ashworth A 1997 *The construction industry of Great Britain* Butterworth-Heinemann, Oxford

HM Treasury 1992 *Guidance note 33: project sponsorship* HM Treasury, Central Unit on Purchasing, London

HM Treasury 1993 *Guidance note 36: contract strategy, selection for major projects* HM Treasury, Central Unit on Purchasing, London

Joint Contracts Tribunal 1998 *The standard form of building contract* JCT, London

Kelly J, MacPerson S and Males S 1992 *The briefing process: a review and critique* Royal Institution of Chartered Surveyors, London

Latham M 1994 *Constructing the Team* HMSO, London

Masterman J W E 1992 *An introduction to building procurement systems* E & F N Spon, London

Morledge R 1996 *The procurement guide* Royal Institution of Chartered Surveyors, London

National Economic Development Office 1998 *Faster building for commerce* HMSO, London

Ndekugri I and Rycroft M 2000 *The JCT 98 building contract law and administration* Edward Arnold, London

Porter M 1980 *Competitive strategy: techniques for analysing industries and competitors* John Wiley, Chichester

Royal Institute of British Architects 1973 *Plan of work for design team operations* RIBA, London

Royal Institution of Chartered Surveyors 1990 *A client's guide to successful building: the role of the chartered quantity surveyor* RICS, London

Skitmore R M and Marsden D E 1988 Which procurement system? Towards a universal procurement selection technique *Construction Management and Economics* 6, 25–34

Turner D 1994 *Building contracts: a practical guide* Longman, Harlow

Walker A 1989 *Project management in construction* Blackwell Science, Oxford

Womack J *et al.* 1990 *The machine that changed the world* Harper Perennial, New York

Womack J and Jones D 1996 *Lean thinking* Harper Perennial, New York

INDEX